Student Solutions Manual

for

Yoshiwara and Yoshiwara's

Intermediate Algebra:
Functions and Graphs

Kirby Bunas
Santa Rosa Junior College

THOMSON

BROOKS/COLE

Australia • Canada • Mexico • Singapore • Spain • United Kingdom • United States

Printed in Canada
    4 5 6 7 07 06 05

Printer: Webcom

ISBN: 0-534-38610-5

For more information about our products, contact us at:
Thomson Learning Academic Resource Center
1-800-423-0563

For permission to use material from this text,
contact us by:
Phone: 1-800-730-2214
Fax: 1-800-730-2215
Web: http://www.thomsonrights.com

Brooks/Cole—Thomson Learning
10 Davis Drive
Belmont, CA 94002-3098
USA

Asia
Thomson Learning
5 Shenton Way #01-01
UIC Building
Singapore 068808

Australia/New Zealand
Thomson Learning
102 Dodds Street
Southbank, Victoria 3006
Australia

Canada
Nelson
1120 Birchmount Road
Toronto, Ontario M1K 5G4
Canada

Europe/Middle East/South Africa
Thomson Learning
High Holborn House
50/51 Bedford Row
London WC1R 4LR
United Kingdom

Latin America
Thomson Learning
Seneca, 53
Colonia Polanco
11560 Mexico D.F.
Mexico

Spain/Portugal
Paraninfo
Calle/Magallanes, 25
28015 Madrid, Spain

Preface

This manual contains solutions to all of the odd-numbered exercises in the text. We refer to the art in the back of the textbook, where applicable, rather than reproducing it here.

Table of Contents

Chapter One: Linear Models

Homework 1.1

1.

w	0	4	8	12	16
A	250	190	130	70	10

  a. $A = 250 - 15w$

  b. Refer to the graph in the back of the textbook.

  c. From the graph, we see that when $w = 3$, $A = 205$, and when $w = 8$, $A = 130$. The decrease in oil is the difference of the two A-values:

$$\text{decrease in fuel oil} = 205 - 130$$
$$= 75 \text{ gallons}$$

The decrease is illustrated by the vertical arrow in the graph.

  d. Using the graph: when $A = 175$, $w = 5$. Therefore, the tank will contain more than 175 gallons up to the fifth week. See the graph in part (b) for illustration.

3.

t	0	5	10	15	20
P	−800	−600	−400	−200	0

  a. $P = 40t - 800$

  b. Set $t = 0$.
$$P = 40(0) - 800 = -800$$
The P-intercept is −800.
Set $P = 0$.
$$0 = 40t - 800$$
$$800 = 40t$$
$$20 = t$$
The t-intercept is 20. Refer to the graph in the back of the textbook.

  c. The P-intercept, −800, is the initial ($t = 0$) value of the profit. Phil and Ernie start out $800 in debt. The t-intercept, 20, is the number of hours required for Phil and Ernie to break even.

5. a. Set $x = 0$.
$$0 + 2y = 8$$
$$y = 4$$
The y-intercept is the point (0, 4).
Set $y = 0$.
$$x + 2(0) = 8$$
$$x = 8$$
The x-intercept is the point (8, 0).

  b. Refer to the graph in the back of the textbook.

7. a. Set $x = 0$.

$$3(0) - 4y = 12$$
$$-4y = 12$$
$$y = -3$$

The y-intercept is the point $(0, -3)$.

Set $y = 0$.

$$3x - 4(0) = 12$$
$$3x = 12$$
$$x = 4$$

The x-intercept is the point $(4, 0)$.

b. Refer to the graph in the back of the textbook.

9. a. Set $x = 0$.

$$\frac{0}{9} - \frac{y}{4} = 1$$
$$\frac{-y}{4} = 1$$
$$y = -4$$

The y-intercept is the point $(0, -4)$.

Set $y = 0$.

$$\frac{x}{9} - \frac{0}{4} = 1$$
$$\frac{x}{9} = 1$$
$$x = 9$$

The x-intercept is the point $(9, 0)$.

b. Refer to the graph in the back of the textbook.

11. a. Set $x = 0$.

$$\frac{2(0)}{8} + \frac{3y}{11} = 1$$
$$\frac{3y}{11} = 1$$
$$y = \frac{11}{3}$$

The y-intercept is $\left(0, \frac{11}{3}\right)$.

Set $y = 0$.

$$\frac{2x}{3} + \frac{3(0)}{11} = 1$$
$$\frac{2x}{3} = 1$$
$$x = \frac{3}{2}$$

The x-intercept is $\left(\frac{3}{2}, 0\right)$.

b. Refer to the graph in the back of the textbook.

13. a. Set $x = 0$.

$$20(0) = 30y - 45000$$
$$0 = 30y - 45000$$
$$45000 = 30y$$
$$1500 = y$$

The y-intercept is the point $(0, 1500)$.

Set $y = 0$.

$$20x = 30(0) - 45,000$$
$$20x = -45,000$$
$$x = -2250$$

The x-intercept is the point $(-2250, 0)$.

b. Refer to the graph in the back of the textbook.

15. a. Set $x = 0$.
$$0.4(0) + 1.2y = 4.8$$
$$1.2y = 4.8$$
$$y = 4$$
The y-intercept is the point $(0, 4)$.
Set $y = 0$.
$$0.4x + 1.2(0) = 4.8$$
$$0.4x = 4.8$$
$$x = 12$$
The x-intercept is the point $(12, 0)$.

b. Refer to the graph in the back of the textbook.

17. a. The owner spends $\$0.6x$ on regular unleaded and $\$0.8y$ on premium unleaded.

b. $0.6x + 0.8y = 4800$

c. Set $x = 0$.
$$0.6(0) + 0.8y = 4800$$
$$0.8y = 4800$$
$$y = 6000$$
The y-intercept is $(0, 6000)$.
Set $y = 0$
$$0.6x + 0.8(0) = 4800$$
$$0.6x = 4800$$
$$x = 8000$$
The x-intercept is $(8000, 0)$.
Refer to the graph in the back of the textbook.

d. The y-intercept, 6000, is the amount of premium the gas station owner can buy if he buys no regular. The x-intercept, 8000, is the amount of regular he can buy if he buys no premium.

19.

s	200	400	500	800	1000
I	16	22	25	34	40

a. $I = 0.03s + 10$

b. Refer to the graph in the back of the textbook.

c. Substitute $I = 16$ into the equation and solve for s.
$$16 = 0.03s + 10$$
$$6 = 0.03s$$
$$200 = s$$
Substitute $I = 22$ into the equation and solve for s.
$$22 = 0.03s + 10$$
$$12 = 0.03s$$
$$400 = s$$
Hence, annual sales are between $\$200,000$ and $\$400,000$, as illustrated on the graph in part (b).

d. Using the graph: When the sales is 500,000, the income is $\$25,000$, and when the sales is 700,000, the income is $\$31,000$. The increase in salary is the difference of the two incomes. Therefore, the increase of salary is $\$31,000 - \$25,000 = \$6000$. This is illustrated in the graph in part (b).

3

Homework 1.2

1. a. $2x + y = 6$
$y = 6 - 2x$

b. Refer to the graph in the back of the textbook.

c. Set $x = 0$.
$2(0) + y = 6$
$y = 6$
The y-intercept is $(0, 6)$.
Set $y = 0$.
$2x + 0 = 6$
$2x = 6$
$x = 3$
The x-intercept is $(3, 0)$. Refer to the graph in the back of the textbook.

3. a. $3x - 4y = 1200$
$-4y = -3x + 1200$
$y = \dfrac{3}{4}x - 300$

b. Refer to the graph in the back of the textbook.

c. Set $x = 0$.
$3(0) - 4y = 1200$
$-4y = 1200$
$y = -300$
The y-intercept is $(0, -300)$.
Set $y = 0$.
$3x - 4(0) = 1200$
$3x = 1200$
$x = 400$
The x-intercept is $(400, 0)$.
Refer to the graph in the back of the textbook.

5. a. $0.2x + 5y = 0.1$
$5y = -0.2x + 0.1$
$y = -0.04x + 0.02$

b. Refer to the graph in the back of the textbook.

c. Set $x = 0$.
$0.2(0) + 5y = 0.1$
$5y = 0.1$
$y = 0.02$
The y-intercept is $(0, 0.02)$.
Set $y = 0$.
$0.2x + 5(0) = 0.1$
$0.2x = 0.1$
$x = 0.5$
The x-intercept is $(0.5, 0)$.
Refer to the graph in the back of the textbook.

7. a. $70x + 3y = y + 420$
$70x + 2y = 420$
$2y = -70x + 420$
$y = -35x + 210$

b. Refer to the graph in the back of the textbook.

c. Set $x = 0$.
$70(0) + 2y = 420$
$2y = 420$
$y = 210$
The y-intercept is $(0, 210)$.
Set $y = 0$.
$70x + 2(0) = 420$
$70x = 420$
$x = 6$
The x-intercept is $(6, 0)$.
Refer to the graph in the back of the textbook.

9. a. Set $x = 0$.

$0 + y = 100$

$y = 100$

The y-intercept is the point $(0, 100)$.

Set $y = 0$.

$x + 0 = 100$

$x = 100$

The x-intercept is the point $(100, 0)$.

b. $x + y = 100$

$y = -x + 100$

c. Xmin $= -20$, Xmax $= 120$, Ymin $= -20$, and Ymax $= 120$.

d. Refer to the graph in the back of the textbook.

11. a. Set $x = 0$.

$25(0) - 36y = 1$

$-36y = 1$

$y = -\dfrac{1}{36}$

The y-intercept is the point

$\left(0, -\dfrac{1}{36}\right) \approx (0, -0.028)$.

Set $y = 0$.

$25x - 36(0) = 1$

$25x = 1$

$x = \dfrac{1}{25}$

The x-intercept is the point

$\left(\dfrac{1}{25}, 0\right) = (0.04, 0)$.

b. $25x - 36y = 1$

$-36y = -25x + 1$

$y = \dfrac{25}{36}x - \dfrac{1}{36}$

c. Xmin $= -0.1$, Xmax $= 0.1$, Ymin $= -0.1$, and Ymax $= 0.1$.

d. Refer to the graph in the back of the textbook.

13. a. Set $x = 0$.

$\dfrac{y}{12} - \dfrac{0}{47} = 1$

$\dfrac{y}{12} = 1$

$y = 12$

The y-intercept is the point $(0, 12)$.

Set $y = 0$.

$\dfrac{0}{12} - \dfrac{x}{47} = 1$

$\dfrac{-x}{47} = 1$

$x = -47$

The x-intercept is the point $(-47, 0)$.

b. $\dfrac{y}{12} - \dfrac{x}{47} = 1$

$\dfrac{y}{12} = \dfrac{x}{47} + 1$

$y = \dfrac{12}{47}x + 12$

c. Xmin $= -50$, Xmax $= 10$, Ymin $= -5$, and Ymax $= 15$.

d. Refer to the graph in the back of the textbook.

Homework 1.2

15. a. Set $x = 0$.

$$-2(0) = 3y + 84$$
$$0 = 3y + 84$$
$$-3y = 84$$
$$y = -28$$

The y-intercept is $(0, -28)$.
Set $y = 0$.

$$-2x = 3(0) + 84$$
$$-2x = 84$$
$$x = -42$$

The x-intercept is $(-42, 0)$.

b. $-2x = 3y + 84$
$$-3y = 2x + 84$$
$$y = -\frac{2}{3}x - 28$$

c. Xmin $= -50$, Xmax $= 10$,
Ymin $= -35$, and Ymax $= 5$.

d. Refer to the graph in the back of
the textbook.

17. a. (i) $x = -3$ since the coordinate
$(-3, 12)$ is on the graph.

(ii) $x < -3$ since for these values of
x, all of the y-coordinates on the
line are greater than 12.

(iii) $x > -3$ since for these values
of x, all of the y-coordinates on the
line are less than 12.

b. The answers to (i), (ii), and (iii)
are the same as in part (a).

c. For values of y on the graph, we
can replace y with $-2x + 6$ since
$y = -2x + 6$.

19. a. $x = 0.6$
Solve algebraically:
$$1.4x - 0.64 = 0.2$$
$$1.4x = 0.84$$
$$x = 0.6$$

b. $x = -0.4$
Solve algebraically:
$$1.4x - 0.64 = -1.2$$
$$1.4x = -0.56$$
$$x = -0.4$$

c. $x > 0.6$
Solve algebraically:
$$1.4x - 0.64 > 0.2$$
$$1.4x > 0.84$$
$$x > 0.6$$

d. $x < -0.4$
Solve algebraically:
$$-1.2 > 1.4x - 0.64$$
$$-0.56 > 1.4x$$
$$-0.4 > x$$

21.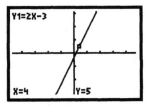

a. $x = 4$
Solve algebraically:
$2x - 3 = 5$

$2x = 8$

$x = 4$

b. $x = -5$
Solve algebraically:
$2x - 3 = -13$

$2x = -10$

$x = -5$

c. $x > 1$
Solve algebraically:
$2x - 3 > -1$

$2x > 2$

$x > 1$

d. $x < 14$
Solve algebraically:
$2x - 3 < 25$

$2x < 28$

$x < 14$

23.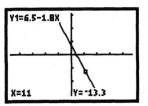

a. $x = 11$
Solve algebraically:
$6.5 - 1.8x = -13.3$

$-1.8x = -19.8$

$x = 11$

b. $x = -10$
Solve algebraically:
$6.5 - 1.8x = 24.5$

$-1.8x = 18$

$x = -10$

c. $x \geq -5$
Solve algebraically:
$6.5 - 1.8x \leq 15.5$

$-1.8x \leq 9$

$x \geq -5$

Note: The inequality sign changes direction when each side is multiplied or divided by a negative number.

d. $x \leq 8$
Solve algebraically:
$6.5 - 1.8x \geq -7.9$

$-1.8x \geq -14.4$

$x \leq 8$

Note: The inequality sign changes direction when each side is multiplied or divided by a negative number.

25.

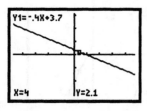

a. $x = 4$
Solve algebraically:
$$-0.4x + 3.7 = 2.1$$
$$-0.4x = -1.6$$
$$x = 4$$

b. $x < 22$
Solve algebraically:
$$-0.4x + 3.7 > -5.1$$
$$-0.4x > -8.8$$
$$x < 22$$

27.

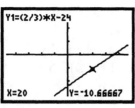

a. $x = 20$
Solve algebraically:
$$\frac{2}{3}x - 24 = -10\frac{2}{3}$$
$$\frac{2}{3}x - 24 = -\frac{32}{3}$$
$$3\left(\frac{2}{3}x - 24\right) = 3\left(-\frac{32}{3}\right)$$
$$2x - 72 = -32$$
$$2x = 40$$
$$x = 20$$

b. $x \le 7$
Solve algebraically:
$$\frac{2}{3}x - 24 \le -19\frac{1}{3}$$
$$\frac{2}{3}x - 24 \le -\frac{58}{3}$$
$$3\left(\frac{2}{3}x - 24\right) \le 3\left(-\frac{58}{3}\right)$$
$$2x - 72 \le -58$$
$$2x \le 14$$
$$x \le 7$$

29. a.

t	0	1	2	5	10
p	120	126	132	150	180

b. $p = 120 + 6t$

c. Refer to the graph in the back of the textbook.

d. Let $t = 3.5$.
$p = 120 + 6(3.5) = 141$
After 3.5 minutes, Kieran's blood pressure is 141 mm Hg.

e. Let $p = 165$.
$$165 = 120 + 6t$$
$$45 = 6t$$
$$t = \frac{45}{6} = 7\frac{1}{2}$$
The treadmill should stop increasing his exercise intensity after 7.5 minutes.

Homework 1.3

1. Carl's average speed is
$$\frac{100 \text{ meters}}{10 \text{ seconds}} = 10 \text{m/s}$$
and Anthony's average speed is
$$\frac{200 \text{ meters}}{19.6 \text{ seconds}} \approx 10.2 \text{m/s}$$
Anthony has the faster average speed.

3. Slope of Grimy Gulch pass:
$$\frac{0.6}{26} = \frac{3}{130} \approx 0.0231$$
Slope of Bob's driveway:
$$\frac{12}{150} = \frac{2}{25} = 0.08$$
Bob's driveway is steeper than Grimy Gulch pass.

5. $m = \dfrac{\text{change in } y \text{ coordinate}}{\text{change in } x \text{ coordinate}}$

$$= \frac{\Delta y}{\Delta x} = \frac{-4}{4} = -1$$

7. $m = \dfrac{\text{change in } y \text{ coordinate}}{\text{change in } x \text{ coordinate}}$

$$= \frac{\Delta y}{\Delta x} = \frac{-4}{6} = -\frac{2}{3}$$

9. a. Notice that each tick mark on the y-axis represents 2 units, while every tick mark on the x-axis represents 1 unit. For example, using the points (−4, −4) and (0, 2), $m = \dfrac{\Delta y}{\Delta x} = \dfrac{6}{4} = \dfrac{3}{2}$

b. If you move 4 units in the positive x-direction starting from the point (0, 2), then you must move 6 units in the y-direction to get back to the line. Therefore, once again,
$$m = \frac{\Delta y}{\Delta x} = \frac{6}{4} = \frac{3}{2}$$

c. Yes, the answers are the same. The ratio $\dfrac{3}{2}$ does not change no matter which x-value we start with.

d. Moving −6 units in the x-direction means moving 6 units in the negative x-direction. If you start from the point (0, 2) and move 6 units to the left, then you must move −9 units in the y-direction to get back to the line. Therefore,
$$m = \frac{\Delta y}{\Delta x} = \frac{-9}{-6} = \frac{3}{2}$$

e. $m = \dfrac{\Delta y}{\Delta x} = \dfrac{3}{2}$ and $\Delta x = 18$.
$$\frac{\Delta y}{18} = \frac{3}{2}$$
$$\Delta y = \frac{3}{2} \cdot 18 = 27$$
You would have to move 27 units in the positive y-direction.

11. $m = \dfrac{\Delta y}{\Delta x} = -\dfrac{4}{5}$

a. $\dfrac{-4}{\Delta x} = -\dfrac{4}{5}$

$\Delta x = 5$

b. $\dfrac{2}{\Delta x} = -\dfrac{4}{5}$

$10 = -4\Delta x$

$\dfrac{10}{-4} = \Delta x$

$\Delta x = -\dfrac{5}{2}$

c. $\dfrac{-12}{\Delta x} = -\dfrac{4}{5}$

$-60 = -4\Delta x$

$\dfrac{-60}{-4} = \Delta x$

$\Delta x = 15$

d. $\dfrac{5}{\Delta x} = -\dfrac{4}{5}$

$25 = -4\Delta x$

$\dfrac{25}{-4} = \Delta x$

$\Delta x = -\dfrac{25}{4} = -6\dfrac{1}{4}$

13. $\dfrac{\text{vertical distance}}{\text{horizontal space}} = \text{slope}$

$\dfrac{10\text{ ft}}{\text{horizontal space}} = \dfrac{7}{10}$

$\text{horizontal space} = \dfrac{100\text{ ft}}{7}$

$\approx 14.29\text{ ft}$

15. If the echo returns in 4.5 seconds, then it took $\dfrac{4.5}{2} = 2.25$ seconds to reach the floor of the ocean. The depth of the ocean is then $1500\text{ m / s} \times 2.25\text{ s} = 3375$ meters.

17. a. Set $x = 0$.

$3(0) - 4y = 12$

$-4y = 12$

$y = -3$

The y-intercept is the point $(0, -3)$.

Set $y = 0$.

$3x - 4(0) = 12$

$3x = 12$

$x = 4$

The x-intercept is the point $(4, 0)$.

Refer to the graph in the back of the textbook.

b. $m = \dfrac{\Delta y}{\Delta x} = \dfrac{3}{4}$

19. a. Set $x = 0$.

$2y + 6(0) = -18$

$2y = -18$

$y = -9$

The y-intercept is $(0, -9)$.

Set $y = 0$.

$2(0) + 6x = -18$

$6x = -18$

$x = -3$

The x-intercept is $(-3, 0)$.

Refer to the graph in the back of the textbook.

b. $m = \dfrac{\Delta y}{\Delta x} = \dfrac{9}{-3} = -3$

21. a. Set $x = 0$.

$$\frac{0}{5} - \frac{y}{8} = 1$$

$$\frac{-y}{8} = 1$$

$$y = -8$$

The y-intercept is the point $(0, -8)$.
Set $y = 0$.

$$\frac{x}{5} - \frac{0}{8} = 1$$

$$\frac{x}{5} = 1$$

$$x = 5$$

The x-intercept is the point $(5, 0)$.
Refer to the graph in the back of the textbook.

b. $m = \dfrac{\Delta y}{\Delta x} = \dfrac{8}{5}$

23. a. The points will lie on a straight line. To prove that this is true, calculate the slope between each pair of consecutive points and show that you always get the same result:

$$\frac{34.0 - 33.0}{12 - 10} = \frac{1}{2},$$

$$\frac{35.5 - 34.0}{15 - 12} = \frac{1}{2}, \text{ etc.}$$

b. The rate of change of salt dissolved with respect to temperature is 0.5 grams per $°C$.

25. If the tables represent variables that are related by a linear equation, the slopes from one point to the next must be constant.

a. $m = \dfrac{17 - 12}{3 - 2} = 5$

$$m = \frac{22 - 17}{4 - 3} = 5$$

$$m = \frac{27 - 22}{5 - 4} = 5$$

$$m = \frac{32 - 27}{6 - 5} = 5$$

The table represents variables that are related by a linear equation.

b. $m = \dfrac{9 - 4}{3 - 2} = 5$

$$m = \frac{16 - 9}{4 - 3} = 7$$

$$m = \frac{25 - 16}{5 - 4} = 9$$

$$m = \frac{36 - 25}{6 - 5} = 11$$

The table does not represent variables that are related by a linear equation.

27. a.

t	4	8	20	40
S	32	64	160	320

b. Refer to the graph in the back of the textbook.

c. For example, use $(4, 32)$ and $(8, 64)$.

$$m = \frac{\Delta s}{\Delta t} = \frac{64 - 32}{8 - 4} = \frac{32}{4} = 8$$

dollars per hour.

d. The slope gives the typist's rate of pay, in dollars per hour.

29. a. For example, use (0, 2000) and (4, 7000).

$$m = \frac{7000 - 2000}{4 - 0} = \frac{5000}{4}$$
$$= 1250 \text{ barrels per day.}$$

b. The slope gives the rate of pumping.

31. a. For example, use (0, 48) and (8, 0).

$$m = \frac{0 - 48}{8 - 0} = \frac{-48}{8}$$
$$= -6 \text{ L per day.}$$

b. The slope gives the rate of water consumption.

33. a. For example, use (0,0) and (5, 60).

$$m = \frac{60 - 0}{5 - 0} = \frac{60}{5}$$
$$= 12 \text{ in. per ft.}$$

b. The slope gives the conversion rate from feet to inches.

35. a. For example, use (0, 0) and (5, 20).

$$m = \frac{20 - 0}{5 - 0} = \frac{20}{5}$$
$$= 4 \text{ dollars per kg.}$$

b. The slope gives the unit cost of beans, in dollars per kilogram

37. a. The points will lie on a straight line since the slopes from one point to the next are constant. (In this case, the slopes are the same when rounded to 3 decimal place accuracy.)

$$\frac{37.699 - 25.133}{6 - 4} = 6.283$$

$$\frac{62.832 - 37.699}{10 - 6} = 6.28325$$

$$\frac{94.248 - 62.832}{15 - 10} = 6.2832$$

b. The slope is approximately 6.283, which is equal to 2π rounded to three decimal places. Note that the formula for the circumference of a circle is $C = 2\pi r$, so $\frac{C}{r} = 2\pi$, and we have found that from part (a), $\frac{\Delta C}{\Delta r} = 2\pi$.

39. a. The graph is plotted with distance as the independent variable because the travel time for P-waves from a shallow earthquake depends on the distance from the epicenter.

b. Using the points (58, 10) and (160, 28),

$$m = \frac{28 - 10}{160 - 58} = \frac{18}{102} = \frac{9}{51}$$

seconds per kilometer. The speed of the wave is then $\frac{51}{9} \approx 5.7$ kilometers per second.

b. 582 meters = 0.582 kilometers. The equivalent horizontal distance is $0.582 \times 7.92 + 3.75 \approx 8.35$ km.

c. $(8.35 \text{ km})(15 \text{ min} / \text{km}) = 125.25$ minutes, or about 2 hours and 5 minutes.

Midchapter 1 Review

1.

h	0	3	6	9	10
T	65	80	95	110	115

  a. $T = 65 + 5h$

  b. Refer to the graph in the back of the textbook.

  c. Noon is 6 hours past 6 a.m., so $h = 6$ at noon. The temperature at noon is $T = 5(6) + 65 = 95°F$.

  d. Let $T = 110$.

$$110 = 5h + 65$$
$$45 = 5h$$
$$9 = h$$

The temperature will be 110 degrees nine hours after 6 a.m., or at 3 p.m.

3. a.

t	0	10	20	30	40
A	300	250	200	150	100

Refer to the graph in the back of the textbook.

  b. $A = 300 - 5t$

  c. Note from the table that the A-intercept is (0, 300). To find the t-intercept, set $A = 0$.

$$0 = 300 - 5t$$
$$5t = 300$$
$$t = 60$$

The t-intercept is (60, 0).

The t-intercept shows that Delbert reached the ground after 60 minutes; the A-intercept shows that Delbert began at an altitude of 300 meters.

5. a. Set $x = 0$.

$$36(0) - 24y = 7200$$
$$-24y = 7200$$
$$y = -300$$

The y-intercept is (0, –300).

Set $y = 0$.

$$36x - 24(0) = 7200$$
$$36x = 7200$$
$$x = 200$$

The x-intercept is (200, 0).

  b. Refer to the graph in the back of the textbook.

  c. $36x - 24y = 7200$

$$-24y = -36x + 7200$$
$$y = \frac{3}{2}x - 300$$

  d.

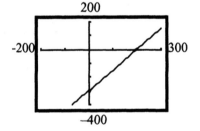

13

7. a. Set $x = 0$.
$$0.3y - 56(0) = 84$$
$$0.3y = 84$$
$$y = 280$$
The y-intercept is $(0, 280)$.

Set $y = 0$.
$$0.3(0) - 56x = 84$$
$$-56x = 84$$
$$x = -1.5$$
The x-intercept is $(-1.5, 0)$.

b. Refer to the graph in the back of the textbook.

c. $0.3y - 56x = 84$
$$0.3y = 56x + 84$$
$$y = \frac{560}{3}x + 280$$

d.

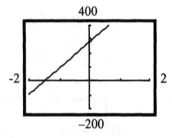

9. a. $x = 6$ since there is a point $(6, -2)$ on the graph.

b. verify algebraically:
$$0.24x - 3.44 = -2$$
$$0.24x = 1.44$$
$$x = 6$$

c. $x > 6$ since for these values of x, the y-coordinates on the line are greater than -2.

d. verify algebraically:
$$0.24x - 3.44 > -2$$
$$0.24x > 1.44$$
$$x > 6$$

11.

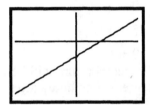

a. Using TRACE, we see that $x = 4$ when $y = 8.9$.

b. Using TRACE, we see that $x = -4$ when $y = -25.5$. The solution to the inequality is $x < -4$ since for these values of x, $y < -25.5$.

c. Using TRACE, we see that $x = 2$ when $y = 0.3$. The solution to the inequality is $x \geq 2$ since for these values of x, $y \geq 0.3$.

d. Using TRACE, we see that $x = -1$ when $y = -12.6$. The solution to the inequality is $x \leq -1$ since for these values of x, $y \leq -12.6$.

13. $m = \dfrac{-400}{2} = -200$

15. $m = \dfrac{1}{20}$

17. a. Set $x = 0$.
$$3y - 5(0) = 30$$
$$3y = 30$$
$$y = 10$$
The y-intercept is $(0, 10)$.

Set $y = 0$.
$$3(0) - 5x = 30$$
$$-5x = 30$$
$$y = -6$$
The x-intercept is $(-6, 0)$.

b. $m = \dfrac{10 - 0}{0 - (-6)} = \dfrac{10}{6} = \dfrac{5}{3}$

19. a. Set $x = 0$.

$$y = 40(0) - 120$$
$$y = -120$$
The y-intercept is $(0, -120)$.

Set $y = 0$.
$$0 = 40x - 120$$
$$120 = 40x$$
$$3 = x$$
The x-intercept is $(3, 0)$.

b. $m = \dfrac{-120 - 0}{0 - 3} = \dfrac{-120}{-3} = 40$

21. a. For example, choose the points $(10, 20)$ and $(20, 35)$.
$$m = \frac{35 - 20}{20 - 10} = \frac{15}{10} = \frac{3}{2} \text{ ft/sec}$$

b. The elevator is rising at a rate of $\dfrac{3}{2}$ feet per second.

Homework 1.4

1. a. $3x + 2y = 1$

$$2y = -3x + 1$$

$$y = -\frac{3}{2}x + \frac{1}{2}$$

b. $m = -\frac{3}{2}, \ b = \frac{1}{2}$

3. a. $\frac{1}{4}x + \frac{3}{2}y = \frac{1}{6}$

$$\frac{3}{2}y = -\frac{1}{4}x + \frac{1}{6}$$

$$y = -\frac{1}{6}x + \frac{1}{9}$$

b. $m = -\frac{1}{6}, \ b = \frac{1}{9}$

5. a. $4.2x - 0.3y = 6.6$

$$-0.3y = -4.2x + 6.6$$

$$y = 14x - 22$$

b. $m = 14, \ b = -22$

7. a. $y + 29 = 0$

$$y = -29$$

b. $m = 0, \ b = -29$

9. a. $250x + 150y = 2450$

$$150y = -250x + 2450$$

$$y = -\frac{5}{3}x + \frac{49}{3}$$

b. $m = -\frac{5}{3}, \ b = \frac{49}{3} = 16\frac{1}{3}$

11. a. Plot the y-intercept $(0, -2)$. Move 3 units in the y-direction and 1 unit in the x-direction to arrive at $(1, 1)$. Draw the line through the two points. Refer to the graph in the back of the textbook.

b. $y = 3x - 2$

c. Set $y = 0$.

$$0 = 3x - 2$$

$$\frac{2}{3} = x$$

The x-intercept is $\left(\frac{2}{3}, 0\right)$.

13. a. Plot the y-intercept $(0, -6)$. Move 5 units up (the positive y-direction) and 3 units left (the negative x-direction) to arrive at $(-3, -1)$. Draw the line through the two points. Refer to the graph in the back of the textbook.

b. $y = -\frac{5}{3}x - 6$

c. Set $y = 0$.

$$0 = -\frac{5}{3}x - 6$$

$$6 = -\frac{5}{3}x$$

$$-\frac{18}{5} = x$$

The x-intercept is $\left(-\frac{18}{5}, 0\right)$.

15. a. Using the points (0, 100) and (4, 700), $m = \dfrac{\Delta a}{\Delta t} = \dfrac{600}{4} = 150$.
The a-intercept is (0, 100). So the equation of the line is
$a = 150t + 100$.

b. a-intercept = (0, 100), $m = 150$. The skier's starting altitude is 100 feet and she rises at 150 feet per minute.

17. a. Using the points (0, 25) and (10, 150), $m = \dfrac{\Delta G}{\Delta t} = \dfrac{125}{10} = 12.5$.
The equation of the line is
$G = 12.5t + 25$.

b. G-intercept = (0, 25), $m = 12.5$. There were 25 tons in the dump before the regulations. The dump is filling at 12.5 tons per year.

19. a. Using the points (0, 7000) and (10, 3000),
$m = \dfrac{\Delta M}{\Delta w} = \dfrac{-4000}{10} = -400$. The equation is $M = -400w + 7000$.

b. M-intercept = (0, 7000), $m = -400$. Tammy has \$7000 in her account before losing all sources of income, and she then loses \$400 per week from this account.

21. a. Set $C = 10$:
$$F = \frac{9}{5}(10) + 32$$
$$F = 50$$
Hence, $10°C = 50°F$.

b. Set $F = -4$.
$$-4 = \frac{9}{5}C + 32$$
$$-20 = 9C + 160$$
$$-180 = 9C$$
$$-20 = C$$
Hence, $-4°F = -20°C$.

c. Refer to the graph in the back of the textbook.

d. $m = \dfrac{9}{5}$. The Fahrenheit temperature increases $\dfrac{9}{5}$ of a degree for each degree increase in Celsius temperature.

e. Set $F = 0$.
$$0 = \frac{9}{5}C + 32$$
$$-\frac{9}{5}C = 32$$
$$C = -\frac{160}{9}$$
Hence the C-intercept is
$\left(-\dfrac{160}{9}, 0\right) \approx (-17.8, 0)$.
The C-intercept gives the Celsius temperature at zero degrees Fahrenheit.

Set $C = 0$.
$$F = \frac{9}{5}(0) + 32$$
$$F = 32$$
The F-intercept is $(0, 32)$.
The F-intercept gives the Fahrenheit temperature at zero degrees Celsius.

23. a. Let G equal the gas mark and let d equal the temperature in degrees Fahrenheit. Refer to the graph at the back of the textbook.

b. $m = \dfrac{475 - 325}{9 - 3} = 25$

$b = 250$

c. $d = 25G + 250$

25. a. Refer to the graph in the back of the textbook. It appears that the y-intercept is $(0, 25.4)$.

b. Use two of the data points to calculate the slope. For example, using $(3, 25.76)$ and $(22, 28.04)$,

$m = \dfrac{28.04 - 25.76}{22 - 3} = 0.12$. The equation is $y = 0.12x + 25.4$.

c. $27.56 = 0.12x + 25.4$

$2.16 = 0.12x$

$18 = x$

The attached weight is 18 kg.

27. Note: $2\dfrac{1}{3} = \dfrac{7}{3}$ and $4\dfrac{1}{3} = \dfrac{13}{3}$.

$m = \dfrac{y_2 - y_1}{x_2 - x_1} = \dfrac{\dfrac{7}{3} - \left(-\dfrac{2}{3}\right)}{-2 - \dfrac{13}{3}}$

$= \dfrac{\dfrac{9}{3}}{-\dfrac{19}{3}} = -\dfrac{9}{19}$

29. $m = \dfrac{y_2 - y_1}{x_2 - x_1} = \dfrac{-1.2 - (-3.6)}{1.4 - (-4.8)}$

$= \dfrac{2.4}{6.2} = \dfrac{24}{62} = \dfrac{12}{31} \approx 0.39$

31. $m = \dfrac{y_2 - y_1}{x_2 - x_1} = \dfrac{7000 - (-2000)}{5000 - 5000}$

$= \dfrac{9000}{0} =$ undefined

33. a. Refer to the graph in the back of the textbook.

b. $y - (-5) = -3(x - 2)$ or

$y + 5 = -3(x - 2)$ (point-slope form)

c. Solve the equation for y:

$y + 5 = -3x + 6$

$y = -3x + 1$ (slope-int form)

35. a. Refer to the graph in the back of the textbook.

b. $y - (-1) = \dfrac{5}{3}(x - 2)$ or

$y + 1 = \dfrac{5}{3}(x - 2)$ (point-slope form)

c. Solve the equation for y:

$y + 1 = \dfrac{5}{3}x - \dfrac{10}{3}$

$y = \dfrac{5}{3}x - \dfrac{13}{3}$ (slope-int. form)

37. a. $y = \frac{3}{4}x + 2$ has a positive slope and positive y-intercept, so the answer is II.

b. $y = -\frac{3}{4}x + 2$ has a negative slope and a positive y-intercept, so the answer is III.

c. $y = \frac{3}{4}x - 2$ has a positive slope and a negative y-intercept, so the answer is I.

d. $y = -\frac{3}{4}x - 2$ has a negative slope and a negative y-intercept, so the answer is IV.

39. $m = 2, \quad (6, -1)$

41. $m = -\frac{4}{3}, \quad (-5, 3)$

(Multiply both sides by $(x + 5)$.)

43. a. $y - (-3.5) = -0.25(x - (-6.4))$ or $y + 3.5 = -0.25(x + 6.4)$

b. Solve the equation for y:
$y + 3.5 = -0.25x - 1.6$
$y = -0.25x - 5.1$ (slope-int. form)

c. Refer to the graph at the back of the textbook.

45. a. $y - (-250) = 2.4(x - 80)$ or $y + 250 = 2.4(x - 80)$

b. Solve the equation for y:
$y + 250 = 2.4x - 192$
$y = 2.4x - 442$ (slope-int. form)

c. Refer to the graph at the back of the textbook.

47. a. $m = 4, b = 40$

b. $y = 4x + 40$

49. a. $m = -80, b = -2000$

b. $P = -80t - 2000$

51. a. $m = \frac{1}{4}, \quad b = 0$

b. $V = \frac{1}{4}d$

53. $y = 2.5x + 6.25$

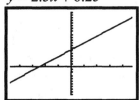

Xscl = 1, Yscl = 1

a. Using points (0, 6.25) and (1, 8.75),
$$m = \frac{y_2 - y_1}{x_2 - x_1} = \frac{8.75 - 6.25}{1 - 0} = 2.5$$

(You should get a slope of 2.5 no matter which two points you use on the line.)

b. y-intercept: (0, 6.25)

55. $y = -8.4x + 63$

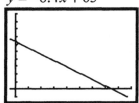

Xscl = 1, Yscl = 10

a. Using points (4, 29.4) and (3, 37.8),
$$m = \frac{y_2 - y_1}{x_2 - x_1} = \frac{37.8 - 29.4}{3 - 4} = -8.4$$

(You should get a slope of -8.4 no matter which two points you use on the line.)

b. y-intercept: (0, 63)

Homework 1.5

1. *Note: Answers may vary slightly in this problem due to the nature of the estimations.*

a. There is a point (25, 12) on the graph, so the 25-year-old drill sergeant would have a time of 12 seconds.

b. There is a point on the graph at about (39, 12.6). Hence the drill sergeant whose time was 12.6 seconds was 39 years old.

c. Draw a line such that approximately half of the points lie above the line and approximately half of the points lie below the line. Refer to the graph in the back of your textbook.

d. Using the line drawn in part (c), when the age of the drill sergeant is 28 years, the hundred-meter time is approximately 11.6 seconds.

e. For example, choose the points (28, 11.6) and (50, 13.6). Then

$$m = \frac{13.6 - 11.6}{50 - 28} = \frac{2}{22} \approx 0.09$$

$$y - y_1 = m(x - x_1)$$

$$y - 11.6 = 0.09(x - 28)$$

$$y - 11.6 = 0.09x - 2.52$$

$$y = 0.09x + 9.1$$

f. When $x = 40$,
$y = 0.09(40) + 9.1 = 12.7$.
According to the equation from part (e), a 40 year-old drill sergeant would have a hundred-meter time of approximately 12.7 seconds. This time seems reasonable when compared to the actual data points on the graph. (For example, we know that a 39 year-old drill sergeant would have a time of about 12.6 according to our observations in part (b).)

When $x = 12$,
$y = 0.09(12) + 9.1 = 10.18$.
According to the equation in part (e), a 12 year-old drill sergeant would have a time of 10.18 seconds. This answer does not seem reasonable. Not only would a 12 year-old be too young to be a drill sergeant, a 12-year old could probably not run faster than, say, an 18 year-old.

3. *Note: Answers may vary slightly in this problem due to the nature of the estimations.*

  a. Draw a line such that approximately half of the points lie above the line and approximately half of the points lie below the line. Refer to the graph in the back of your textbook.

  b. According to the line drawn in part (a), a 65-inch-tall runner has a weight of 129 pounds and a 71-inch-tall runner has a weight of 145 pounds.

  c. Use the points (65, 129) and (71, 145).

$$m = \frac{145 - 129}{71 - 65} = \frac{16}{6} = \frac{8}{3} \approx 2.67.$$

$$y - y_1 = m(x - x_1)$$

$$y - 129 = \frac{8}{3}(x - 65)$$

$$y - 129 = \frac{8}{3}x - \frac{520}{3}$$

$$y = \frac{8}{3}x - \frac{133}{3}$$

or approximately,
$$y = 2.67x - 44.33.$$

  d. When $x = 68$,
$y = 2.67(68) - 44.33 = 137.23$.
A runner who is 68 inches tall would have a weight of 137.23 pounds.

5. a. $y = 2.84x - 55.74$

  b. When $x = 68$,
$y = 2.84(68) - 55.74 = 137.38$.
According to the regression line, a runner who is 68 inches tall has a weight of 137.38 pounds, which is very close to the estimation made in part (d) of problem 3.

7. *Note: Answers may vary slightly in this problem due to the nature of the estimations.*

  a. Choose two points on the line and calculate the slope. For example, choose (0, 0) and (250, 85). Then

$$m = \frac{85}{250} = \frac{17}{50} = 0.34.$$

The rate of growth of a bracken colony is about 0.34 meters per year.

  b. The vertical intercept is (0, 0) so the equation is $y = 0.34x$.

  c. Let $y = 450$ meters. Then
$$450 = 0.34x$$
$$1324 \approx x$$

The bracken colonies over 450 meters in diameter are over 1324 years old.

9. a. Refer to the graph in the back of the textbook.

  b. Using the calculator to find the regression line, we have
$y = 0.69 + 0.16x$. (If you used your line of best fit to find the equation, your answer may differ slightly.)

  c. The problem gives us the model
$m = m_0 + kx$, where k is the extinction coefficient and m_0 is the apparent magnitude. Hence for our data the extinction coefficient is $k = 0.16$ and the apparent magnitude is $m_0 = 0.69$.

Homework 1.5

11. *Note: Answers may vary slightly in this problem due to the nature of the estimations.*

a. Refer to the graph in the back of the textbook. Student E made a mistake in the experiment since for a loss in mass of 88 mg, according to a line passing through the other points, the volume should have been about 68 cm^3 instead of 76 cm^3.

b. Refer to the graph in the back of the textbook.

c. The line of best fit passes through points (0, 0) and (80, 60). The slope is $m = \dfrac{60}{80} = \dfrac{3}{4} = 0.75.$

Since the y-intercept is (0, 0), the equation is $y = 0.75x$.

d. Let $y = 1000$.

$1000 = 0.75x$

$1333\dfrac{1}{3} = x$

The mass of 1000 cm^3 of the gas is about 1333 mg.

e. The density of the unknown gas is

$\dfrac{1333 \text{ mg}}{1000 \text{ cm}^3} = \dfrac{1333 \text{ mg}}{1000 \text{ milliliters}}$

$= \dfrac{1333 \text{ mg}}{1 \text{ liter}}$

$= 1333 \text{ mg / liter}$

Since oxygen has a density of 1330 mg/liter, oxygen is the most likely gas.

13. a.

x	50	125
C	$9000	$15,000

b. $m = \dfrac{15,000 - 9000}{125 - 50} = \dfrac{6000}{75} = 80$

$C - 9000 = 80(x - 50)$

$C - 9000 = 80x - 4000$

$C = 80x + 5000$

c. Plot the points (50, 9000) and (125, 15,000). Draw the line through the two points. Refer to the graph in the back of the textbook.

d. $m = 80$ dollars/bike is the cost of making each bike.

15. a.

g	12	5
d	312	130

b. $m = \dfrac{312 - 130}{12 - 5} = \dfrac{182}{7} = 26$

$d - 130 = 26(g - 5)$

$d - 130 = 26g - 130$

$d = 26g$

c. Plot the points (12, 312) and (5,130). Draw a line through the points. Refer to the graph in the back of the textbook.

d. $m = 26$ miles per gallon is the ideal gas mileage of Andrea's Porsche.

17. a.

C	15	−5
F	59	23

b. $m = \dfrac{59 - 23}{15 - (-5)} = \dfrac{36}{20} = \dfrac{9}{5}$

$F - 59 = \dfrac{9}{5}(C - 15)$

$F - 59 = \dfrac{9}{5}C - 27$

$F = \dfrac{9}{5}C + 32$

c. Plot the points (15, 59) and (−5, 23). Draw a line through the points. Refer to the graph in the back of the textbook.

d. $m = \dfrac{9}{5}$ degrees Fahrenheit per degree Celsius is the amount of degree Fahrenheit increase for every degree increase in Celsius.

19. a. Let x equal the number of hours since midnight and let y equal the temperature. Form the ordered pairs (1, 81) and (5, 73).

$m = \dfrac{73 - 81}{5 - 1} = \dfrac{-8}{4} = -2$

$y - 81 = -2(x - 1)$

$y - 81 = -2x + 2$

$y = -2x + 83$

At 4 a.m., the temperature is $y = -2(4) + 83 = 75$ degrees.

b. The slope −2 gives the rate of change of the temperature in degrees per hour.

21. a. Let x equal the number of seconds after the car starts and let y equal the speed of the car. Form the ordered pairs (0, 0) and (6, 60).

$m = \dfrac{60}{6} = 10$ and $y = 10x$. The car's speed 2 seconds after the car starts accelerating is $y = 10(2) = 20$ miles per hour.

b. $m = 10$ is the acceleration of the car in miles per hour per second. That is, $m = 10$ is the rate at which the speed increases.

23. Let x be the time in minutes since the engine starts. Let y be the temperature of the engine. Form the ordered pairs (0, 9) and (7, 51). Then $m = \dfrac{51 - 9}{7 - 0} = \dfrac{42}{7} = 6$.

The y-intercept is (0, 9) so $y = 6x + 9$. At two minutes, the engine temperature is $y = 6(2) + 9 = 21$ degrees. This is a reasonable answer. At two hours, the engine temperature, according to the model, is $6(120) + 9 = 729$ degrees which is much too hot to be a reasonable answer.

25. Let x equal Ben's age in years and let y equal Ben's weight in pounds. Form the ordered pairs (0, 8) and (1, 20). Then $m = \dfrac{20 - 8}{1 - 0} = 12$ and since the y-intercept is (0, 8), the equation is $y = 12x + 8$. According to the model, at age ten Ben will weigh $y = 12(10) + 8 = 128$ pounds. This is very high for a ten year old, and therefore a child's weight does not increase at a constant rate throughout childhood.

27. a. The least squares regression line given by the calculator is $y = -0.54x + 58.7$. Refer to the graph in the back of the textbook.

b. The probability of predators finding a nest if its nearest neighbor is 50 meters away is $y = -0.54(50) + 58.7 = 31.7\%$.

c. Let $y = 10$.
$$10 = -0.54x + 58.7$$
$$-48.7 = -0.54x$$
$$90.2 \approx x$$
Its nearest neighbor is about 90 meters away.

d. According to the model, if its nearest neighbor is 120 meters away, the probability of predators finding a nest is $y = -0.54(120) + 58.7 = -6.1\%$. This is not reasonable since a probability cannot be negative.

Homework 1.6

1. a. $y = -3$
Refer to the graph in the back of the textbook.

b. $m = 0$

3. a. $2x = 8$
$x = 4$
Refer to the graph in the back of the textbook.

b. Undefined

5. a. $x = 0$
Refer to the graph in the back of the textbook.

b. Undefined

7. $x = -5$

9. $y = 0$

11. $y = 9$

13. a. l_1 negative, l_2 negative, l_3 positive, l_4 zero

b. l_1, l_2, l_4, l_3

15. a.

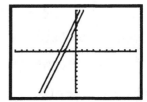

The lines appear parallel.

b. The slope of the line $y = 3x + 8$ is 3 and the slope of the line $y = 3.1x + 6$ is 3.1. Since the slopes are different, the lines are not parallel.

c. When $x = 20$,
$y = 3x + 8 = 3(20) + 8 = 68$ and
$y = 3.1x + 6 = 3.1(20) + 6 = 68$.
Therefore, the point (20, 68) is on both lines which means that the lines intersect at this point. This confirms that the lines are not parallel.

17. The line $y = 0.75x + 2$ has slope $m = 0.75 = \frac{3}{4}$. Therefore, a line parallel to it must have a slope of $m = \frac{3}{4}$ and a line perpendicular to it must have a slope of $m = -\frac{4}{3}$.

a. parallel since $m = \frac{3}{4}$

b. neither since $m = \frac{8}{6} = \frac{4}{3}$

c. perpendicular since $m = \frac{-20}{15} = -\frac{4}{3}$

d. neither since $m = \frac{-39}{52} = -\frac{3}{4}$

e. neither since $m = \frac{4}{3}$

f. perpendicular since $m = \frac{-16}{12} = -\frac{4}{3}$

g. parallel since $m = \frac{36}{48} = \frac{3}{4}$

h. parallel since $m = \frac{9}{12} = \frac{3}{4}$

19. a. $y = \frac{3}{5}x - 7$, so $m_1 = \frac{3}{5}$.

$$3x - 5y = 2$$
$$-5y = -3x + 2$$
$$y = \frac{3}{5}x - \frac{2}{5},$$

so $m_2 = \frac{3}{5}$.
given lines are parallel.

b. $y = 4x + 3$, so $m_1 = 4$.
$y = \frac{1}{4}x - 3$, so $m_2 = \frac{1}{4}$.
The given lines are neither parallel nor perpendicular.

c. $6x + 2y = 1$
$$2y = -6x + 1$$
$$y = -3x + \frac{1}{2},$$
so $m_1 = -3$.

$$x = 1 - 3y$$
$$3y = -x + 1$$
$$y = -\frac{1}{3}x + \frac{1}{3},$$
so $m_2 = -\frac{1}{3}$.
The given lines are neither parallel nor perpendicular.

d. $2y = 5$ so $y = \frac{5}{2}$ and $m_1 = 0$.

$5y = -2$ so $y = -\frac{2}{5}$ and $m_2 = 0$.
Notice that these are both horizontal lines. The given lines are parallel.

21. a.

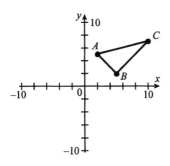

b. Slope $\overline{AB}$: $\dfrac{2-5}{5-2} = \dfrac{-3}{3} = -1$,

Slope $\overline{BC}$: $\dfrac{7-2}{10-5} = \dfrac{5}{5} = 1$,

Slope $\overline{AC}$: $\dfrac{7-5}{10-2} = \dfrac{2}{8} = \dfrac{1}{4}$.

$\overline{AB} \perp \overline{BC}$ since their slopes are negative reciprocals, so the triangle is a right triangle.

23. a.

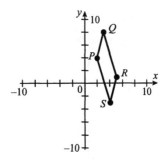

b. Slope $\overline{PQ}$: $\dfrac{8-4}{3-2} = \dfrac{4}{1} = 4$,

Slope $\overline{QR}$: $\dfrac{1-8}{5-3} = \dfrac{-7}{2} = -\dfrac{7}{2}$,

Slope $\overline{RS}$: $\dfrac{-3-1}{4-5} = \dfrac{-4}{-1} = 4$,

Slope $\overline{SP}$: $\dfrac{4-(-3)}{2-4} = \dfrac{7}{-2} = -\dfrac{7}{2}$.

$\overline{PQ} \parallel \overline{RS}$ since their slopes are the same and $\overline{QR} \parallel \overline{SP}$ since their slopes are the same, so the quadrilateral is a parallelogram.

25. The line passes through points A, B, and C if the slope of $\overline{AB}$ is equal to the slope of $\overline{AC}$.

Slope $\overline{AB}$: $\dfrac{\frac{1}{2}-(-3)}{3-0} = \dfrac{\frac{7}{2}}{3} = \dfrac{7}{6}$

Slope $\overline{AC}$:

$\dfrac{-10-(-3)}{-6-0} = \dfrac{-7}{-6} = \dfrac{7}{6}$

Hence the line through points A and B also passes through the point C.

27. a. $x - 2y = 5$
$-2y = -x + 5$
$y = \dfrac{1}{2}x - \dfrac{5}{2}$

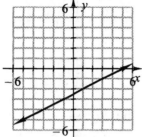

b. Any line parallel to $x - 2y = 5$
has slope $m = \dfrac{1}{2}$

c.

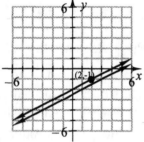

d. $y - (-1) = \dfrac{1}{2}(x - 2)$
$y + 1 = \dfrac{1}{2}x - 1$
$y = \dfrac{1}{2}x - 2$ or $x - 2y = 4$

29. a. $2y - 3x = 5$
$2y = 3x + 5$
$y = \dfrac{3}{2}x + \dfrac{5}{2}$

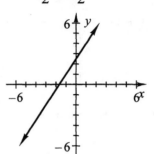

b. Any line perpendicular to
$2y - 3x = 5$ has slope $m = -\dfrac{2}{3}$.

c.

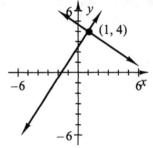

d. $y - 4 = -\dfrac{2}{3}(x - 1)$
$y - 4 = -\dfrac{2}{3}x + \dfrac{2}{3}$
$y = -\dfrac{2}{3}x + \dfrac{14}{3}$ or $2x + 3y = 14$

31. a. Slope $\overline{AB}$: $\dfrac{-4-2}{-2-(-5)} = \dfrac{-6}{3} = -2$

$y - 2 = -2(x - (-5))$

$y - 2 = -2(x + 5)$

$y - 2 = -2x - 10$

$y = -2x - 8$

b. Slope

$\overline{BC} = -\dfrac{1}{\text{slope } \overline{AB}} = -\dfrac{1}{-2} = \dfrac{1}{2}$

$y - (-4) = \dfrac{1}{2}(x - (-2))$

$y + 4 = \dfrac{1}{2}(x + 2)$

$y + 4 = \dfrac{1}{2}x + 1$

$y = \dfrac{1}{2}x - 3$

35. Slope $\overline{CP}$: $\dfrac{4-6}{2-(-1)} = \dfrac{-2}{3} = -\dfrac{2}{3}$

Slope of the tangent line:

$-\dfrac{1}{\text{slope } \overline{CP}} = -\dfrac{1}{-\frac{2}{3}} = \dfrac{3}{2}$

$y - 6 = \dfrac{3}{2}(x - (-1))$

$y - 6 = \dfrac{3}{2}(x + 1)$

$y - 6 = \dfrac{3}{2}x + \dfrac{3}{2}$

$y = \dfrac{3}{2}x + \dfrac{15}{2}$

33. a.

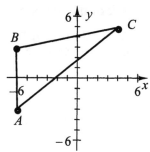

b. Slope $\overline{AC} = \dfrac{5 - (-3)}{4 - (-6)} = \dfrac{8}{10} = \dfrac{4}{5}$

c. The altitude from point B to side $\overline{AC}$ is perpendicular to $\overline{AC}$ and therefore has slope $-\dfrac{5}{4}$.

d. $y - 3 = -\dfrac{5}{4}(x + 6)$

$y - 3 = -\dfrac{5}{4}x - \dfrac{15}{2}$

$y = -\dfrac{5}{4}x - \dfrac{9}{2}$

Chapter 1 Review

1. a.

n	100	500	800	1200	1500
C	4000	12,000	18,000	26,000	32,000

b. $C = 20n + 2000$

c. Refer to the graph in the back of the textbook.

d. Substitute $n = 1000$ into the equation.
$$C = 20(1000) + 2000$$
$$= 20,000 + 2000$$
$$= 22,000$$
The cost is $22,000.
This is the point (1000, 22,000) on the graph.

e. Substitute $C = 10,000$ into the equation.
$$10,000 = 20n + 2000$$
$$8000 = 20n$$
$$400 = n$$
The number of calculators produced is 400. This is the point (400, 10,000) on the graph.

3. a.

t	5	10	15	20	25
R	1560	1460	1360	1260	1160

b. $R = 1660 - 20t$

c. Substitute $t = 0$ into the equation.
$R = 1660 - 20(0) = 1660$
The R-intercept is 1660.
Substitute $R = 0$ into the equation.
$$0 = 1660 - 20t$$
$$20t = 1660$$
$$t = 83$$
The t-intercept is 83.
Plot the points (0, 1660) and (83, 0), then draw a line through the points. Refer to the graph in the back of the textbook.

d. When $t = 0$ (1976) R was 1660.
R will be 0 when $t = 83$ (in 2059).

30

5. a. $5A + 2C = 1000$

b. Substitute $A = 0$ into the equation.
$$5(0) + 2C = 1000$$
$$2C = 1000$$
$$C = 500$$
The C-intercept is 500.

Substitute $C = 0$ into the equation.
$$5A + 2(0) = 1000$$
$$5A = 1000$$
$$A = 200$$
The A-intercept is 200.

For example, let C be the first coordinate. Plot the points $(0, 200)$ and $(500, 0)$, then draw a line through the points. Refer to the graph in the back of the textbook.

c. Substitute $A = 120$ into the equation.
$$5(120) + 2C = 1000$$
$$600 + 2C = 1000$$
$$2C = 400$$
$$C = 200$$
He must sell 200 children's tickets.

d. The A-intercept, 200, is the number of adult tickets that must be sold if no children's tickets are sold.

The C-intercept, 500, is the number of children's tickets that must be sold if no adult tickets are sold.

7. For example, plot the points $(0, -4)$ and $(3, 0)$, then draw a line through the points. Refer to the graph in the back of the textbook.

9. For example, plot the points $(0, 500)$ and $(-400, 0)$, then draw a line through the points. Refer to the graph in the back of the textbook.

11. For example, plot the points $(0, 0)$ and $(4, 3)$, then draw a line through the points. Refer to the graph in the back of the textbook.

13. $4x = -12$
$$x = -3$$
Graph the vertical line through $(-3, 0)$. Refer to the graph in the back of the textbook.

15. For the volleyball, the ratio of feet per second is $\dfrac{6}{0.04} = 150$ ft/sec. For the baseball, the ratio of feet per second is $\dfrac{66}{0.48} = 137.5$ ft/sec. The spiked volleyball travels faster.

17. For Stone Canyon Drive, the ratio of vertical gain to horizontal distance is
$$\frac{840 \text{ feet}}{1500 \text{ feet}} = 0.56,$$
and for Highway 33, the ratio is
$$\frac{1150 \text{ feet}}{2000 \text{ feet}} = 0.575.$$
Highway 33 is steeper.

19. a. $m = \dfrac{750 - 800}{10 - 0} = \dfrac{-50}{10} = -5$
B-intercept $= 800$
$B = 800 - 5t$

b. Refer to the graph in the back of the textbook.

c. $m = -5$ thousands of barrels per minute gives the rate at which the oil is leaking.

21. a. $m = \dfrac{1500-1000}{10,000-5000} = \dfrac{500}{5000} = 0.1$

$F - 1000 = 0.1(C - 5000)$

$F - 1000 = 0.1C - 500$

$F = 0.1C + 500$

b. Refer to the graph in the back of the textbook.

c. $m = 0.1$ gives the percent of her fee with respect to the cost. The decorator charges 10% of the cost of the job (plus a flat $500 fee).

23. $m = \dfrac{-2-4}{3-(-1)} = \dfrac{-6}{4} = -\dfrac{3}{2}$

25. $m = \dfrac{4.8-1.4}{-2.1-6.2} = \dfrac{3.4}{-8.3} \approx -0.4$

27. a. Let m_1 be the slope connecting the points $(1, 5)$ and $(2, 5/2)$ and let m_2 be the slope connecting the points $(2, 5/2)$ and $(3, 5/3)$. Then

$$m_1 = \dfrac{\frac{5}{2}-5}{2-1} = -\dfrac{5}{2}$$

$$m_2 = \dfrac{\frac{5}{3}-\frac{5}{2}}{3-2} = -\dfrac{5}{6}$$

The slopes are different, so the points are not linear.

b. Let m_1 be the slope connecting the points $(10, 6.2)$ and $(20, 9.7)$ and let m_2 be the slope connecting the points $(20, 9.7)$ and $(30, 12.6)$. Then

$$m_1 = \dfrac{9.7-6.2}{20-10} = 0.35$$

$$m_2 = \dfrac{12.6-9.7}{30-20} = 0.29$$

The slopes are different, so the points are not linear.

29. First find the linear equation.

$$m = \dfrac{-4.8-(-3)}{-5-(-2)} = \dfrac{-1.8}{-3} = 0.6$$

$V - (-4.8) = 0.6(d - (-5))$

$V + 4.8 = 0.6d + 3$

$V = 0.6d - 1.8$

Substitute $V = -1.2$ into the equation.

$-1.2 = 0.6d - 1.8$

$0.6 = 0.6d$

$1 = d$

Substitute $d = 10$ into the equation.

$V = 0.6(10) - 1.8 = 6 - 1.8 = 4.2$

The missing values are $d = 1$ and $V = 4.2$.

31. Let x be the distance from the end to the base.

$\dfrac{20}{x} = 0.25$

$20 = 0.25x$

$80 = x$

The end should be 80 ft from the base.

33. $2x - 4y = 5$

$-4y = -2x + 5$

$y = \dfrac{1}{2}x - \dfrac{5}{4}$

$m = \dfrac{1}{2}; \ b = -\dfrac{5}{4}$

35. $8.4x + 2.1y = 6.3$

$2.1y = -8.4x + 6.3$

$y = -4x + 3$

$m = -4; \ b = 3$

37. a. Plot the point (–4, 6). Move 2 units in the *y*-direction and –3 units in the *x*-direction to arrive at (–7, 8). Draw a line through the points. Refer to the graph in the back of the textbook.

b.
$$y - 6 = -\frac{2}{3}(x - (-4))$$
$$y - 6 = -\frac{2}{3}(x + 4)$$
$$y - 6 = -\frac{2}{3}x - \frac{8}{3}$$
$$y = -\frac{2}{3}x + \frac{10}{3}$$

39. a. Since the lapse rate is $3.6°F$ for every 1000 feet, the lapse rate is $0.0036°F$ for every single foot. Hence the equation is $T = 62 - 0.0036h$.

b. Set $h = 30,000$. Assuming the ground temperature is $62°F$ (from (a)), then the temperature outside the airplane is
$$T = 62 - 0.0036(30,000)$$
$$= -46°F$$
This is $0.0036(30,000) = 108°F$ cooler than the ground temperature.

c. At the top of the troposphere, $h = 7$ miles = 36,960 feet (1 mi = 5280 ft). The temperature there is
$$T = 62 - 0.0036(36,960) \approx -71°F$$

41.
$$m = \frac{4 - (-5)}{-2 - 3} = \frac{9}{-5} = -\frac{9}{5}$$
$$y - (-5) = -\frac{9}{5}(x - 3)$$
$$y + 5 = -\frac{9}{5}x + \frac{27}{5}$$
$$y = -\frac{9}{5}x + \frac{2}{5}$$

43. a.

t	0	15
P	4800	6780

b.
$$m = \frac{6780 - 4800}{15 - 0} = \frac{1980}{15} = 132$$
$$P - 4800 = 132(t - 0)$$
$$P = 4800 + 132t$$

c. $m = 132$ people per year gives the rate of population growth.

45. a.
$$m = \frac{1 - 3}{1 - 0} = \frac{-2}{1} = -2;$$
$$b = 3$$

b. $y = -2x + 3$

47.
$$m = \frac{3 - 0}{0 - (-5)} = \frac{3}{5}$$

49. a. $y = 2 + \frac{3}{2}(x - 4)$ written in point slope form is $y - 2 = \frac{3}{2}(x - 4)$ so $m = \frac{3}{2}$.

b. $y = 2 + \frac{3}{2}(4 - 4) = 2$. The point is (4, 2). There can only be one such point since when we substitute 4 for *x*, we have 2 as our only answer.

c.
$$(x_2, y_2) = (x_1 + \Delta x, y_1 + \Delta y)$$
$$= (4 + 2, 2 + 3)$$
$$= (6, 5)$$

51. $y = \frac{1}{2}x + 3$

so $m_1 = \frac{1}{2}$.

$x - 2y = 8$

$-2y = -x + 8$

$y = \frac{1}{2}x - 4$

so $m_2 = \frac{1}{2}$.

Thus, $m_1 = m_2$ and the lines are parallel.

53. $2x + 3y = 6$

$3y = -2x + 6$

$y = -\frac{2}{3}x + 2$

Using $m = -\frac{2}{3}$:

$y - 4 = -\frac{2}{3}(x - 1)$

$y - 4 = -\frac{2}{3}x + \frac{2}{3}$

$y = -\frac{2}{3}x + \frac{14}{3}$

55. slope of $\overline{AB}$: $\frac{-4 - 2}{7 - 3} = \frac{-6}{4} = \frac{-3}{2}$

slope of $\overline{BC}$: $-\frac{1}{\frac{-3}{2}} = \frac{2}{3}$

$y - (-4) = \frac{2}{3}(x - 7)$

$y + 4 = \frac{2}{3}x - \frac{14}{3}$

$y = \frac{2}{3}x - \frac{26}{3}$

57. Let N equal the number of incidents of O-ring failure due to temperature and let T equal the temperature in degrees Fahrenheit. Then the ordered pairs of the form (T, N) are $(70, 1)$ and $(54, 3)$.

$m = \frac{3 - 1}{54 - 70} = \frac{2}{-16} = -\frac{1}{8}$

$N - N_1 = m(T - T_1)$

$N - 1 = -\frac{1}{8}(T - 70)$

$N - 1 = -\frac{1}{8}T + \frac{35}{4}$

$N = -\frac{1}{8}T + \frac{39}{4}$

Let $T = 30$.

$N = -\frac{1}{8}(30) + \frac{39}{4} = 6$. You would expect about 6 O-ring failures when the temperature is $30° F$.

59. *Note: Answers may vary slightly in parts (a) through (d) due to the nature of the estimations.*

a. If we sketch a line so that approximately half of the data points are above the line and approximately half are below the line, the line appears to go through the point (40, 45). Hence an Archaeopteryx whose femur is 40 centimeters has a humerus of length approximately 45 centimeters.

b. Using the line sketched for part (a), we see that when the femur length is 75 centimeters, the corresponding humerus length is approximately 87 centimeters.

c. Using the points (40, 45) and (75, 87),
$$m = \frac{87 - 45}{75 - 40} = \frac{42}{35} = \frac{6}{5} = 1.2.$$
$$y - y_1 = m(x - x_1)$$
$$y - 45 = 1.2(x - 40)$$
$$y - 45 = 1.2x - 48$$
$$y = 1.2x - 3$$

d. Let $x = 60$. Then $y = 1.2(60) - 3 = 69$. According to the equation in part (c), the Archaeopteryx has a humerus of length 69 cm.

e. The calculator gives the regression line $y = 1.197x - 3.660$ (with values rounded to three decimal places). If $x = 60$ cm, then $y = 1.197(60) - 3.660 = 68.16$ cm, which is less than a centimeter different from our answer in part (d).

Chapter Two: Applications of Linear Models

Homework 2.1

1. It appears from the graph that the solution is (3, 0).
   Algebraic verification:
   $2.3(3) - 3.7(0) = 6.9 - 0 = 6.9$
   $1.1(3) + 3.7(0) = 3.3 + 0 = 3.3$

3. It appears from the graph that the solution is (50, 70).
   Algebraic verification:
   $35(50) - 17(70) = 1750 - 1190$
   $$= 560$$
   $24(50) + 15(70) = 1200 + 1050$
   $$= 2250$$

5. **a.** From 1990 to 1998, the rate at which the median age of women increased was
   $$\frac{36.4 - 34.0}{1998 - 1990} = \frac{2.4}{8} = 0.3 \text{ years}$$
   per year and the rate at which the mean age of women increased was $\frac{37.5 - 36.6}{1998 - 1990} = \frac{0.9}{8} = 0.1125$ years per year. Hence the median age is growing more rapidly than the mean age.

   b. Refer to the graph in the back of the textbook.

   c. The median age for women is increasing by 0.3 years per year.

   d. Refer to the graph in the back of the textbook.

   e. Extend the two lines on the graph until they meet. The median age will be the same as the mean age in about 2004.

   f. There are more young women than old women, with some women much older than the mean.

7. It appears from the graph that the solution is (–2, 3).

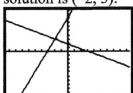

Algebraic verification:
$2.6(-2) + 8.2 = -5.2 + 8.2 = 3$
$1.8 - 0.6(-2) = 1.8 - (-1.2) = 3$

9. To graph the second equation on the calculator, first solve for y:
$$-2.8x + 3.7y = 5.5$$
$$3.7y = 2.8x + 5.5$$
$$y = \frac{2.8}{3.7}x + \frac{5.5}{3.7}$$
$$y = \frac{28}{37}x + \frac{55}{37}$$
It appears from the graph that the solution is (2, 3).

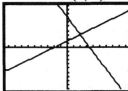

Algebraic verification:
$7.2 - 2.1(2) = 7.2 - 4.2 = 3$
$-2.8(2) + 3.7(3) = -5.6 + 11.1 = 5.5$

11. The graph of $2x = y + 4$ has y-intercept –4 and x-intercept 2. The graph of $8x - 4y = 8$ has y-intercept –2 and x-intercept 1.

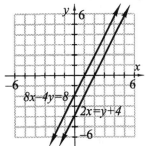

If we solve each equation for y, we have $y = -2x + 4$ and $y = -2x - 2$. The slopes are both –2, and so the lines are parallel. The system is inconsistent.

13. The graph of $w - 3z = 6$ has z-intercept –2 and w-intercept 6. The graph of $2w + z = 8$ has z-intercept 8 and w-intercept 4.

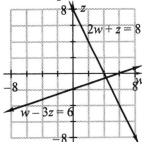

The lines intersect and they are different lines so the system is consistent and independent.

15. The graph of $2L - 5W = 6$ has
W-intercept $-\frac{6}{5}$ and L-intercept 3.

The graph of $\frac{15W}{2} + 9 = 3L$ has

W-intercept $-\frac{6}{5}$ and L-intercept 3.

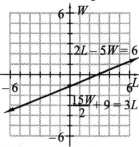

The equations represent the same line and so the system is dependent.

17. a. Let D equal the monthly bill for Dash Phone Company. Then
$D = 0.09x + 10$.

b. Let F equal the monthly bill for Friendly Phone Company. Then
$F = 0.05x + 15$.

c.

x	0	30	60	90	120	150
D	10.00	12.70	15.40	18.10	20.80	23.50
F	15.00	16.50	18.00	19.50	21.00	22.50

According to the table, Dash's bill becomes larger than Friendly's somewhere between the 120th and 150th minute.

d. Xmin = 0, Xmax = 200, Ymin = 0, Ymax = 30

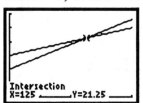

e. Using the *intersect* feature of the calculator, we find that the graphs intersect at the point (125, 21.25).
To verify algebraically:
$0.09(125) + 10 = \$21.25$
$0.05(125) + 15 = \$21.25$

Hence the bills would be equal for 125 minutes of talking time.

19. a. $y = 50x$

  b. $y = 2100 - 20x$

  c. The graph of $y = 50x$ passes through the points $(0, 0)$ and $(1, 50)$.
  The graph of $y = 2100 - 20x$ has y-intercept $(0, 2100)$ and x-intercept $(105, 0)$.

  Xmin = 0, Xmax = 120
  Ymin = 0, Ymax = 2500

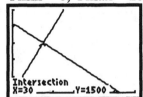

  d. The equilibrium price occurs at the intersection point $(30, 1500)$ in the above graph.

  To verify:
  $y = 50(30) = 1500$
  $y = 2100 - 20(30) = 1500$

  Yasuo should sell the wheat at 30 cents per bushel and produce 1500 bushels.

21. a. The cost is the initial cost ($200) plus the cost per pendant ($4), so the cost of production is $C = 200 + 4x$.

  b. The revenue is $12 per pendant sold, so $R = 12x$.

  c.

x	5	10	15	20	25	30
C	220	240	260	280	300	320
R	60	120	180	240	300	360

  d.

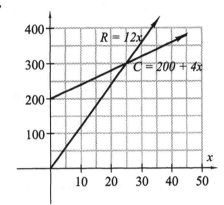

  e. From the graph, it appears that the solution is $x = 25$.

  To verify algebraically:
  $R = 12(25) = 300$
  $C = 200 + 4(25) = 300$

  Hence the Aquarius company needs to sell 25 pendants to break even.

23. a.

	Number of tickets	Cost per ticket	Revenue
Adults	x	7.50	$7.50x$
Students	y	4.25	$4.25y$
Total	82		465.50

b. $x + y = 82$

c. $7.50x + 4.25y = 465.50$

d. To graph the equations, solve each for y:

$x + y = 82$

$y = -x + 82$

and

$7.50x + 4.25y = 465.50$

$4.25y = -7.50x + 465.50$

$y = -\dfrac{7.50}{4.25}x + \dfrac{465.50}{4.25}$

$y = -\dfrac{30}{17}x + \dfrac{1862}{17}$

Xmin = 0, Xmax = 100,
Ymin = 0, Ymax = 125

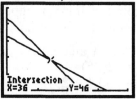

From the graph, we find the solution (36, 46). To verify algebraically:

$36 + 46 = 72$

and

$7.50(36) + 4.25(46) = 465.50$

There were 36 adults and 46 students in attendance.

25. Graph $Y_1 = \dfrac{-55.2 - 38x}{2.3}$ and $Y_2 = 15x + 121$ using Xmin = −10, Xmax = 1, Ymin = −10, Ymax = 100. Then use the intersection feature to find the solution (−4.6, 52).

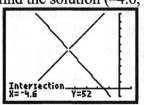

27. Graph $Y_1 = \dfrac{707 - 64x}{58}$ and $Y_2 = \dfrac{496 - 82x}{-21}$ using Xmin = −1, Xmax = 10, Ymin = −1, Ymax = 10. Then use the intersection feature to find the solution (7.15, 4.3).

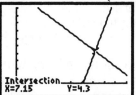

Homework 2.2

1. Given $3m + n = 7$ and $2m = 5n - 1$, solve the first equation for n to get $n = 7 - 3m$. Substitute this into the second equation:

   $2m = 5(7 - 3m) - 1$

   $2m = 35 - 15m - 1$

   $17m = 35 - 1$

   $17m = 34$

   $m = 2$

   Substitute this value for m into the first equation:

   $3(2) + n = 7$

   $6 + n = 7$

   $n = 1$

   The solution is $m = 2$, $n = 1$.

3. Given $2u - 3v = -4$ and $5u + 2v = 9$, multiply the first equation by 2 and the second equation by 3:

   $4u - 6v = -8$

   $15u + 6v = 27$

   Add these equations:

   $19u = 19$

   $u = 1$

   Put this value into the first equation:

   $2(1) - 3v = -4$

   $2 - 3v = -4$

   $-3v = -6$

   $v = 2$

   The solution is $u = 1$, $v = 2$.

5. Given $3y = 2x - 8$ and $4y + 11 = 3x$, rewrite the equations in standard form:

   $-2x + 3y = -8$

   $3x - 4y = 11$

   Multiply the first equation by 3 and the second equation by 2:

   $-6x + 9y = -24$

   $6x - 8y = 22$

   Add these equations:

   $y = -2$

   Substitute this value into the first equation:

   $3(-2) = 2x - 8$

   $-6 = 2x - 8$

   $2 = 2x$

   $1 = x$

   The solution is $x = 1$, $y = -2$.

7. Given $\dfrac{2}{3}A - B = 4$ and $A - \dfrac{3}{4}B = 6$, multiply the first equation by -3 and the second equation by 4:

   $-2A + 3B = -12$

   $4A - 3B = 24$

   Add these equations:

   $2A = 12$

   $A = 6$

   Put this value into the first equation:

   $\dfrac{2}{3}(6) - B = 4$

   $4 - B = 4$

   $-B = 0$

   $B = 0$

   The solution is $A = 6$, $B = 0$.

9. Given

$$\frac{M}{4} = \frac{N}{3} - \frac{5}{12} \quad \text{and}$$

$$\frac{N}{5} = \frac{1}{2} - \frac{M}{10}$$

Multiply the first equation by 12 and the second equation by 10:

$$3M = 4N - 5$$

$$2N = 5 - M$$

We can solve the second of these equations for M to get $M = 5 - 2N$. Put this into the first of the new equations:

$$3(5 - 2N) = 4N - 5$$

$$15 - 6N = 4N - 5$$

$$20 - 6N = 4N$$

$$20 = 10N$$

$$2 = N$$

Thus, $M = 5 - 2(2) = 5 - 4 = 1$ and the solution is $M = 1$, $N = 2$.

11. Given

$$\frac{s}{2} = \frac{7}{6} - \frac{t}{3} \quad \text{and}$$

$$\frac{s}{4} = \frac{3}{4} - \frac{t}{4}$$

multiply the first equation by 6 and the second equation by 4:

$$3s = 7 - 2t \quad (1)$$

$$s = 3 - t \quad (2)$$

Put equation (2) into equation (1):

$$3(3 - t) = 7 - 2t$$

$$9 - 3t = 7 - 2t$$

$$2 - 3t = -2t$$

$$2 = t$$

Put this value into (2):

$$s = 3 - 2 = 1.$$

The solution is $s = 1$, $t = 2$.

13. Given

$$2m = n + 1 \quad \text{and}$$

$$8m - 4n = 3$$

write the equations in standard form:

$$2m - n = 1$$

$$8m - 4n = 3$$

Multiply the first equation by -4:

$$-8m + 4n = -4$$

$$8m - 4n = 3$$

Add the equations:

$$0 = -1$$

The system is inconsistent.

15. Given

$$r - 3s = 4 \quad \text{and}$$

$$2r + s = 6$$

multiply the first equation by -2:

$$-2r + 6s = -8$$

$$2r + s = 6$$

Add the equations:

$$7s = -2$$

$$s = -\frac{2}{7}$$

The system is consistent and independent.

17. Given

$$3L + 4W = 18 \quad \text{and}$$

$$\frac{2W}{3} = 3 - \frac{L}{2}$$

multiply the second equation by 6:

$$3L + 4W = 18$$

$$4W = 18 - 3L$$

Write the equations in standard form:

$$3L + 4W = 18$$

$$3L + 4W = 18$$

Multiply the first equation by –1:

$$-3L - 4W = -18$$

$$3L + 4W = 18$$

Add the equations:

$$0 = 0$$

The system is dependent.

19. a. In problem 5 from Section 2.1, the rate of increase of median age was found to be 0.3. So $m = 0.3$. Let x be the number of years since 1990. Hence $x = 0$ corresponds to the year 1990 which means that the y-intercept must be 34.0. The equation is then $y = 0.3x + 34$.

b. Let x be the number of years since 1990. The data points for the mean age do not lie on a straight line. Use the regression feature on your calculator to find the equation that best fits the data: $y = 0.115x + 36.58$.

c. Set the y-values of the two equations equal to one another:

$$0.3x + 34 = 0.115x + 36.58$$

$$0.185x = 2.58$$

$$x \approx 13.95$$

The year in which the mean age and the median age are the same is 13.95 years after 1990, which means at the very end of the year 2003. The year 2004 was a good estimation.

21. a.

	Principal	Interest rate	Interest
Bonds	x	0.10	$0.10x$
Certificate	y	0.08	$0.08y$
Total	2000	----	184

b. $x + y = 2000$

c. $0.1x + 0.08y = 184$

d. Solve $x + y = 2000$ for x so $x = 2000 - y$. Substitute this into the equation from part (c).
$$0.1(2000 - y) + 0.08y = 184$$
$$200 - 0.1y + 0.08y = 184$$
$$200 - 0.02y = 184$$
$$-0.02y = -16$$
$$y = 800$$
Thus Francine has $800 invested at 8% and $2000 - $800 = $1200 invested at 10%.

23. a.

	Pounds	% silver	Amt. silver
First alloy	x	0.45	$0.45x$
Second alloy	y	0.60	$0.60y$
Mixture	40	0.48	$0.48(40)$

b. $x + y = 40$

c. $0.45x + 0.60y = 0.48(40)$ or $0.45x + 0.60y = 19.2$

d. Solve $x + y = 40$ for y so that $y = 40 - x$. Substitute into (c):
$$0.45x + 0.60(40 - x) = 19.2$$
$$0.45x + 24 - 0.60x = 19.2$$
$$-0.15x + 24 = 19.2$$
$$-0.15x = -4.8$$
$$x = 32$$
Paul needs 32 pounds of the 45% alloy.

25. Let $x =$ the number points a true-false question were worth and $y =$ the number of points the fill-in questions were worth. Then
$13x + 9y = 71$ and
$9x + 13y = 83$.

Multiply the first equation by -9 and the second equation by 13:
$$-117x - 81y = -639$$
$$117x + 169y = 1079$$

Add the two equations:
$$88y = 440$$
$$y = 5$$

Substitute this value into the first equation:
$$13x + 9(5) = 71$$
$$13x + 45 = 71$$
$$13x = 26$$
$$x = 2$$

The true-false questions are worth 2 points each, the fill-in questions are worth 5 points each.

27. a. Let x = the speed of the airplane and let y = the speed of the wind.

	Rate	Time	Distance
Detroit to Denver	$x - y$	4	1120
Denver to Detroit	$x + y$	3.5	1120

b. (rate) × (time) = distance:
$$4(x - y) = 1120$$

c. (rate) × (time) = distance:
$$3.5(x + y) = 1120$$

d. Use the distributive property on the two equations:

$$4x - 4y = 1120 \quad (1)$$

$$3.5x + 3.5y = 1120 \quad (2)$$

Divide (1) by 4 and move the y term to the other side:
$x = 280 + y$.
Put this into (2):
$$3.5(280 + y) + 3.5y = 1120$$

$$980 + 3.5y + 3.5y = 1120$$

$$980 + 7y = 1120$$

$$7y = 140$$

$$y = 20$$

The speed of the wind is 20 mph, while the speed of the airplane is $280 + 20 = 300$ mph.

29. a. Supply equation:
$y = 35x$
Demand equation:
$y = 1700 - 15x$

b. The equilibrium price occurs when
$35x = 1700 - 15x$. Solving this equation:
$$35x = 1700 - 15x$$

$$50x = 1700$$

$$x = 34$$

The equilibrium price is $34 per pair, and Sanaz will produce and sell $35(34) = 1190$ pairs at that price.

31. a. Let x = the amount of oats and y = the amount of wheat flakes in one cup of the new cereal.

	Cups	Calories per cup	Calories
Oat flakes	x	310	$310x$
Wheat flakes	y	290	$290y$
Mixture	1	----	302

b. $x + y = 1$

c. $310x + 290y = 302$

d. Solve $x + y = 1$ for x so that $x = 1 - y$. Put this into the equation from part (c):
$$310(1 - y) + 290y = 302$$

$$310 - 310y + 290y = 302$$

$$310 - 20y = 302$$

$$-20y = -8$$

$$y = 0.4$$

Thus, the new cereal uses 0.4 cups of wheat flakes and $1 - 0.4 = 0.6$ cups of oats.

33. Given

$4.8x - 3.5y = 5.44$ and

$2.7x + 1.3y = 8.29$

first multiply both equations by 10 to better see which variable to eliminate.

$48x - 35y = 54.4$ (1)

$27x + 13y = 82.9$ (2)

Multiply (1) by –9 and (2) by 16:

$-432x + 315y = -489.6$

$432x + 208y = 1326.4$

Add these equations:

$523y = 836.8$

$y = 1.6$

Put this value into the first equation given:

$4.8x - 3.5(1.6) = 5.44$

$4.8x - 5.6 = 5.44$

$4.8x = 11.04$

$x = 2.3$

The solution is (2.3, 1.6).

35. Given

$0.9x = 25.78 + 1.03y$

$0.25x + 0.3y = 85.7$

Put the first equation into standard form by subtracting $1.03y$ from both sides:

$0.9x - 1.03y = 25.78$ (1)

$0.25x + 0.3y = 85.7$ (2)

Multiply (1) by –0.25 and (2) by .9:

$-0.225x + 0.2575y = -6.445$

$0.225x + 0.27y = 77.13$

Add these equations:

$0.5275y = 70.685$

$y = 134$

Put this value into the first equation:

$0.9x = 25.78 + 1.03(134)$

$0.9x = 25.78 + 138.02$

$0.9x = 163.8$

$x = 182$

The solution is (182, 134).

Homework 2.3

1. Solve the third equation for y to get $y = 2$. Put this value into the second equation:
$$3(2) + z = 5$$
$$6 + z = 5$$
$$z = -1$$
Now put the values for y and z into the first equation:
$$x + 2 + (-1) = 2$$
$$x + 1 = 2$$
$$x = 1$$
The solution is $(1, 2, -1)$.

3. Solve the third equation for y to get $y = -1$. Put this value into the second equation:
$$5(-1) + 3z = -8$$
$$-5 + 3z = -8$$
$$3z = -3$$
$$z = -1$$
Now put the values for y and z into the first equation:
$$2x - (-1) - (-1) = 6$$
$$2x + 2 = 6$$
$$2x = 4$$
$$x = 2$$
The solution is $(2, -1, -1)$.

5. Solve the third equation for x to get $x = 4$. Put this value for x into the first equation:
$$2(4) + z = 5$$
$$8 + z = 5$$
$$z = -3$$
Put this value for z into the second equation:
$$3y + 2(-3) = 6$$
$$3y - 6 = 6$$
$$3y = 12$$
$$y = 4$$
The solution is $(4, 4, -3)$.

7.
$$x + y + z = 0 \quad (1)$$
$$2x - 2y + z = 8 \quad (2)$$
$$3x + 2y + z = 2 \quad (3)$$
Multiply (1) by 2 and add the result to (2) to obtain (4)
$$2x + 2y + 2z = 0 \quad (1a)$$
$$\underline{2x - 2y + z = 8 \quad (2)}$$
$$4x + 3z = 8 \quad (4)$$
Now add (2) and (3):
$$2x - 2y + z = 8 \quad (2)$$
$$\underline{3x + 2y + z = 2 \quad (3)}$$
$$5x + 2z = 10 \quad (5)$$
Multiply (4) by -2 and (5) by 3 and add the result:
$$-8x - 6z = -16 \quad (4a)$$
$$\underline{15x + 6z = 30 \quad (5a)}$$
$$7x = 14$$
$$x = 2 \quad (6)$$
We form a new system consisting of (1), (4), and (6):
$$x + y + z = 0 \quad (1)$$
$$4x + 3z = 8 \quad (4)$$
$$x = 2 \quad (6)$$
Put the value for x from (6) into (4):
$$4(2) + 3z = 8$$
$$8 + 3z = 8$$
$$3z = 0$$
$$z = 0$$
Put these values for x and z into (1):
$$2 + y + 0 = 0$$
$$y = -2$$
The solution is $(2, -2, 0)$.

47

9. $4x + z = 3$ (1)

$2x - y = 2$ (2)

$3y + 2z = 0$ (3)

Multiply (2) by –2 and add the result to (1). The result will be an equation in y and z that can be combined with (3):

$4x + z = 3$ (1)

$-4x + 2y = -4$ (2a)

$2y + z = -1$ (4)

Multiply (4) by –2, add the result to (3) and simplify:

$-4y - 2z = 2$ (4a)

$3y + 2z = 0$ (3)

$-y = 2$

$y = -2$

Put this value for y into (4):

$2(-2) + z = -1$

$-4 + z = -1$

$z = 3$

Now put this value for z into (1):

$4x + 3 = 3$

$4x = 0$

$x = 0$

The solution is (0, –2, 3).

11. $2x + 3y - 2z = 5$ (1)

$3x - 2y - 5z = 5$ (2)

$5x + 2y + 3z = -9$ (3)

Multiply (1) by 2, (2) by 3, and add the results:

$4x + 6y - 4z = 10$ (1a)

$9x - 6y - 15z = 15$ (2a)

$13x - 19z = 25$ (4)

Now, add (2) and (3), then simplifying the result:

$3x - 2y - 5z = 5$ (2)

$5x + 2y + 3z = -9$ (3)

$8x - 2z = -4$

Divide each side of this result by 2:

$4x - z = -2$ (5)

Multiply (5) by –19, add the result to (4), and simplify:

$13x - 19z = 25$ (4)

$-76x + 19z = 38$ (5a)

$-63x = 63$

$x = -1$

Put this value for x into (5):

$4(-1) - z = -2$

$-z = 2$

$z = -2$

Put the values for x and z into (1):

$2(-1) + 3y - 2(-2) = 5$

$-2 + 3y + 4 = 5$

$3y = 3$

$y = 1$

The solution is (–1, 1, –2).

13. $4x + 6y + 3z = -3$ (1)
$2x - 3y - 2z = 5$ (2)
$-6x + 6y + 2z = -5$ (3)
Multiply (2) by –2 and add the result to (1):
$4x + 6y + 3z = -3$ (1)
$-4x + 6y + 4z = -10$ (2a)
_____
$12y + 7z = -13$ (4)
Now multiply (2) by 3 and add the result to (3):
$6x - 9y - 6z = 15$ (2b)
$-6x + 6y + 2z = -5$ (3)
_____
$-3y - 4z = 10$ (5)
Multiply (5) by 4, add the result to (4) and simplify:
$12y + 7z = -13$ (4)
$-12y - 16z = 40$ (5a)
_____
$-9z = 27$

$z = -3$
Put this value for z into (5):
$-3y - 4(-3) = 10$

$-3y + 12 = 10$

$-3y = -2$

$y = \dfrac{2}{3}$
Put the values for y and z into (1):
$4x + 6\left(\dfrac{2}{3}\right) + 3(-3) = -3$

$4x + 4 - 9 = -3$

$4x = 2$

$x = \dfrac{1}{2}$

The solution is $\left(\dfrac{1}{2},\ \dfrac{2}{3},\ -3\right)$.

15. $x - \dfrac{1}{2}y - \dfrac{1}{2}z = 4$
$x - \dfrac{3}{2}y - 2z = 3$
$\dfrac{1}{4}x + \dfrac{1}{4}y - \dfrac{1}{4}z = 0$
First, clear all the denominators by multiplying the first and second equations by 2 and the third equation by 4.
$2x - y - z = 8$ (1)
$2x - 3y - 4z = 6$ (2)
$x + y - z = 0$ (3)
Multiply (1) by –1 and add the result to (2):
$-2x + y + z = -8$ (1a)
$2x - 3y - 4z = 6$ (2)
_____
$-2y - 3z = -2$ (4)
Multiply (3) by –2 and add the result to (2):
$2x - 3y - 4z = 6$ (2)
$-2x - 2y + 2z = 0$ (3a)
_____
$-5y - 2z = 6$ (5)
Now, multiply (4) by –2, (5) by 3, add and simplify the result.
$4y + 6z = 4$ (4a)
$-15y - 6z = 18$ (5a)
_____
$-11y = 22$

$y = -2$
Put this value for y into (5):
$-5(-2) - 2z = 6$

$-2z = -4$

$z = 2$
Put the values for y and z into (3):
$x + (-2) - 2 = 0$

$x - 4 = 0$

$x = 4$
The solution is $(4, -2, 2)$.

17.
$$x + y - z = 2$$
$$\frac{1}{2}x - y + \frac{1}{2}z = -\frac{1}{2}$$
$$x + \frac{1}{3}y - \frac{2}{3}z = \frac{4}{3}$$

First, clear all denominators by multiplying the second equation by 2 and the third equation by 3.

$$x + y - z = 2 \quad (1)$$
$$x - 2y + z = -1 \quad (2)$$
$$3x + y - 2z = 4 \quad (3)$$

Multiply (2) by –1 and add the result to (1):

$$x + y - z = 2 \quad (1)$$
$$\underline{-x + 2y - z = 1 \quad (2a)}$$
$$3y - 2z = 3 \quad (4)$$

Multiply (2) by –3 and add the result to (3):

$$-3x + 6y - 3z = 3 \quad (2b)$$
$$\underline{3x + y - 2z = 4 \quad (3)}$$
$$7y - 5z = 7 \quad (5)$$

Multiply (4) by –5, (5) by 2, add and simplify the result:

$$-15y + 10z = -15 \quad (4a)$$
$$\underline{14y - 10z = 14 \quad (5a)}$$
$$-y = -1$$
$$y = 1$$

Put this value for *y* into (4):

$$3(1) - 2z = 3$$
$$3 - 2z = 3$$
$$-2z = 0$$
$$z = 0$$

Put the values for *y* and *z* into (1):

$$x + 1 - 0 = 2$$
$$x = 1$$

The solution is (1, 1, 0).

19.
$$x = -y$$
$$x + z = \frac{5}{6}$$
$$y - 2z = \frac{-7}{6}$$

First, put the system into standard form by moving the –*y* term to the left in the first equation and multiplying the second and third equations by 6.

$$x + y = 0 \quad (1)$$
$$6x + 6z = 5 \quad (2)$$
$$6y - 12z = -7 \quad (3)$$

Multiply (1) by –6 and add to (2):

$$-6x - 6y = 0 \quad (1a)$$
$$\underline{6x + 6z = 5 \quad (2)}$$
$$-6y + 6z = 5 \quad (4)$$

Add (3) and (4):

$$6y - 12z = -7 \quad (3)$$
$$\underline{-6y + 6z = 5 \quad (4)}$$
$$-6z = -2$$
$$z = \frac{1}{3}$$

Put this value for *z* into (4):

$$-6y + 6\left(\frac{1}{3}\right) = 5$$
$$-6y + 2 = 5$$
$$-6y = 3$$
$$y = -\frac{1}{2}$$

Put this value for *y* into (1):

$$x + \left(-\frac{1}{2}\right) = 0$$
$$x = \frac{1}{2}$$

The solution is $\left(\frac{1}{2}, -\frac{1}{2}, \frac{1}{3}\right)$.

21.
$$3x - 2y + z = 6 \quad (1)$$
$$2x + y - z = 2 \quad (2)$$
$$4x + 2y - 2z = 3 \quad (3)$$

Multiply (2) by –2 and add the result to (3):

$$-4x - 2y + 2z = -4 \quad (2a)$$
$$\underline{4x + 2y - 2z = 3 \quad (3)}$$
$$0 = -1 \quad (4)$$

Equation (4) can also be written as $0x + 0y + 0z = -1$, so the system is inconsistent.

23.
$$2x + 3y - z = -2$$
$$x - y + \frac{1}{2}z = 2$$
$$4x - \frac{1}{3}y + 2z = 8$$

Write the system in standard form by multiplying the second equation by 2 and the third equation by 3:

$$2x + 3y - z = -2 \quad (1)$$
$$2x - 2y + z = 4 \quad (2)$$
$$12x - y + 6z = 24 \quad (3)$$

Multiply (2) by –1 and add the result to (1):

$$2x + 3y - z = -2 \quad (1)$$
$$\underline{-2x + 2y - z = -4 \quad (2a)}$$
$$5y - 2z = -6 \quad (4)$$

Multiply (2) by –6 and add the result to (3):

$$-12x + 12y - 6z = -24 \quad (2b)$$
$$\underline{12x - y + 6z = 24 \quad (3)}$$
$$11y = 0$$
$$y = 0$$

Put this value for y into (4):
$$5(0) - 2z = -6$$
$$-2z = -6$$
$$z = 3$$

Put the values for y and z into (1)
$$2x + 3(0) - 3 = -2$$
$$2x = 1$$
$$x = \frac{1}{2}$$

The solution is $\left(\dfrac{1}{2},\ 0,\ 3\right)$.

25.
$$x = 2y - 7$$
$$y = 4z + 3$$
$$z = 3x + y$$

The first step is to write the system in standard form by moving the variable terms to the left sides of the equations.

$$x - 2y = -7 \quad (1)$$
$$y - 4z = 3 \quad (2)$$
$$-3x - y + z = 0 \quad (3)$$

Multiply (1) by 3 and add the result to (3):

$$3x - 6y \quad\;\; = -21 \quad (1a)$$
$$\underline{-3x - y + z = \quad 0 \quad (3)}$$
$$-7y + z = -21 \quad (4)$$

Multiply (2) by 7, add the result to (4) and simplify:

$$7y - 28z = 21 \quad (2a)$$
$$\underline{-7y + z = -21 \quad (4)}$$
$$-27z = 0$$
$$z = 0$$

Put this value for z into (2) and simplify:

$$y - 4(0) = 3$$
$$y = 3$$

Put this value for y into (1):

$$x - 2(3) = -7$$
$$x - 6 = -7$$
$$x = -1$$

The solution is $(-1, 3, 0)$.

27.
$$\frac{1}{2}x + y = \frac{1}{2}z$$
$$x - y = -z - 2$$
$$-x - 2y = -z + \frac{4}{3}$$

Multiply the first equation by 2, the third equation by 3, and move the variable terms to the left sides of the equations:

$$x + 2y - z = 0 \quad (1)$$
$$x - y + z = -2 \quad (2)$$
$$-3x - 6y + 3z = 4 \quad (3)$$

Multiply (1) by 3 and add the result to (3):

$$3x + 6y - 3z = 0 \quad (1a)$$
$$\underline{-3x - 6y + 3z = 4 \quad (3)}$$
$$0 = 4 \quad (4)$$

Equation (4) can also be written $0x + 0y + 0z = 4$, so the system is inconsistent.

29.

$$x - y = 0$$

$$2x + 2y + z = 5$$

$$2x + y - \frac{1}{2}z = 0$$

We put this system into standard form by multiplying the third equation by 2.

$$x - y = 0 \quad (1)$$

$$2x + 2y + z = 5 \quad (2)$$

$$4x + 2y - z = 0 \quad (3)$$

Multiply (1) by –2 and add the result to (2):

$$-2x + 2y = 0 \quad (1a)$$

$$\underline{2x + 2y + z = 5 \quad (2)}$$

$$4y + z = 5 \quad (4)$$

Multiply (2) by –2 and add the result to (3):

$$-4x - 4y - 2z = -10 \quad (2a)$$

$$\underline{4x + 2y - z = 0 \quad (3)}$$

$$-2y - 3z = -10 \quad (5)$$

Multiply (5) by 2, add the result to (4), and simplify:

$$-4y - 6z = -20 \quad (5a)$$

$$\underline{4y + z = 5 \quad (4)}$$

$$-5z = -15$$

$$z = 3$$

Put this value for z into (4):

$$4y + 3 = 5$$

$$4y = 2$$

$$y = \frac{1}{2}$$

Put this value for y into (1):

$$x - \frac{1}{2} = 0$$

$$x = \frac{1}{2}$$

The solution is $\left(\frac{1}{2}, \frac{1}{2}, 3\right)$.

31. a. Let n = the number of nickels, d = the number of dimes, and q = the number or quarters in the box.

b.

	number of coins	value per coin (in cents)	total value (in cents)
nickels	n	5	$5n$
dimes	d	10	$10d$
quarters	q	25	$25q$
total	$n+d+q$	------	$5n+10d+25q$

c. Since there are 85 coins,
$n + d + q = 85$.
The value of the coins is $6.25, so
$5n + 10d + 25q = 625$
(converting the values to cents).
Three times as many nickels as dimes means that $n = 3d$ or
$n - 3d = 0$
Thus we have the system:

$$n + d + q = 85 \quad (1)$$

$$5n + 10d + 25q = 625 \quad (2)$$

$$n - 3d = 0 \quad (3)$$

Multiply (1) by –25, add the result to (2), and simplify:

$$-25n - 25d - 25q = -2125 \quad (1a)$$

$$\underline{5n + 10d + 25q = 625 \quad (2)}$$

$$-20n - 15d = -1500$$

$$4n + 3d = 300 \quad (4)$$

Add (3) to (4) and simplify the result:

$$n - 3d = \quad 0 \quad (3)$$

$$\underline{4n + 3d = 300} \quad (4)$$

$$5n = 300$$

$$n = 60$$

Put this value for n into (3):

$$60 - 3d = 0$$

$$60 = 3d$$

$$20 = d$$

Put the values for n and d into (1):

$$60 + 20 + q = 85$$

$$80 + q = 85$$

$$q = 5$$

The box contains 60 nickels, 20 dimes, and 5 quarters.

33. a. The variables x, y, and z are the lengths of the sides of the triangle as described by the problem.

b. No table is used for this problem.

c. From the perimeter, $x + y + z = 155$. The relationships between the sides are $x + 20 = y$ or $x - y = -20$ and $y - 5 = z$ or $y - z = 5$. The system is

$$x + y + z = 155 \quad (1)$$

$$x - y = -20 \quad (2)$$

$$y - z = 5 \quad (3)$$

Add (1) and (3):

$$x + y + z = 155 \quad (1)$$

$$\underline{y - z = \quad 5} \quad (3)$$

$$x + 2y = 160 \quad (4)$$

Multiply (2) by 2, add the result to (4), and simplify:

$$2x - 2y = -40 \quad (2a)$$

$$\underline{x + 2y = 160 \quad\quad (4)}$$

$$3x = 120$$

$$x = 40$$

Put this value for x into (2):

$$40 - y = -20$$

$$-y = -60$$

$$y = 60$$

Put this value for y into (3):

$$60 - z = 5$$

$$-z = -55$$

$$z = 55$$

The sides of the triangle are: $x = 40$ in., $y = 60$ in., and $z = 55$ in.

35. a. Let x = the amount of carrots,
y = the amount of green beans,
and z = the amount of
cauliflower in 1 cup of Vegetable
Medley.

b.

	number of cups	Vit. C per cup	total mg. Vitamin C
carrots	x	9	$9x$
green beans	y	15	$15y$
cauliflower	z	69	$69z$
total	$x+y+z$	------	$9x+15y+69z$

	number of cups	calcium per cup	total mg. calcium
carrots	x	48	$48x$
green beans	y	63	$63y$
cauliflower	z	26	$26z$
total	$x+y+z$	------	$48x+63y+26z$

c. In one cup of Vegetable Medley,
$x + y + z = 1$.
The information on Vitamin C
gives the equation:
$9x + 15y + 69z = 29.4$.
The information on calcium
gives the equation:
$48x + 63y + 26z = 47.4$.
The system is:

$$x + y + z = 1 \quad (1)$$
$$9x + 15y + 69z = 29.4 \quad (2)$$
$$48x + 63y + 26z = 47.4 \quad (3)$$

Multiply (1) by -9, add the result
to (2), and simplify:

$$-9x - 9y - 9z = -9 \quad (1a)$$
$$9x + 15y + 69z = 29.4 \quad (2)$$
$$\overline{6y + 60z = 20.4}$$
$$y + 10z = 3.4 \quad (4)$$

Multiply (1) by -48 and add the
result to (3):
$$-48x - 48y - 48z = -48 \quad (1b)$$
$$48x + 63y + 26z = 47.4 \quad (3)$$
$$\overline{15y - 22z = -0.6 \quad (5)}$$
Multiply (4) by -15, add the
result to (5), and simplify the
result:
$$-15y - 150z = -51 \quad (4a)$$
$$15y - 22z = -0.6 \quad (5)$$
$$\overline{-172z = -51.6}$$

$$z = 0.3$$
Put this value for z into (5):
$$15y - 22(0.3) = -0.6$$
$$15y - 6.6 = -0.6$$
$$15y = 6$$
$$y = 0.4$$
Put the values for z and y
into (1):
$$x + 0.4 + 0.3 = 1$$
$$x + 0.7 = 1$$
$$x = 0.3$$
Each cup of Vegetable Medley
contains 0.3 cups of carrots, 0.4
cups of green beans, and 0.3 cups
of cauliflower.

37. a. Let s = the number of score only, e = the number of evaluation, and n = the number of narrative reports processed each day.

b. No table is used for this problem.

c. Converting all times to minutes, we have, for the optical scanner:
$3s + 3e + 3n = 420$.
For the analysis:
$0s + 4e + 5n = 480$.
For the printer:
$1s + 2e + 8n = 720$.
After reducing the equation which corresponds to the scanner, the system is:

$$s + e + n = 140 \quad (1)$$

$$4e + 5n = 480 \quad (2)$$

$$s + 2e + 8n = 720 \quad (3)$$

Multiply (1) by -1 and add the result to (3):

$$-s - e - n = -140 \quad (1a)$$

$$\underline{s + 2e + 8n = 720} \quad (3)$$

$$e + 7n = 580 \quad (4)$$

Multiply (4) by -4, add the result to (2), and simplify:

$$4e + 5n = 480 \quad (2)$$

$$\underline{-4e - 28n = -2320} \quad (4a)$$

$$-23n = -1840$$

$$n = 80$$

Put this value for n into (2):

$$4e + 5(80) = 480$$

$$4e + 400 = 480$$

$$4e = 80$$

$$e = 20$$

Put the values for e and n into (1):

$$s + 20 + 80 = 140$$

$$s + 100 = 140$$

$$s = 40$$

Thus, when using all of its resources, ABC can complete 40 score only, 20 evaluation, and 80 narrative reports.

39. a. Let t = the number of tennis rackets, p = the number of Ping-Pong paddles, and s = the number of squash rackets.

b. No table is used for this problem.

c. The time available for gluing gives the equation:
$3t + 1p + 2s = 95$.
The time available for sanding gives the equation:
$2t + 1p + 2s = 75$.
The time available for finishing gives the equation:
$3t + 1p + 2.5s = 100$.
The system is

$$3t + p + 2s = 95 \quad (1)$$

$$2t + p + 2s = 75 \quad (2)$$

$$3t + p + 2.5s = 100 \quad (3)$$

Multiply (1) by -1, add the result to (3), and simplify:

$$-3t - p - 2s = -95 \quad (1a)$$

$$3t + p + 2.5s = 100 \quad (3)$$

$$0.5s = 5$$

$$s = 10 \quad (4)$$

Multiply (2) by -1 and add the result to (1):

$$3t + p + 2s = 95 \quad (1)$$

$$-2t - p - 2s = -75 \quad (2a)$$

$$t = 20 \quad (5)$$

Put the values for t and s into (1):
$3(20) + p + 2(10) = 95$

$$60 + p + 20 = 95$$

$$p = 15$$

Ace can make 20 tennis rackets, 15 Ping-Pong paddles, and 10 squash rackets every day.

Midchapter 2 Review

1. a.

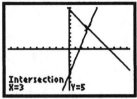

The solution is (3, 5).

b. Verify algebraically:

$y = 3.7(3) - 6.1 = 5.$

$y = -1.6(3) + 9.8 = 5.$

3. Graph the equations:
$Y_1 = 2630 + 32x$ and
$Y_2 = -21x - 1610$ in the window
Xmin = −100, Xmax = 10,
Ymin = −300, Ymax = 300.

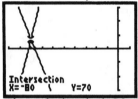

The solution is (−80, 70).

(Note that the above window does not include all intercepts. This is because when including all intercepts in the graphing window for this system, for example by using YMIN = −2000 and YMAX = 3000, the intersection point appears to be on the *x*-axis!)

5. a. The equation $3a - b = 6$ has *a*-intercept $a = 2$ and *b*-intercept $b = -6$. The equation $6a - 12 = 2b$ has *a*-intercept $a = 2$ and *b*-intercept $b = -6$. Use the intercepts to graph the lines. In this case the equations represent the same line.

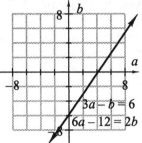

b. The system is dependent.

7. a. $460x + 120y = 7520$

b. The number of chairs is four times the number of tables so the equation is $y = 4x$.

c. The equation in part (a) has *x*-intercept at approximately (16.3, 0) and *y*-intercept at approximately (0, 62.7). The line in part (b) passes through the origin and has slope 4.

Graph $Y_1 = \dfrac{7520 - 460x}{120}$ and $Y_2 = 4x$ in the window
XMIN = −10, XMAX = 20
YMIN = −10, YMAX = 80.

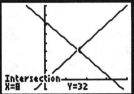

Etienne should buy 8 tables and 32 chairs.

9. Solve the first equation for a to get $a = -6 - 2b$. Substitute this into the second equation:

$$2a - 3b = 16$$
$$2(-6 - 2b) - 3b = 16$$
$$-12 - 4b - 3b = 16$$
$$-12 - 7b = 16$$
$$-7b = 28$$
$$b = -4$$

Put this value back into the first equation: $a = -6 - 2(-4) = 2$. The solution is $a = 2$ and $b = -4$.

11. $5x - 2y = -4$ (1)

$-6x + 3y = 5$ (2)

Multiply (1) by 3 and (2) by 2 and add the results:

$$15x - 6y = -12$$
$$\underline{-12x + 6y = 10}$$
$$3x = -2$$
$$x = -\frac{2}{3}$$

Put this value into (1):

$$5\left(-\frac{2}{3}\right) - 2y = -4$$
$$-\frac{10}{3} - 2y = -4$$
$$-10 - 6y = -12$$
$$-6y = -2$$
$$y = \frac{1}{3}$$

The solution is $\left(-\frac{2}{3}, \frac{1}{3}\right)$.

13. Write the equations in standard form:

$$3a - 2b = 6 \quad (1)$$
$$-6a + 4b = -8 \quad (2)$$

Multiply (1) by 2 and add the result to (2).

$$6a - 4b = 12$$
$$\underline{-6a + 4b = -8}$$
$$0 = 4$$

The system is inconsistent.

15. **a.**

	Rate	Time	Distance
P waves	5.4	x	y
S waves	3	$x + 90$	y

b. Using rate $\times$ time = distance, $y = 3(x + 90)$.

c. Using the formula rate $\times$ time = distance, $y = 5.4x$.

d. Substitute $y = 5.4x$ into the equation in part (b):

$$5.4x = 3(x + 90)$$
$$5.4x = 3x + 270$$
$$2.4x = 270$$
$$x = 112.5$$

Put this value into $y = 5.4x$:
$y = 5.4(112.5) = 607.5$.
The solution is $(112.5, 607.5)$.
The seismograph is 607.5 miles from the earthquake.

17. Solve the third equation for z to get $z = -3$. Substitute this value into the second equation:

$$4x - 3 = 5$$
$$4x = 8$$
$$x = 2$$

Put the values of x and z into the first equation:

$$-3(2) + 2y + 3(-3) = -3$$
$$-6 + 2y - 9 = -3$$
$$-15 + 2y = -3$$
$$2y = 12$$
$$y = 6$$

The solution is $(2, 6, -3)$.

19.
$$x - 5y - z = 2 \quad (1)$$
$$3x - 9y + 3z = 6 \quad (2)$$
$$x - 3y - z = -6 \quad (3)$$

Multiply (1) by -3, add the result to (2), and reduce:

$$-3x + 15y + 3z = -6 \quad (1a)$$
$$\underline{3x - 9y + 3z = 6 \quad (2)}$$
$$6y + 6z = 0$$
$$y + z = 0 \quad (4)$$

Multiply (1) by -1 and add the result to (3):

$$-x + 5y + z = -2 \quad (1b)$$
$$\underline{x - 3y - z = -6 \quad (3)}$$
$$2y = -8$$
$$y = -4$$

Substitute $y = -4$ into (4):

$$-4 + z = 0$$
$$z = 4$$

Put the values of y and z into (1):

$$x - 5(-4) - 4 = 2$$
$$x + 20 - 4 = 2$$
$$x + 16 = 2$$
$$x = -14$$

The solution is $(-14, -4, 4)$.

21.

$$2x + y = 6 \quad (1)$$
$$x - z = 4 \quad (2)$$
$$3x + y - z = 10 \quad (3)$$

Add (1) and (2):

$$2x + y = 6 \quad (1)$$
$$\underline{x - z = 4 \quad (2)}$$
$$3x + y - z = 10 \quad (4)$$

Multiply (4) by –1 and add the result to (3):

$$3x + y - z = 10 \quad (3)$$
$$\underline{-3x - y + z = -10 \quad (4a)}$$
$$0 = 0 \quad (5)$$

The system is dependent.

23. Let x = the amount of cranberry juice, let y = the amount of apricot nectar, and let z = the amount of club soda per quart. Then $x + y + z = 1$. The information about the cost gives the equation $1.25x + 1.00y + 0.25z = 0.80$. Multiply by 100 to and then reduce.

$$125x + 100y + 25z = 80$$
$$25x + 20y + 5z = 16$$

The information about the calories gives $1200x + 1600y = 800$. Reduce by dividing each term by 400: $3x + 4y = 2$. The system is:

$$x + y + z = 1 \quad (1)$$
$$25x + 20y + 5z = 16 \quad (2)$$
$$3x + 4y = 2 \quad (3)$$

Multiply (1) by –5 and add the result to (2):

$$-5x - 5y - 5z = -5 \quad (1a)$$
$$\underline{25x + 20y + 5z = 16 \quad (2)}$$
$$20x + 15y = 11 \quad (4)$$

Multiply (3) by –20, multiply (4) by 3 and add the results:

$$-60x - 80y = -40 \quad (3a)$$
$$\underline{60x + 45y = 33 \quad (4a)}$$
$$-35y = -7$$
$$y = 0.2$$

Substitute $y = 0.2$ into (3):

$$3x + 4(0.2) = 2$$
$$3x = 1.2$$
$$x = 0.4$$

Put the values of x and y into (1):

$$0.4 + 0.2 + z = 1$$
$$z = 0.4$$

Solution: 0.4 quarts of cranberry juice, 0.2 quarts of apricot nectar, and 0.4 quarts of club soda.

Homework 2.4

1. $\begin{bmatrix} -2 & 1 & 0 \\ 3 & -1 & 2 \end{bmatrix} \rightarrow \begin{bmatrix} -2 & 1 & 0 \\ -9 & 3 & -6 \end{bmatrix}$

3. $\begin{bmatrix} 1 & -3 & 6 \\ -2 & 4 & -1 \end{bmatrix} \rightarrow \begin{bmatrix} 1 & -3 & 6 \\ 0 & -2 & 11 \end{bmatrix}$

5. $\begin{bmatrix} 0 & -3 & 2 & -3 \\ 2 & 6 & -1 & 4 \\ 1 & 0 & -2 & 5 \end{bmatrix} \rightarrow \begin{bmatrix} 1 & 0 & -2 & 5 \\ 2 & 6 & -1 & 4 \\ 0 & -3 & 2 & -3 \end{bmatrix}$

7. $\begin{bmatrix} 1 & 2 & 1 & -5 \\ 0 & 4 & -2 & 3 \\ 4 & -1 & 6 & -8 \end{bmatrix} \rightarrow \begin{bmatrix} 1 & 2 & 1 & -5 \\ 0 & 4 & -2 & 3 \\ 0 & -9 & 2 & 12 \end{bmatrix}$

9. To obtain 0 as the first entry in the second row, add -2(row 1) to row 2. (Let R_1 denote row 1 and let R_2 denote row 2.)

$$\begin{bmatrix} 1 & -3 & 2 \\ 2 & 1 & 4 \end{bmatrix} \xrightarrow{-2R_1 + R_2} \begin{bmatrix} 1 & -3 & 2 \\ 0 & 7 & 0 \end{bmatrix}$$

11. To obtain 0 as the second entry in the second row, add $-\frac{1}{2}$(row 1) to row 2. (Let R_1 denote row 1 and let R_2 denote row 2.)

$$\begin{bmatrix} 2 & 6 & -4 \\ 5 & 3 & 1 \end{bmatrix} \xrightarrow{-\frac{1}{2}R_1 + R_2} \begin{bmatrix} 2 & 6 & -4 \\ 4 & 0 & 3 \end{bmatrix}$$

13. To obtain 0 as the first entry in the second and third rows, add -2(row 1) to row 2 and -4(row 1) to row 3: (Let R_1 denote row 1, R_2 denote row 2, and R_3 denote row 3.)

$$\begin{bmatrix} 1 & -2 & 2 & 1 \\ 2 & 3 & -1 & 6 \\ 4 & 1 & -3 & 3 \end{bmatrix} \xrightarrow[\substack{-2R_1 + R_2 \\ -4R_1 + R_3}]{} \begin{bmatrix} 1 & -2 & 2 & 1 \\ 0 & 7 & -5 & 4 \\ 0 & 9 & -11 & -1 \end{bmatrix}$$

15. To obtain 0 as the second entry in the second and third rows, add row 1 to 2(row 2) add row 1 to –2(row 3):

$$\begin{bmatrix} -1 & 4 & 3 & | & 2 \\ 2 & -2 & -4 & | & 6 \\ 1 & 2 & 3 & | & -3 \end{bmatrix} \xrightarrow[R_1+(-2R_3)]{R_1+2R_2} \begin{bmatrix} -1 & 4 & 3 & | & 2 \\ 3 & 0 & -5 & | & 14 \\ -3 & 0 & -3 & | & 8 \end{bmatrix}$$

17. <u>Step 1</u>: To obtain 0 as the first entry in the second and third rows, add 2(row 1) to row 2 and 3(row 1) to row 3.

<u>Step 2</u>: To obtain 0 as the second entry in the third row, add $-\dfrac{1}{2}$(row 2) to row 3.

$$\begin{bmatrix} -2 & 1 & -3 & | & -2 \\ 4 & 2 & 0 & | & 2 \\ 6 & -1 & 2 & | & 0 \end{bmatrix} \xrightarrow[3R_1+R_3]{2R_1+R_2} \begin{bmatrix} -2 & 1 & -3 & | & -2 \\ 0 & 4 & -6 & | & -2 \\ 0 & 2 & -7 & | & -6 \end{bmatrix} \xrightarrow[-\frac{1}{2}R_2+R_3]{} \begin{bmatrix} -2 & 1 & -3 & | & -2 \\ 0 & 4 & -6 & | & -2 \\ 0 & 0 & -4 & | & -5 \end{bmatrix}$$

19. The augmented matrix is $\begin{bmatrix} 1 & 3 & | & 11 \\ 2 & -1 & | & 1 \end{bmatrix}$. Perform matrix reduction:

$$\begin{bmatrix} 1 & 3 & | & 11 \\ 2 & -1 & | & 1 \end{bmatrix} \xrightarrow{-2R_1+R_2} \begin{bmatrix} 1 & 3 & | & 11 \\ 0 & -7 & | & -21 \end{bmatrix}$$

Which corresponds to the system
$x + 3y = 11$

$-7y = -21.$
Therefore, $y = 3$. Substitute 3 for y to find
$x + 3(3) = 11$

$x = 2$
The solution is the ordered pair (2, 3).

21. The augmented matrix is $\begin{bmatrix} 1 & -4 & | & -6 \\ 3 & 1 & | & -5 \end{bmatrix}$. Perform matrix reduction:

$$\begin{bmatrix} 1 & -4 & | & -6 \\ 3 & 1 & | & -5 \end{bmatrix} \xrightarrow{-3R_1+R_2} \begin{bmatrix} 1 & -4 & | & -6 \\ 0 & 13 & | & 13 \end{bmatrix}$$

Which corresponds to the system
$x - 4y = -6$

$13y = 13.$
Therefore, $y = 1$. Substitute 1 for y to find
$x - 4(1) = -6$

$x = -2$
The solution is the ordered pair (–2, 1).

23. The augmented matrix is $\begin{bmatrix} 2 & 1 & 5 \\ 3 & -5 & 14 \end{bmatrix}$. Perform matrix reduction:

$$\begin{bmatrix} 2 & 1 & 5 \\ 3 & -5 & 14 \end{bmatrix} \xrightarrow{-3R_1+2R_2} \begin{bmatrix} 2 & 1 & 5 \\ 0 & -13 & 13 \end{bmatrix}$$

Which corresponds to the system
$2x + y = 5$

$-13y = 13$.

Therefore, $y = -1$. Substitute -1 for y to find
$2x + (-1) = 5$

$x = 3$.

The solution is the ordered pair $(3, -1)$.

25. The augmented matrix is $\begin{bmatrix} 1 & -1 & -8 \\ 1 & 2 & 9 \end{bmatrix}$. Perform matrix reduction:

$$\begin{bmatrix} 1 & -1 & -8 \\ 1 & 2 & 9 \end{bmatrix} \xrightarrow{-R_1+R_2} \begin{bmatrix} 1 & -1 & -8 \\ 0 & 3 & 17 \end{bmatrix}$$

which corresponds to the system
$x - y = -8$

$3y = 17$.

Therefore, $y = \dfrac{17}{3}$. Substitute $\dfrac{17}{3}$ for y to find

$x - \dfrac{17}{3} = -8$

$x = -\dfrac{7}{3}$.

The solution is the ordered pair $\left(-\dfrac{7}{3}, \dfrac{17}{3} \right)$.

27. Write the augmented matrix and perform matrix reduction:

$$\begin{bmatrix} 1 & 3 & -1 & | & 5 \\ 3 & -1 & 2 & | & 5 \\ 1 & 1 & 2 & | & 7 \end{bmatrix} \xrightarrow[\substack{-3R_1+R_2 \\ -R_1+R_3}]{} \begin{bmatrix} 1 & 3 & -1 & | & 5 \\ 0 & -10 & 5 & | & -10 \\ 0 & -2 & 3 & | & 2 \end{bmatrix} \xrightarrow[-\frac{1}{5}R_2+R_3]{} \begin{bmatrix} 1 & 3 & -1 & | & 5 \\ 0 & -10 & 5 & | & -10 \\ 0 & 0 & 2 & | & 4 \end{bmatrix}$$

which corresponds to the system

$$x + 3y - z = 5$$
$$-10y + 5z = -10$$
$$2z = 4$$

Therefore, $z = 2$. Substitute 2 for z to find

$$-10y + 5(2) = -10$$
$$y = 2$$

Substitute 2 for z and 2 for y to find

$$x + 3(2) - 2 = 5$$
$$x = 1$$

The solution is the ordered triplet $(1, 2, 2)$.

29. Write the augmented matrix and perform matrix reduction:

$$\begin{bmatrix} 2 & -1 & 1 & | & 5 \\ 1 & -2 & -2 & | & 2 \\ 3 & 3 & -1 & | & 4 \end{bmatrix} \xrightarrow{\text{Swap } R_1 \text{ and } R_2} \begin{bmatrix} 1 & -2 & -2 & | & 2 \\ 2 & -1 & 1 & | & 5 \\ 3 & 3 & -1 & | & 4 \end{bmatrix} \xrightarrow[\substack{-2R_1+R_2 \\ -3R_1+R_3}]{} \begin{bmatrix} 1 & -2 & -2 & | & 2 \\ 0 & 3 & 5 & | & 1 \\ 0 & 9 & 5 & | & -2 \end{bmatrix}$$

$$\xrightarrow[-3R_2+R_3]{} \begin{bmatrix} 1 & -2 & -2 & | & 2 \\ 0 & 3 & 5 & | & 1 \\ 0 & 0 & -10 & | & -5 \end{bmatrix}$$

which corresponds to the system

$$x - 2y - 2z = 2$$
$$3y + 5z = 1$$
$$-10z = -5$$

Therefore, $z = \frac{1}{2}$. Substitute $\frac{1}{2}$ for z to find

$$3y + 5\left(\frac{1}{2}\right) = 1$$
$$y = -\frac{1}{2}$$

Substitute $\frac{1}{2}$ for z and $-\frac{1}{2}$ for y to find

$$x - 2\left(-\frac{1}{2}\right) - 2\left(\frac{1}{2}\right) = 2$$
$$x = 2$$

The solution is the ordered triplet $\left(2, -\frac{1}{2}, \frac{1}{2}\right)$.

31. Write the augmented matrix and perform matrix reduction:

$$\begin{bmatrix} 2 & -1 & -1 & | & -4 \\ 1 & 1 & 1 & | & -5 \\ 1 & 3 & -4 & | & 12 \end{bmatrix} \xrightarrow{\text{Swap } R_1 \text{ and } R_2} \begin{bmatrix} 1 & 1 & 1 & | & -5 \\ 2 & -1 & -1 & | & -4 \\ 1 & 3 & -4 & | & 12 \end{bmatrix} \xrightarrow[-R_1+R_3]{-2R_1+R_2} \begin{bmatrix} 1 & 1 & 1 & | & -5 \\ 0 & -3 & -3 & | & 6 \\ 0 & 2 & -5 & | & 17 \end{bmatrix}$$

$$\xrightarrow{\frac{2}{3}R_2+R_3} \begin{bmatrix} 1 & 1 & 1 & | & -5 \\ 0 & -3 & -3 & | & 6 \\ 0 & 0 & -7 & | & 21 \end{bmatrix}$$

which corresponds to the system

$x+y+z=-5$

$-3y-3z=6$

$-7z=21.$

Therefore, $z=-3$. Substitute -3 for z to find

$-3y-3(-3)=6$

$y=1.$

Substitute -3 for z and 1 for y to find

$x+1+(-3)=-5$

$x=-3.$

The solution is the ordered triplet $(-3, 1, -3)$.

33. Write the augmented matrix and perform matrix reduction:

$$\begin{bmatrix} 2 & -1 & 0 & | & 0 \\ 0 & 3 & 1 & | & 7 \\ 2 & 0 & 3 & | & 1 \end{bmatrix} \xrightarrow[-R_1+R_3]{} \begin{bmatrix} 2 & -1 & 0 & | & 0 \\ 0 & 3 & 1 & | & 7 \\ 0 & 1 & 3 & | & 1 \end{bmatrix} \xrightarrow[\text{Swap } R_2 \text{ and } R_3]{} \begin{bmatrix} 2 & -1 & 0 & | & 0 \\ 0 & 1 & 3 & | & 1 \\ 0 & 3 & 1 & | & 7 \end{bmatrix}$$

$$\xrightarrow[-3R_2+R_3]{} \begin{bmatrix} 2 & -1 & 0 & | & 0 \\ 0 & 1 & 3 & | & 1 \\ 0 & 0 & -8 & | & 4 \end{bmatrix}$$

which corresponds to the system

$$2x - y = 0$$
$$y + 3z = 1$$
$$-8z = 4$$

Therefore, $z = -\dfrac{1}{2}$. Substitute this value of z to find

$$y + 3\left(-\frac{1}{2}\right) = 1$$
$$y = \frac{5}{2}$$

Substitute $\dfrac{5}{2}$ for y to find

$$2x - \frac{5}{2} = 0$$
$$x = \frac{5}{4}$$

The solution is the ordered triplet $\left(\dfrac{5}{4}, \dfrac{5}{2}, -\dfrac{1}{2}\right)$.

Homework 2.5

1. Refer to the graph in the back of the textbook. Graph the line $y = 2x + 4$. Since the inequality is $y > 2x + 4$, the region above the line is shaded and the line is dashed.

3. Refer to the graph in the back of the textbook. Graph the line $3x - 2y = 12$. Since the inequality can be written as $y \geq -6 + \frac{3}{2}x$, the region above the line is shaded and the line is solid.

5. Refer to the graph in the back of the textbook. Graph the line $x + 4y = -6$. Since the inequality can be written as $y \geq -\frac{3}{2} - \frac{x}{4}$, the region above the line is shaded and the line is solid.

7. Refer to the graph in the back of the textbook. Graph the line $x = -3y + 1$. Since the inequality can be written as $y > -\frac{x}{3} + \frac{1}{3}$, the region above the line is shaded and the line is dashed.

9. Refer to the graph in the back of the textbook. Graph the line $x = -3$. Since the inequality is $x \geq -3$, the region to the right of the line is shaded and the line is solid.

11. Refer to the graph in the back of the textbook. Graph the line $y = \frac{1}{2}x$. Since the inequality is $y < \frac{1}{2}x$, the region below the line is shaded and the line is dashed.

13. Refer to the graph in the back of the textbook. Graph the line $0 = x - y$. Since the inequality can be written as $y \geq x$, the region above the line is shaded and the line is solid.

15. Refer to the graph in the back of the textbook. Graph the lines $y = -1$ and $y = 4$. The line $y = -1$ should be dashed and the line $y = 4$ should be solid. Since the inequality is $-1 < y \leq 4$, the region between the two lines is shaded.

17. Refer to the graph in the back of the textbook.
For $y > 2$, graph the line $y = 2$ with a dashed line and shade the region above it.
For $x \geq -2$, graph the line $x = -2$ with a solid line and shade the region to the right of the line.

19. Refer to the graph in the back of the textbook.
For $y < x$, graph the line $y = x$ with a dashed line and shade the region below the line.
For $y \geq -3$, graph the line $y = -3$ with a solid line and shade the region above the line.

21. Refer to the graph in the back of the textbook.
For $x + y \leq 6$, graph the line $x + y = 6$ with a solid line. Shade the region below this line, since the inequality can be written as $y \leq 6 - x$.
For $x + y \geq 4$, graph the line $x + y = 4$ with a solid line. Shade the region above this line, since the inequality can be written as $y \geq 4 - x$.

23. Refer to the graph in the back of the textbook. For $2x - y \le 4$, graph the line $2x - y = 4$ with a solid line. Shade the region above this line since the inequality can be written as $y \ge -4 + 2x$. For $x + 2y > 6$, graph $x + 2y = 6$ with a dashed line. Shade the region above this line, since the inequality can be written as

$$y > -\frac{1}{2}x + 3.$$

25. Refer to the graph in the back of the textbook.
For $3y - 2x < 2$, graph the line $3y - 2x = 2$ with a dashed line. Shade the region below this line since the inequality can be written

as $y < \frac{2}{3}x + \frac{2}{3}$.

For $y > x - 1$, graph the line $y = x - 1$ with a dashed line and shade the region above this line.

27. Refer to the graph in the back of the textbook.
For $2x + 3y - 6 < 0$, graph the line $2x + 3y - 6 = 0$ with a dashed line. Shade the region below this line, since the inequality can be written

as $y < 2 - \frac{2}{3}x$.

For $x \ge 0$, graph the line $x = 0$ (the y-axis) with a solid line and shade the region to the right of this line.
For $y \ge 0$, graph the line $y = 0$ (the x-axis) with a solid line and shade the region above this line.
The vertices are the points of intersection of the lines $2x + 3y - 6 = 0$, $x = 0$, and $y = 0$.
The lines $2x + 3y - 6 = 0$ and $x = 0$ intersect at the point $(0, 2)$, $2x + 3y - 6 = 0$ and $y = 0$ intersect at the point $(3, 0)$, while $x = 0$ and $y = 0$ intersect at the origin $(0, 0)$.

29. Refer to the graph in the back of the textbook.
For $5y - 3x \le 15$, graph the line $5y - 3x = 15$ with a solid line. Shade the region below this line since the inequality can be written as

$$y \le 3 + \frac{3}{5}x.$$

For $x + y \le 11$, graph the line $x + y = 11$ with a solid line. Shade the region below this line since the inequality can be written as
$y \le 11 - x$.
For graphing $x \ge 0$ and $y \ge 0$, see 27.
The vertices are $(0, 0)$, the intersection of the lines $x = 0$ and $y = 0$; $(0, 3)$, the intersection of the lines $x = 0$ and $5y - 3x = 15$; $(5, 6)$, the intersection of the lines $5y - 3x = 15$ and $x + y = 11$; and $(11, 0)$, the intersection of $y = 0$ and $x + y = 11$.

31. Refer to the graph in the back of the textbook.
For $2y \le x$, graph the line $2y = x$ with a solid line. Shade the region below this line since the inequality

can be written as $y \le \dfrac{x}{2}$.

For $2x \le y + 12$, graph the line $2x = y + 12$ with a solid line. Shade the region above this line since the inequality can be written as
$y \ge 2x - 12$.
For graphing $x \ge 0$ and $y \ge 0$, see 27.
The vertices are $(0, 0)$, the intersection of the lines $x = 0$ and $y = 0$; $(8, 4)$, the intersection of the lines $2y = x$ and $2x = y + 12$; and $(6, 0)$, the intersection of the lines $2x = y + 12$ and $y = 0$.

33. Refer to the graph in the back of the textbook.

For $x + y \geq 3$, graph the line $x + y = 3$ with a solid line. Shade the region above this line since the inequality can be written as $y \geq 3 - x$.
For $2y \leq x + 8$, graph the line $2y = x + 8$ with a solid line. Shade the region below this line since the inequality can be written as $y \leq \frac{x}{2} + 4$.
For $2y + 3x \leq 24$, graph the line $2y + 3x = 24$ with a solid line. Shade the region below this line since the inequality can be written as $y \leq 12 - \frac{3}{2}x$.
For graphing $x \geq 0$ and $y \geq 0$, see 27. The vertices are (0, 3), the intersection of the lines $x = 0$ and $x + y = 3$; (0, 4), the intersection of the lines $x = 0$ and $2y = x + 8$; (4, 6), the intersection of the lines $2y = x + 8$ and $2y + 3x = 24$; (8, 0), the intersection of the lines $2y + 3x = 24$ and $y = 0$; and (3, 0) the intersection of the lines $y = 0$ and $x + y = 3$.

35. Refer to the graph in the back of the textbook.
For $3y - x \geq 3$, graph the line $3y - x = 3$ with a solid line. Shade the region above this line since the inequality can be written as $y \geq 1 + \frac{x}{3}$.

For $y - 4x \geq -10$, graph the line $y - 4x = -10$ with a solid line. Shade the region above this line since the inequality can be written as $y \geq 4x - 10$.
For $y - 2 \leq x$, graph the line $y - 2 = x$ with a solid line. Shade the region below this line since the inequality can be written as $y \leq x + 2$.
For graphing $x \geq 0$ and $y \geq 0$, see 27. The vertices are (0, 1), the intersection of the lines $x = 0$ and $3y - x = 3$; (0, 2), the intersection of the lines $x = 0$ and $y - 2 = x$; (4,6), the intersection of the lines $y - 2 = x$ and $y - 4x = -10$; and (3, 2), the intersection of the lines $y - 4x = -10$ and $3y - x = 3$.

37. Let x represent the number of student tickets sold, and y represent the number of faculty tickets sold. The information that student tickets cost $1, faculty tickets cost $2, and the receipts must be at least $250, can be stated in the inequality $x + 2y \geq 250$. The fact that only positive numbers of tickets will be sold can be stated in the inequalities $x \geq 0$ and $y \geq 0$. The system of inequalities is
$x + 2y \geq 250$
$x \geq 0, y \geq 0$
Refer to the graph in the back of the textbook.

39. Let x represent the amount that Vassilis invests at 6% interest and let y represent the amount that he invests at 5%. Vassilis will invest at most $10,000, so $x + y \leq 10,000$. Earning at least $540 from these investments requires that $0.06x + 0.05y \geq 540$. It is impossible for Vassilis to invest a negative amount with either bank, so $x \geq 0$ and $y \geq 0$. The system of inequalities is
$x + y \leq 10,000$
$0.06x + 0.05y \geq 540$
$x \geq 0, y \geq 0$
Refer to the graph in the back of the textbook.

41. Let x represent the amount of corn meal and let y represent the amount of whole wheat flour that Gary uses. Since he cannot use more than 3 cups of the two ingredients combined, $x + y \leq 3$. The amount of linoleic acid in x cups of corn meal is $2.4x$ grams, while the amount of linoleic acid in y cups of whole wheat flour is $0.8y$ grams. To provide at least 3.2 grams in the combined mixture requires that $2.4x + 0.8y \geq 3.2$. Similarly, to makes sure that the mixture contains at least 10 milligrams of niacin, $2.5x + 5y \geq 10$. Since using a negative amount of either ingredient does not make sense, $x \geq 0$ and $y \geq 0$. The system of inequalities is
$x + y \leq 3$
$2.4x + 0.8y \geq 3.2$
$2.5x + 5y \geq 10$
$x \geq 0, y \geq 0$
Refer to the graph in the back of the textbook.

Chapter 2 Review

1. Refer to the graph in the back of the textbook. The solution is $(-1, 2)$.

3. $x + 5y = 18$ (1)

$x - y = -3$ (2)

Add -1 times (2) to (1):

$x + 5y = 18$ (1)

$\underline{-x + y = 3}$ (2a)

$6y = 21$

$y = \dfrac{21}{6}$

$y = \dfrac{7}{2}$

Put this value for y into (2):

$x - \dfrac{7}{2} = -3$

$x = -3 + \dfrac{7}{2}$

$x = \dfrac{1}{2}$

The solution is $\left(\dfrac{1}{2}, \dfrac{7}{2}\right)$.

5. $\dfrac{2}{3}x - 3y = 8$ (1)

$x + \dfrac{3}{4}y = 12$ (2)

Add 4 times (2) to (1):

$\dfrac{2}{3}x - 3y = 8$ (1)

$\underline{4x + 3y = 48}$ (2a)

$\dfrac{14}{3}x = 56$

$x = 12$

Put this value for x into (1):

$\dfrac{2}{3}(12) - 3y = 8$

$8 - 3y = 8$

$-3y = 0$

$y = 0$

The solution is $(12, 0)$.

7. $2x - 3y = 4$ (1)

$x + 2y = 7$ (2)

Add (1) to -2 times (2):

$2x - 3y = 4$ (1)

$\underline{-2x - 4y = -14}$ (2a)

$-7y = -10$

$y = \dfrac{10}{7}$

Put this value of y into (2):

$x + 2\left(\dfrac{10}{7}\right) = 7$

$x + \dfrac{20}{7} = 7$

$x = \dfrac{29}{7}$

The system has one solution, so the system is consistent and independent.

9. $2x - 3y = 4$ (1)

$6x - 9y = 12$ (2)

Multiply (1) by 3:

$6x - 9y = 12$ (1a)

$6x - 9y = 12$ (2)

These equations are identical, so the system is dependent.

72

11.
$$x + 3y - z = 3 \quad (1)$$
$$2x - y + 3z = 1 \quad (2)$$
$$3x + 2y + z = 5 \quad (3)$$
To eliminate x, multiply (1) by -2 and add the result to (2):
$$-2x - 6y + 2z = -6 \quad (1a)$$
$$\underline{2x - y + 3z = 1 \quad (2)}$$
$$-7y + 5z = -5 \quad (4)$$
Multiply (1) by -3 and add the result to (3):
$$-3x - 9y + 3z = -9 \quad (1b)$$
$$\underline{3x + 2y + z = 5 \quad (3)}$$
$$-7y + 4z = -4 \quad (5)$$
Now consider the system formed by (4) and (5):
$$-7y + 5z = -5 \quad (4)$$
$$-7y + 4z = -4 \quad (5)$$
Multiply (5) by -1 and add the result to (4):
$$-7y + 5z = -5 \quad (4)$$
$$\underline{7y - 4z = 4 \quad (5a)}$$
$$z = -1$$
Put this value for z into (4):
$$-7y + 5(-1) = -5$$
$$-7y - 5 = -5$$
$$-7y = 0$$
$$y = 0$$
Put the values for y and z into (1):
$$x + 3(0) - (-1) = 3$$
$$x + 1 = 3$$
$$x = 2$$
The solution is $(2, 0, -1)$.

13.
$$x + z = 5 \quad (1)$$
$$y - z = -8 \quad (2)$$
$$2x + z = 7 \quad (3)$$
To eliminate z, multiply (1) by -1 and add the result to (3):
$$-x - z = -5 \quad (1a)$$
$$\underline{2x + z = 7 \quad (3)}$$
$$x = 2$$
Put this value for x into (1):
$$2 + z = 5$$
$$z = 3$$
Put this value for z into (2):
$$y - 3 = -8$$
$$y = -5$$
The solution is $(2, -5, 3)$.

15. $\frac{1}{2}x + y + z = 3$

$x - 2y - \frac{1}{3}z = -5$

$\frac{1}{2}x - 3y - \frac{2}{3}z = -6$

Put the system into standard form by multiplying the first equation by 2, the second equation by 3, and the third equation by 6.

$x + 2y + 2z = 6$ (1)

$3x - 6y - z = -15$ (2)

$3x - 18y - 4z = -36$ (3)

To eliminate x, multiply (1) by -3 and add the result to (2):

$-3x - 6y - 6z = -18$ (1a)

$\underline{3x - 6y - z = -15}$ (2)

$-12y - 7z = -33$ (4)

Add -1 times (2) to (3):

$-3x + 6y + z = 15$ (2a)

$\underline{3x - 18y - 4z = -36}$ (3)

$-12y - 3z = -21$ (5)

Add -1 times (4) to (5):

$12y + 7z = 33$ (4a)

$\underline{-12y - 3z = -21}$ (5)

$4z = 12$

$z = 3$

Put this value for z into (4):

$-12y - 7(3) = -33$

$-12y - 21 = -33$

$-12y = -12$

$y = 1$

Put these values for y and z into (1):

$x + 2(1) + 2(3) = 6$

$x + 8 = 6$

$x = -2$

The solution is $(-2, 1, 3)$.

17. The augmented matrix is $\begin{bmatrix} 1 & -2 & | & 5 \\ 2 & 1 & | & 5 \end{bmatrix}$.

Let row 1 and row 2 be denoted by R_1 and R_2. Perform matrix reduction:

$\begin{bmatrix} 1 & -2 & | & 5 \\ 2 & 1 & | & 5 \end{bmatrix} \xrightarrow{-2R_1 + R_2} \begin{bmatrix} 1 & -2 & | & 5 \\ 0 & 5 & | & -5 \end{bmatrix}$

which corresponds to the system

$x - 2y = 5$

$5y = -5$

Therefore $y = -1$. Substitute -1 for y to find

$x - 2(-1) = 5$

$x = 3$

The solution is the ordered pair $(3, -1)$.

19. The augmented matrix is

$\begin{bmatrix} 2 & -1 & | & 7 \\ 3 & 2 & | & 14 \end{bmatrix}$.

Let row 1 and row 2 be denoted by R_1 and R_2. Perform matrix reduction:

$\begin{bmatrix} 2 & -1 & | & 7 \\ 3 & 2 & | & 14 \end{bmatrix} \xrightarrow{-\frac{3}{2}R_1 + R_2} \begin{bmatrix} 2 & -1 & | & 7 \\ 0 & \frac{7}{2} & | & \frac{7}{2} \end{bmatrix}$

which corresponds to the system

$2x - y = 7$

$\frac{7}{2}y = \frac{7}{2}$

Therefore, $y = 1$. Substitute 1 for y to find

$2x - 1 = 7$

$x = 4$

The solution is the ordered pair $(4, 1)$.

21. The augmented matrix is $\begin{bmatrix} 1 & 2 & -1 & -3 \\ 2 & -3 & 2 & 2 \\ 1 & -1 & 4 & 7 \end{bmatrix}$. Let row 1, row 2, and row 3 be denoted

by R_1, R_2, and R_3. Perform matrix reduction:

$$\begin{bmatrix} 1 & 2 & -1 & -3 \\ 2 & -3 & 2 & 2 \\ 1 & -1 & 4 & 7 \end{bmatrix} \xrightarrow[\substack{-2R_1+R_2 \\ -R_1+R_3}]{} \begin{bmatrix} 1 & 2 & -1 & -3 \\ 0 & -7 & 4 & 8 \\ 0 & -3 & 5 & 10 \end{bmatrix} \xrightarrow[-\frac{3}{7}R_2+R_3]{} \begin{bmatrix} 1 & 2 & -1 & -3 \\ 0 & -7 & 4 & 8 \\ 0 & 0 & \frac{23}{7} & \frac{46}{7} \end{bmatrix}$$

which corresponds to the system
$x + 2y - z = -3$

$-7y + 4z = 8$

$\dfrac{23}{7} z = \dfrac{46}{7}.$

Therefore, $z = 2$. Substitute 2 for z to find
$-7y + 4(2) = 8$
$y = 0$.
Substitute 2 for z and 0 for y to find
$x + 2(0) - 2 = -3$
$x = -1$.
The solution is the ordered triplet $(-1, 0, 2)$.

23. Let x = the number of questions Lupe answered correctly and y = the number of questions she answered incorrectly. Since there are 40 questions on the exam, $x + y = 40$. The scoring system is $5x - 2y$. Since Lupe got a score of 102, $5x - 2y = 102$.
The system of equations is

$$x + y = 40 \quad (1)$$
$$5x - 2y = 102 \quad (2)$$

We want to know how many questions Lupe answered correctly, i.e., we want to find x. To eliminate y, multiply (1) by 2 and add the result to (2):

$$2x + 2y = 80 \quad (1a)$$
$$\underline{5x - 2y = 102 \quad (2)}$$
$$7x = 182$$
$$x = 26$$

Lupe answered 26 questions correctly.

25. Let x = the amount that Barbara invests at 8% and y = the amount that she invests at 13.5%. Since she is investing \$5000, $x + y = 5000$. To earn \$500 per year in interest, she needs $0.08x + 0.135y = 500$.
The system of equations is

$$x + y = 5000 \quad (1)$$
$$0.08x + 0.135y = 500 \quad (2)$$

To eliminate x, multiply (1) by -0.08 and add the result to (2).

$$-0.08x - 0.08y = -400 \quad (1a)$$
$$\underline{0.08x + 0.135y = 500 \quad (2)}$$
$$0.055y = 100$$
$$y = 1818.\overline{18}$$

Put $y = 1818.18$ into (1):
$$x + 1818.18 = 5000$$
$$x = 3181.82$$

Barbara needs to invest \$3181.82 at 8% and \$1818.18 at 13.5%.

27. Let x, y, and z represent the lengths of the sides of the triangle. Since the perimeter of the triangle is 30 cm, $x + y + z = 30$. One side is 7 cm shorter than the second side, say the side of length x is 7 cm shorter than the side of length y, or $x = y - 7$. The third side is 1 cm longer than the second side, so $z = y + 1$.
The system of equations is

$$x + y + z = 30 \quad (1)$$
$$x = y - 7 \quad (2)$$
$$z = y + 1 \quad (3)$$

(2) and (3) give x and z in terms of y, so substitute these into (1):

$$y - 7 + y + y + 1 = 30$$
$$3y - 6 = 30$$
$$3y = 36$$
$$y = 12$$

Put this value for y into (2):
$$x = 12 - 7 = 5$$
Put the value for y into (3):
$$z = 12 + 1 = 13$$
The sides of the triangle are 5 cm, 12 cm, and 13 cm long.

29. Graph the line $3x - 4y = 12$ with a dashed line. The inequality can be written as $y > \dfrac{3}{4}x - 3$, so the region above the line is shaded. Refer to the graph in the back of the textbook.

31. Graph the line $y = -\dfrac{1}{2}$ with a dashed line. Because the inequality is $y < -\dfrac{1}{2}$, the region below the line is shaded. Refer to the graph in the back of the textbook.

33. Graph the line $y = 3$ with a dashed line and the line $x = 2$ with a solid line. The region above the line $y = 3$ and the region to the left of $x = 2$ are shaded. The solution set is where the shaded regions overlap. Refer to the graph in the back of the textbook.

35. Graph the lines $3x - y = 6$ and $x + 2y = 6$ with dashed lines. The first inequality can be written as $y > 3x - 6$, so the region above the line $3x - y = 6$ is shaded. The second inequality can be written as

$$y > -\frac{x}{2} + 3,$$ so the region above the

line $x + 2y = 6$ is shaded. The solution set is where the shaded regions overlap. Refer to the graph in the back of the textbook.

37. Graph the lines $3x - 4y = 12, x = 0$ (the y-axis), and $y = 0$ (the x-axis). The first inequality can be written as

$$y \geq \frac{3}{4}x - 3,$$ so the region that is

shaded is above the line $3x - 4y = 12$, to the right of the y-axis, and below the x-axis. The vertices are
$(0, 0)$, the intersection of the lines $x = 0$ and $y = 0$;
$(4, 0)$, the intersection of $y = 0$ and $3x - 4y = 12$; and
$(0, -3)$, the intersection of $x = 0$ and $3x - 4y = 12$.
Refer to the graph in the back of the textbook.

39. Graph the lines $x + y = 5, y = x,$ $y = 2,$ and $x = 0$. The first inequality can be written as $y \leq 5 - x$, so the solution set is the region below the line $x + y = 5$, above the line $y = x$, above the line $y = 2$, and to the right of the y-axis. The vertices are $(0, 2)$, the intersection of the lines $x = 0$ and $y = 2$; $(0, 5)$, the intersection of $x = 0$ and $x + y = 5$; $\left(\frac{5}{2}, \frac{5}{2}\right)$, the intersection of $y = x$ and $x + y = 5$; and $(2, 2)$, the intersection of $y = x$ and $y = 2$. Refer to the graph in the back of the textbook.

41. Let $p =$ the number of batches of peanut butter cookies and $g =$ the number of batches of granola cookies. From the mixing considerations, $20p + 8g \leq 120$, or $5p + 2g \leq 30$, with time in minutes. From the baking considerations, $10p + 10g \leq 120$, or $p + g \leq 12$. Since Ruth cannot make a negative number of cookies, $p \geq 0$ and $g \geq 0$. The system of inequalities is
$5p + 2g \leq 30$
$p + g \leq 12$
$p \geq 0, g \geq 0$
Graph the lines $5p + 2g = 30,$ $p + g = 12, p = 0,$ and $g = 0$, with p along the horizontal axis and g along the vertical axis. The first inequality can be written as

$$g \leq 15 - \frac{5}{2}p$$ and the second as

$g \leq 12 - p$, so the solution set is the region below the lines $5p + 2g = 30$ and $p + g = 12$, above the p-axis, and to the right of the g-axis. Refer to the graph in the back of the textbook.

Chapter Three: Quadratic Models

Homework 3.1

1. $9x^2 = 25$

$$x^2 = \frac{25}{9}$$

$$x = \pm\sqrt{\frac{25}{9}}$$

$$x = \pm\frac{5}{3}$$

The solutions are $x = -\frac{5}{3}$ and $x = \frac{5}{3}$.

3. $4x^2 - 24 = 0$

$$4x^2 = 24$$

$$x^2 = 6$$

$$x = \pm\sqrt{6}$$

The solutions are $x = -\sqrt{6}$ and $x = \sqrt{6}$.

5. $\frac{2x^2}{3} = 4$

$$x^2 = 6$$

$$x = \pm\sqrt{6}$$

The solutions are $x = -\sqrt{6}$ and $x = \sqrt{6}$.

7. $2x^2 = 14$

$$x^2 = 7$$

$$x = \pm\sqrt{7}$$

$$x \approx \pm 2.65$$

The solutions are $x = -2.65$ and $x = 2.65$.

9. $1.5x^2 = 0.7x^2 + 26.2$

$$0.8x^2 = 26.2$$

$$x^2 = 32.75$$

$$x = \pm\sqrt{32.75}$$

$$x \approx \pm 5.72$$

The solutions are $x \approx -5.72$ and $x \approx 5.72$.

11. $5x^2 - 97 = 3.2x^2 - 38$

$$1.8x^2 = 59$$

$$x^2 = \frac{59}{1.8}$$

$$x = \pm\sqrt{\frac{59}{1.8}}$$

$$x \approx \pm 5.73$$

The solutions are $x \approx -5.73$ and $x \approx 5.73$.

13. Solve $F = \frac{mv^2}{r}$, for v

$$Fr = mv^2$$

$$\frac{Fr}{m} = v^2$$

$$\pm\sqrt{\frac{Fr}{m}} = v$$

The solutions are $v = -\sqrt{\frac{Fr}{m}}$ and $v = \sqrt{\frac{Fr}{m}}$.

15. Solve $s = \frac{1}{2}gt^2$, for t

$$2s = gt^2$$
$$\frac{2s}{g} = t^2$$
$$\pm\sqrt{\frac{2s}{g}} = t$$

The solutions are $t = -\sqrt{\frac{2s}{g}}$ and $t = \sqrt{\frac{2s}{g}}$.

17. a. The volume of a cone is
$V = \frac{1}{3}\pi r^2 h$, where r is the radius and h is the height. Substitute $h = 8.4$ into the formula:
$$V = \frac{1}{3}\pi r^2(8.4) \approx 8.8r^2$$

b.

r	1	2	3	4	5	6	7	8
V	8.8	35.2	79.2	140.8	220	316.8	431.2	563.2

As the values of r double, the values of V increase by a factor of 4.

c. Let $V = 302.4$.
$$302.4 = 8.8r^2$$
$$\frac{302.4}{8.8} = r^2$$
$$\sqrt{\frac{302.4}{8.8}} = r$$
$$5.86 \approx r$$
The radius is approximately 5.86 centimeters. (Note that r cannot be negative.)

d. Draw the graphs of $Y_1 = 8.8x^2$ and $Y_2 = 302.4$. Use the *intersect* feature of your calculator to find that they intersect where $x = 5.86$. Refer to the graph in the back of the textbook.

19. a. Let h represent the height of the TV screen.

b. $28^2 + h^2 = 35^2$
$$784 + h^2 = 1225$$
$$h^2 = 441$$
$$h = \sqrt{441} = 21$$
The height of the TV screen is 21 inches. (Note that h cannot be negative.)

21. a. Let d represent the distance from the tip of the shadow to the top of the tree.

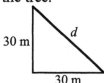

b. $30^2 + 30^2 = d^2$
$$900 + 900 = d^2$$
$$1800 = d^2$$
$$d = \sqrt{1800} \approx 42.4$$
The distance from the tip of the shadow to the top of the tree is approximately 42.4 meters. (Note that d cannot be negative.)

23. a. Let s represent the length of a side of the square.

b. $s^2 + s^2 = 16^2$
$$2s^2 = 256$$
$$s^2 = 128$$
$$s = \sqrt{128}$$
$$\approx 11.3$$

The length of a side of the square is approximately 11.3 inches. (Note that s cannot be negative.)

25. a. Refer to the graph in the back of the textbook.

b. Trace along the curve on your graphing calculator until the y-value is 36. There are two x-values where this occurs: $x = -12$ and $x = 12$. Alternatively, graph the line $y = 36$ in the same window as your graph from part (a) and use the *intersect* feature of your calculator.

c. Check algebraically:
$\frac{1}{4}(-12)^2 = 36$ and $\frac{1}{4}(12)^2 = 36$.

27. a. Refer to the graph in the back of the textbook.

b. Trace along the curve on your graphing calculator until the y-value is 16. There are two x-values where this occurs: $x = 1$ and $x = 9$. Alternatively, graph the line $y = 16$ in the same window as your graph from part (a) and use the *intersect* feature of your calculator.

c. Check algebraically:
$(1 - 5)^2 = 16$ and $(9 - 5)^2 = 16$.

29. a. Refer to the graph in the back of the textbook.

b. Trace along the curve on your graphing calculator until the y-value is 108. There are two x-values where this occurs: $x = -2$ and $x = 10$. Alternatively, graph the line $y = 108$ in the same window as your graph from part (a) and use the *intersect* feature of your calculator.

c. Check algebraically:
$3(-2 - 4)^2 = 108$ and
$3(10 - 4)^2 = 108$.

31. $(x - 2)^2 = 9$
$$x - 2 = \pm\sqrt{9}$$
$$x - 2 = \pm 3$$
$$x - 2 = 3 \quad \text{or} \quad x - 2 = -3$$
$$x = 5 \qquad\qquad x = -1$$

The solutions are $x = 5$ and $x = -1$.

33. $(2x-1)^2 = 16$

$\qquad 2x-1 = \pm\sqrt{16}$

$\qquad 2x-1 = \pm 4$

$\qquad 2x-1 = 4 \quad$ or $\quad 2x-1 = -4$

$\qquad\qquad 2x = 5 \qquad\qquad 2x = -3$

$\qquad\qquad x = \dfrac{5}{2} \qquad\qquad x = \dfrac{-3}{2}$

The solutions are $x = \dfrac{5}{2}$ and $x = \dfrac{-3}{2}$.

35. $4(x+2)^2 = 12$

$\qquad (x+2)^2 = 3$

$\qquad\quad x+2 = \pm\sqrt{3}$

$\qquad\qquad x = -2 \pm \sqrt{3}$

The solutions are $x = -2 + \sqrt{3}$ and $x = -2 - \sqrt{3}$.

37. $\left(x-\dfrac{1}{2}\right)^2 = \dfrac{3}{4}$

$\qquad x - \dfrac{1}{2} = \pm\sqrt{\dfrac{3}{4}}$

$\qquad\quad x = \dfrac{1}{2} \pm \dfrac{\sqrt{3}}{\sqrt{4}}$

$\qquad\quad x = \dfrac{1}{2} \pm \dfrac{\sqrt{3}}{2}$

The solutions are $x = \dfrac{1}{2} + \dfrac{\sqrt{3}}{2}$ and $x = \dfrac{1}{2} - \dfrac{\sqrt{3}}{2}$.

39. $81\left(x+\dfrac{1}{3}\right)^2 = 1$

$\qquad \left(x+\dfrac{1}{3}\right)^2 = \dfrac{1}{81}$

$\qquad\quad x + \dfrac{1}{3} = \pm\sqrt{\dfrac{1}{81}}$

$\qquad\quad x + \dfrac{1}{3} = \pm\dfrac{1}{9}$

$\quad x + \dfrac{1}{3} = \dfrac{1}{9} \quad$ or $\quad x + \dfrac{1}{3} = \dfrac{-1}{9}$

$\qquad\quad x = \dfrac{-2}{9} \qquad\qquad x = \dfrac{-4}{9}$

The solutions are $x = \dfrac{-2}{9}$ and $x = \dfrac{-4}{9}$.

41. $3(8x-7)^2 = 24$

$\qquad (8x-7)^2 = 8$

$\qquad\quad 8x-7 = \pm\sqrt{8}$

$\qquad\qquad 8x = 7 \pm \sqrt{8}$

$\qquad\qquad x = \dfrac{7}{8} \pm \dfrac{\sqrt{8}}{8}$

The solutions are $x = \dfrac{7}{8} + \dfrac{\sqrt{8}}{8}$ and $x = \dfrac{7}{8} - \dfrac{\sqrt{8}}{8}$.

Note: These answers can be simplified to $x = \dfrac{7}{8} + \dfrac{\sqrt{2}}{4}$ and $x = \dfrac{7}{8} - \dfrac{\sqrt{2}}{4}$.

43. a. $B = P(1+r)^n$ We are given that $B = 5000$ and $n = 2$. The equation is $B = 5000(1+r)^2$.

b.

r	0.02	0.04	0.06	0.08
B	5202	5408	5618	5832

c. Put $B = 6250$ into the equation:
$$6250 = 5000(1+r)^2$$
$$1.25 = (1+r)^2$$
$$\pm\sqrt{1.25} = 1+r$$
$$-1 \pm \sqrt{1.25} = r$$
The interest rate must be positive, so
$$r = -1 + \sqrt{1.25} \approx 0.118 \text{ or } 11.8\%.$$

d. Graph $Y_1 = 5000(1+x)^2$ on your calculator. Trace along the curve on your graphing calculator until the *y*-value is 6250. The positive *x*-value where this occurs is $x = 0.118$. Alternatively, graph the line $y = 6250$ in the same window as your graph from part (a) and use the *intersect* feature of your calculator. Refer to the graph in the back of the textbook.

45. Let $P = \$1200$ and let $A = \$1400$ in the formula $A = P(1+r)^n$. Then:
$$1400 = 1200(1+r)^2$$
$$\frac{1400}{1200} = (1+r)^2$$
$$\frac{7}{6} = (1+r)^2$$
$$\pm\sqrt{\frac{7}{6}} = 1+r$$
$$-1 \pm \sqrt{\frac{7}{6}} = r$$
The inflation rate must be positive, so $r = -1 + \sqrt{\frac{7}{6}} \approx 0.08$ or 8%.

47. We are given that $h = 80$ mm and $V = 9000 \ mm^3$:
$$9000 = \pi(r-2)^2 80$$
$$\frac{9000}{80\pi} = (r-2)^2$$
$$\pm\sqrt{\frac{9000}{80\pi}} = r-2$$
$$2 \pm 5.98 \approx r$$
The radius must be positive, so $r \approx 2 + 5.98 = 7.98$ mm.

49. a. The area of an equilateral triangle with sides of length 2 cm is

$$A = \frac{\sqrt{3}}{4}(2)^2 = \sqrt{3} \approx 1.73 \text{ cm}^2.$$

The area of an equilateral triangle with sides of length 4 cm is

$$A = \frac{\sqrt{3}}{4}(4)^2 = 4\sqrt{3} \approx 6.93 \text{ cm}^2.$$

The area of an equilateral triangle with sides of length 10 cm is

$$A = \frac{\sqrt{3}}{4}(10)^2 = 25\sqrt{3} \approx 43.3 \text{ cm}^2.$$

b. Refer to the graph in the back of the textbook.

c. An equilateral triangle with side 5.1 cm has an area of 11.26 cm^2.

d. Trace along the curve until the y-value, which represents the area A, is approximately 20. The first coordinate, which represents the length of a side, is about 6.8 cm.

e.
$$\frac{\sqrt{3}}{4}s^2 = 20$$
$$s^2 = \frac{80}{\sqrt{3}}$$
$$s^2 \approx 46.188$$
$$s \approx 6.8 \text{ cm}$$
Each side is approximately 6.8 cm long.

f.
$$\frac{\sqrt{3}}{4}s^2 = 100\sqrt{3}$$
$$s^2 = 400$$
$$s = 20$$
Each side is approximately 20 cm long.

51.
$$\frac{ax^2}{b} = c$$
$$x^2 = \frac{bc}{a}$$
$$x = \pm\sqrt{\frac{bc}{a}}$$

53.
$$(x-a)^2 = 16$$
$$x - a = \pm\sqrt{16}$$
$$x - a = \pm 4$$
$$x = a \pm 4$$

55.
$$(ax+b)^2 = 9$$
$$ax + b = \pm\sqrt{9}$$
$$ax + b = \pm 3$$
$$ax = -b \pm 3$$
$$x = \frac{-b \pm 3}{a}$$

Homework 3.2

1. a.

Height	Base	Area
1	34	34
2	32	64
3	30	90
4	28	112
5	26	130
6	24	144
7	22	154
8	20	160
9	18	162
10	16	160
11	14	154
12	12	144
13	10	130
14	8	112
15	6	90
16	4	64
17	2	34
18	0	0

b. To draw the graph, use the height and area values from the above table for your ordered pairs, with height as the first coordinate and area as the second coordinate. Refer to the graph in the back of the textbook.

c. The largest value for the area in the table is 162 square feet. This also corresponds to the highest point on the graph in part (b). Therefore, the maximum area is 162 square feet and from the table, we find that this occurs when the height is 9 feet and the base is 18 feet.

d. Let x represent the height of the rectangle and b the length of the base. The area of the region is
$$A = bx = (36 - 2x)x = 36x - 2x^2.$$

e. Enter $Y_1 = 36x - 2x^2$ and go to your *table* feature on your calculator. The x and Y_1 values should correspond to the heights and areas in the table from part (a).

f. Refer to the graph in the back of the textbook.

g. Graph the line $Y_2 = 149.5$ in the same window as your equation from part (e). Use the *intersect* feature to find the x-values of the intersection.

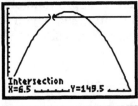

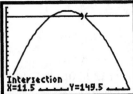

There are two possible rectangles, one with height 6.5 feet and the other with height 11.5 feet.

3. a. Enter the equation $Y_1 = -16x^2 + 20x + 300$ into your calculator. In the table setup menu, enter the table start value of 0 and increment value (ΔTbl) of 0.5. Then go to your table feature on your calculator. You should have

X	Y₁
0	300
.5	306
1	304
1.5	294
2	276
2.5	250
3	216

X=0

b. Refer to the graph in the back of the textbook.

c. Tracing along the curves shows that Delbert's book reaches an altitude of about 306 feet and this occurs after about 0.6 seconds.

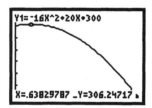

To use your table feature on your calculator to improve this estimate, change the table start value to 0.575 and increment value (ΔTbl) to 0.025.

X	Y₁
.575	306.21
.6	306.24
.625	306.25
.65	306.24
.675	306.21
.7	306.16
.725	306.09

X=.625

The highest altitude reached is about 306.25 feet after about 0.625 seconds.

d. Graph the line $Y_2 = 300$ in the same window. Use the *intersect* feature to find the x-values of the intersection.

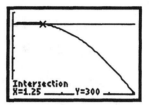

Delbert's book will pass him after 1.25 seconds.

e. It appears from the graph that $h = 0$ when $t = 5$. To confirm, substitute $t = 5$ into the equation: $h = -16(5)^2 + 20(5) + 300 = 0$. The book will hit the ground after 5 seconds.

5. $3x(x-5) = 3x^2 - 15x$

7. $(b+6)(2b-3) = b(2b) + b(-3) + 6(2b) + 6(-3) = 2b^2 - 3b + 12b - 18$
$= 2b^2 + 9b - 18$

9. $(4w-3)^2 = (4w-3)(4w-3) = 4w(4w) + 4w(-3) - 3(4w) - 3(-3)$
$= 16w^2 - 12w - 12w + 9 = 16w^2 - 24w + 9$

11. $3p(2p-5)(p-3) = 3p(2p^2 - 6p - 5p + 15) = 3p(2p^2 - 11p + 15)$
$= 6p^3 - 33p^2 + 45p$

13. $-50(1+r)^2 = -50(1+r)(1+r) = -50(1 + r + r + r^2) = -50(1 + 2r + r^2)$
$= -50 - 100r - 50r^2$

15. $3q^2(2q-3)^2 = 3q^2(2q-3)(2q-3) = 3q^2(4q^2 - 6q - 6q + 9)$
$= 3q^2(4q^2 - 12q + 9) = 12q^4 - 36q^3 + 27q^2$

17. $x^2 - 7x + 10 = (x-2)(x-5)$

19. $x^2 - 225 = (x-15)(x+15)$

21. $w^2 - 4w - 32 = (w-8)(w+4)$

23. $2z^2 + 11z - 40 = (2z-5)(z+8)$

25. $9n^2 + 24n + 16 = (3n+4)(3n+4) = (3n+4)^2$

27. $3a^4 + 6a^3 + 3a^2 = 3a^2(a^2 + 2a + 1) = 3a^2(a+1)(a+1) = 3a^2(a+1)^2$

29. $4h^4 - 36h^2 = 4h^2(h^2 - 9) = 4h^2(h-3)(h+3)$

31. $-10u^2 - 100u + 390 = -10(u^2 + 10u - 39) = -10(u-3)(u+13)$

33. $24t^4 + 6t^2 = 6t^2(4t^2 + 1)$ Note that the sum of two squares is not factorable.

Homework 3.3

Note: As suggested in the text's instructions, the graphs in 1 - 8 use Xmax = −9.4 and Xmax = 9.4, the "friendly" window for the TI-83. Other graphing calculators may have different "friendly" windows. In some problems it may be helpful to ZOOM in on the x-intercepts in order to more easily locate them using TRACE.

1. Ymin = −13, Ymax = 13

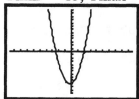

$y = (2x + 5)(x − 2)$, so solve
$(2x + 5)(x − 2) = 0$.

$$2x + 5 = 0 \quad \text{or} \quad x − 2 = 0$$
$$2x = −5 \qquad\qquad x = 2$$
$$x = −\frac{5}{2}$$

The solutions are $x = −\frac{5}{2}$ and $x = 2$.

3. Ymin = −10, Ymax = 10

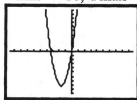

$y = x(3x + 10)$, so solve
$x(3x + 10) = 0$.

$$x = 0 \quad \text{or} \quad 3x + 10 = 0$$
$$3x = −10$$
$$x = −\frac{10}{3}$$

The solutions are $x = 0$ and
$x = −\frac{10}{3}$.

5. Ymin = −60, Ymax = 60

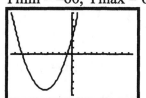

$y = (4x + 3)(x + 8)$, so solve
$(4x + 3)(x + 8) = 0$.

$$4x + 3 = 0 \quad \text{or} \quad x + 8 = 0$$
$$4x = −3 \qquad\qquad x = −8$$
$$x = −\frac{3}{4}$$

The solutions are $x = −\frac{3}{4}$ and
$x = −8$.

7. Ymin = −5, Ymax = 5

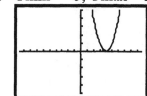

$y = (x − 4)^2$, so solve

$$(x − 4)^2 = 0$$
$$x − 4 = 0$$
$$x = 4$$

The solution is $x = 4$.

9. $\quad 2a^2 + 5a − 3 = 0$
$$(2a − 1)(a + 3) = 0$$
$$2a − 1 = 0 \quad \text{or} \quad a + 3 = 0$$
$$2a = 1 \qquad\qquad a = −3$$
$$a = \frac{1}{2}$$

The solutions are $a = −3$ and $a = \frac{1}{2}$.

11.
$$2x^2 = 6x$$
$$2x^2 - 6x = 0$$
$$2x(x - 3) = 0$$
$$2x = 0 \quad \text{or} \quad x - 3 = 0$$
$$x = 0 \qquad\qquad x = 3$$
The solutions are $x = 0$ and $x = 3$.

13.
$$3y^2 - 6y = -3$$
$$3y^2 - 6y + 3 = 0$$
$$3(y^2 - 2y + 1) = 0$$
$$3(y - 1)(y - 1) = 0$$
$$3(y - 1)^2 = 0$$
$$y - 1 = 0$$
$$y = 1$$
The solution is $y = 1$.

15.
$$x(2x - 3) = -1$$
$$2x^2 - 3x = -1$$
$$2x^2 - 3x + 1 = 0$$
$$(2x - 1)(x - 1) = 0$$
$$2x - 1 = 0 \quad \text{or} \quad x - 1 = 0$$
$$2x = 1 \qquad\qquad x = 1$$
$$x = \frac{1}{2}$$
The solutions are $x = 1$ and $x = \frac{1}{2}$.

17.
$$t(t - 3) = 2(t - 3)$$
$$t(t - 3) - 2(t - 3) = 0$$
$$(t - 3)(t - 2) = 0$$
$$t - 3 = 0 \quad \text{or} \quad t - 2 = 0$$
$$t = 3 \qquad\qquad t = 2$$
The solutions are $t = 3$ and $t = 2$.

19.
$$z(3z + 2) = (z + 2)^2$$
$$3z^2 + 2z = (z + 2)(z + 2)$$
$$3z^2 + 2z = z^2 + 2z + 2z + 4$$
$$3z^2 + 2z = z^2 + 4z + 4$$
$$2z^2 - 2z - 4 = 0$$
$$2(z^2 - z - 2) = 0$$
$$2(z + 1)(z - 2) = 0$$
$$z + 1 = 0 \quad \text{or} \quad z - 2 = 0$$
$$z = -1 \qquad\qquad z = 2$$
The solutions are $z = 2$ and $z = -1$.

21.
$$(v + 2)(v - 5) = 8$$
$$v^2 - 5v + 2v - 10 = 8$$
$$v^2 - 3v - 18 = 0$$
$$(v - 6)(v + 3) = 0$$
$$v - 6 = 0 \quad \text{or} \quad v + 3 = 0$$
$$v = 6 \qquad\qquad v = -3$$
The solutions are $v = 6$ and $v = -3$.

23.

All three graphs have x-intercepts $(-4, 0)$ and $(5, 0)$. In general, the graph of $y = ax^2 + bx + c$ has the same x-intercepts as the graph of $y = k(ax^2 + bx + c)$. To test this generalization, choose any values for k, a, b, and c and draw the graphs of the two equations.

25.

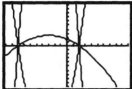

All three graphs have *x*-intercepts
(−8, 0) and (2, 0). See problem 23
for generalization.

27. $[x-(-2)](x-1)=0$
$(x+2)(x-1)=0$
$x^2-x+2x-2=0$
$x^2+x-2=0$

29. $(x-0)[x-(-5)]=0$
$x(x+5)=0$
$x^2+5x=0$

31. $[x-(-3)]\left(x-\frac{1}{2}\right)=0$
$(x+3)\left(x-\frac{1}{2}\right)=0$
$x^2-\frac{1}{2}x+3x-\frac{3}{2}=0$
$x^2+\frac{5}{2}x-\frac{3}{2}=0$
$2\left(x^2+\frac{5}{2}x-\frac{3}{2}\right)=2(0)$
$2x^2+5x-3=0$

33. $\left[x-\left(\frac{-1}{4}\right)\right]\left(x-\frac{3}{2}\right)=0$
$\left(x+\frac{1}{4}\right)\left(x-\frac{3}{2}\right)=0$
$x^2-\frac{3}{2}x+\frac{1}{4}x-\frac{3}{8}=0$
$x^2-\frac{5}{4}x-\frac{3}{8}=0$
$8\left(x^2-\frac{5}{4}x-\frac{3}{8}\right)=8(0)$
$8x^2-10x-3=0$

35.

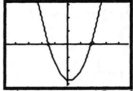

x-intercepts: −15 and 18. So the
equation is $y=0.1(x-18)(x+15)$.

37.

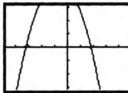

x-intercepts: −32 and 18. So the
equation is $y=-0.08(x+32)(x-18)$

39. a. Let *x* represent the height of the
window. Then the length of the
ladder is *x* + 2. By the
Pythagorean theorem:
$x^2+10^2=(x+2)^2$

b. $x^2+100=x^2+4x+4$
$100=4x+4$
$96=4x$
$24=x$
The window is 24 feet high.

41. a. $h = \frac{-1}{2}(32)t^2 + 16t + 8$

$h = -16t^2 + 16t + 8$

b. When $t = \frac{1}{2}$,

$h = -16\left(\frac{1}{2}\right)^2 + 16\left(\frac{1}{2}\right) + 8$

$h = -16\left(\frac{1}{4}\right) + 8 + 8$

$h = -4 + 16$

$h = 12$

The tennis ball is 12 feet in the air when $t = \frac{1}{2}$ sec.

When $t = 1$,

$h = -16(1)^2 + 16(1) + 8$

$h = -16 + 16 + 8$

$h = 8$

The tennis ball is 8 feet in the air when $t = 1$ sec.

c. Let $h = 11$:

$11 = -16t^2 + 16t + 8$

$16t^2 - 16t + 3 = 0$

$(4t - 1)(4t - 3) = 0$

$4t - 1 = 0 \quad$ or $\quad 4t - 3 = 0$

$\qquad 4t = 1 \qquad\qquad 4t = 3$

$\qquad t = \frac{1}{4} \qquad\qquad t = \frac{3}{4}$

The tennis ball is 11 feet in the air after $\frac{1}{4}$ sec and after $\frac{3}{4}$ sec.

d. Use $\Delta \text{Tbl} = 0.25$ in your *table setup* menu..

e. Refer to the graph in the back of the textbook.

f. By tracing along the curve drawn in part (e), the first coordinate (*t*) is about 1.37 when the second coordinate (*h*) is 0. Therefore, the tennis ball will hit the ground after about 1.37 seconds.

43. a. Since there are 360 yards of fence, the perimeter of the pasture is 360 yards. Let l and w represent the length and width of the pasture, respectively. Then $2l + 2w = 360$, or $l + w = 180$. So $l = 180 - w$.

Width	Length	Area
10	170	1700
20	160	3200
30	150	4500
40	140	5600
50	130	6500
60	120	7200
70	110	7700
80	100	8000

The pasture will contain 8000 square yards when its dimensions are 80 yards by 100 yards.

b. If the width is represented by x, the length is $l = 180 - x$ (from part (a)), so the area is
$A = lw = (180 - x)x = 180x - x^2$.
Use Xmin = 0, Xmax = 200, Ymin = 0, and Ymax = 9000. Graph the area equation as Y_1 and graph the line $Y_2 = 8000$. Use the *intersect* feature to find the two intersection points, which occur at widths $x = 80$ and $x = 100$.

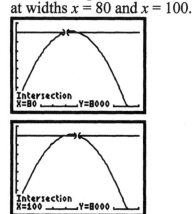

The width is either 80 yards or 100 yards. When the width is 80 yards, the length is $180 - 80 = 100$ yards and vice versa, so the pasture has dimensions 80 yards by 100 yards.

c. From (b), we have
$A = 180x - x^2$. Let $A = 8000$.
$$8000 = 180x - x^2$$
$$x^2 - 180x + 8000 = 0$$
$$(x - 80)(x - 100) = 0$$
As in part (b), $x = 80$ or $x = 100$. The width is either 80 yards or 100 yards. When the width is 80 yards, the length is $180 - 80 = 100$ yards and vice versa, so the pasture has dimensions 80 yards by 100 yards.

There are two solutions because the pasture can be oriented in two directions.

45. a. If the cardboard is x inches long, then the sides (length and width) of the box will be $x - 4$ inches long, and the height of the box is 2 inches. $V = lwh$, so
$$V = (x - 4)(x - 4)2$$
$$= 2(x - 4)^2$$

b. In your graphing calculator, enter $Y_1 = 2(x - 4)^2$, table start (TblStart) as 4, and increment size (ΔTbl) as 1. This gives the table:

c. Use Xmin = 0, Xmax = 10, Ymin = 0, and Ymax = 100.

V increases as x increases for $x > 4$. Note that the problem does not make sense for $x < 4$, since in this case there wouldn't be enough material to cut out 2 inch squares from the corners.

d. From the table in part (b), we find that the cardboard must have side length 9 inches in order to make a box with volume 50 cubic inches. Tracing along the curve in part (c) confirms this result.

e. If $V = 50$ in.^{3}, solve
$$2(x - 4)^2 = 50$$
$$(x - 4)^2 = 25$$
$$x - 4 = \pm\sqrt{25}$$
$$x - 4 = \pm 5$$
$$x - 4 = 5 \quad \text{or} \quad x - 4 = -5$$
$$x = 9 \qquad\qquad x = -1$$

Since the length can never be negative, the piece of cardboard must be 9 inches on a side. This confirms the result from part (d).

Midchapter 3 Review

1. $x^2 + 7 = 13 - 2x^2$
$$3x^2 = 6$$
$$x^2 = 2$$
$$x = \pm\sqrt{2}$$
The solutions are $x = -\sqrt{2}$ and $x = \sqrt{2}$.

3. $3(x+4)^2 = 60$
$$(x+4)^2 = 20$$
$$x + 4 = \pm\sqrt{20}$$
$$x = -4 \pm \sqrt{20}$$
The solutions are
$x = -4 + \sqrt{20} = -4 + 2\sqrt{5}$ and
$x = -4 - \sqrt{20} = -4 - 2\sqrt{5}$.

5. $A = \dfrac{3\sqrt{3}}{2}s^2$, for s:
$$\frac{2A}{3\sqrt{3}} = s^2$$
$$\pm\sqrt{\frac{2A}{3\sqrt{3}}} = s$$
The solutions are $s = -\sqrt{\dfrac{2A}{3\sqrt{3}}}$ and $s = \sqrt{\dfrac{2A}{3\sqrt{3}}}$.

7. The formula for the volume of a cylinder with radius r is $V = \pi r^2 h$. Say that one cylinder has a radius r and another has a radius $R = 2r$. Then the volume of the cylinder with radius R is $V = \pi R^2 h = \pi (2r)^2 h = 4\pi r^2 h$. Hence, if the radius is doubled, the volume quadruples.

9. **a.** $A = lw$ and $P = 2l + 2w$.

   b. Many answers are possible. For example the rectangle could have dimensions 29 inches by 1 inch. In this case, $P = 2(29) + 2(1) = 60$ inches and $A = (29)(1) = 29$ square inches. Or, for example, the rectangle could have dimensions 16 inches by 14 inches. In this case, $P = 2(16) + 2(14) = 60$ inches and $A = (16)(14) = 224$ square inches.

11. $-3b(6b - 2) = -18b^2 + 6b$

13. $2(t - 8)(2t - 1)$
$$= 2(2t^2 - t - 16t + 8)$$
$$= 2(2t^2 - 17t + 8)$$
$$= 4t^2 - 34t + 16$$

15. $2x^2 + 8x - 42$
$$= 2(x^2 + 4x - 21)$$
$$= 2(x + 7)(x - 3)$$

17. $49a^2 + 28ab + 4b^2$
$$= (7a + 2b)(7a + 2b)$$
$$= (7a + 2b)^2$$

19. The product of two factors equals zero if and only if one or both of the factors equals zero.

21. Ymin $= -12$, Ymax $= 12$

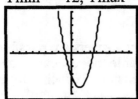

$y = (2x - 7)(x + 1)$, so solve $(2x - 7)(x + 1) = 0$.

$$2x - 7 = 0 \qquad \text{or} \quad x + 1 = 0$$
$$2x = 7 \qquad\qquad x = -1$$
$$x = \frac{7}{2} = 3\frac{1}{2}$$

The solutions are $x = 3\frac{1}{2}$ and $x = -1$.

23. $x^2 + x = 4 - (x + 2)^2$

$$x^2 + x = 4 - \left(x^2 + 4x + 4\right)$$
$$x^2 + x = 4 - x^2 - 4x - 4$$
$$x^2 + x = -x^2 - 4x$$
$$2x^2 + 5x = 0$$
$$x(2x + 5) = 0$$
$$x = 0 \quad \text{or} \quad 2x + 5 = 0$$
$$2x = -5$$
$$x = -\frac{5}{2}$$

The solutions are $x = 0$ and $x = -\frac{5}{2}$.

25. Let d $= 40$:

$$40 = 50t - 10t^2$$
$$10t^2 - 50t + 40 = 0$$
$$10\left(t^2 - 5t + 4\right) = 0$$
$$10(t - 4)(t - 1) = 0$$
$$t - 4 = 0 \quad \text{or} \quad t - 1 = 0$$
$$t = 4 \qquad\qquad t = 1$$

We disregard the answer $t = 4$ since the car has already come to a stop at t $= 2.5$ seconds. Hence it takes 1 second for the car to travel 40 feet.

Homework 3.4

1. Refer to the graph in the back of the textbook. The parabola will open upward and be narrower than the standard parabola.

3. Refer to the graph in the back of the textbook. The parabola will open upward and be wider than the standard parabola.

5. Refer to the graph in the back of the textbook. The parabola will open downward, but will be no wider or narrower than the standard parabola.

7. Refer to the graph in the back of the textbook. The parabola will open downward and be wider than the standard parabola.

9. Refer to the graph in the back of the textbook. This is the standard parabola shifted 2 units upward. The vertex is at $(0, 2)$. There are no x-intercepts.

11. Refer to the graph in the back of the textbook. This is the standard parabola shifted 1 unit downward. The vertex is at $(0, -1)$. To find the x-intercepts, solve the equation:
$$x^2 - 1 = 0$$
$$x^2 = 1$$
$$x = \pm\sqrt{1} = \pm 1$$
The x-intercepts are at $(1, 0)$ and $(-1, 0)$.

13. Refer to the graph in the back of the textbook. This is the standard parabola shifted 5 units downward. The vertex is at $(0, -5)$. To find the x-intercepts, solve the equation:
$$x^2 - 5 = 0$$
$$x^2 = 5$$
$$x = \pm\sqrt{5}$$
The x-intercepts are at $(\sqrt{5}, 0)$ and $(-\sqrt{5}, 0)$.

15. Refer to the graph in the back of the textbook. This is the standard parabola reflected about the x-axis and shifted 100 units upward. The vertex is at $(0, 100)$. To find the x-intercepts, solve the equation:
$$100 - x^2 = 0$$
$$100 = x^2$$
$$x = \pm\sqrt{100} = \pm 10$$
The x-intercepts are at $(10, 0)$ and $(-10, 0)$.

17. Refer to the graph in the back of the textbook. In $y = x^2 - 4x$, $a = 1$, $b = -4$, and $c = 0$. The vertex is where $x = \dfrac{-b}{2a}$, so in this case,
$x = \dfrac{-(-4)}{2(1)} = \dfrac{4}{2} = 2$. When $x = 2$,
$y = (2)^2 - 4(2) = 4 - 8 = -4$, so the vertex is at $(2, -4)$. The x-intercepts occur when $y = 0$, so solve:
$$x^2 - 4x = 0$$
$$x(x - 4) = 0.$$
$$x = 0 \quad \text{or} \quad x - 4 = 0$$
$$x = 4$$
The x-intercepts are $(0, 0)$ and $(4, 0)$.

19. Refer to the graph in the back of the textbook. In $y = x^2 + 2x$, $a = 1$, $b = 2$, and $c = 0$. The vertex is where $x = \frac{-2}{2(1)} = -\frac{2}{2} = -1$. When $x = -1$,

$y = (-1)^2 + 2(-1) = 1 - 2 = -1$, so the vertex is at $(-1, -1)$.

For the x-intercepts, solve
$$0 = x^2 + 2x$$
$$0 = x(x + 2)$$
$$x = 0 \quad \text{or} \quad x + 2 = 0$$
$$x = -2$$
The x-intercepts are at $(0, 0)$ and $(-2, 0)$.

21. Refer to the graph in the back of the textbook. In $y = 3x^2 + 6x$, $a = 3$, $b = 6$, and $c = 0$. The vertex is where $x = \frac{-6}{2(3)} = \frac{-6}{6} = -1$. When $x = -1$,

$y = 3(-1)^2 + 6(-1) = 3 - 6 = -3$, so the vertex is at $(-1, -3)$.

For the x-intercepts, solve
$$3x^2 + 6x = 0$$
$$3x(x + 2) = 0.$$
$$3x = 0 \quad \text{or} \quad x + 2 = 0$$
$$x = 0 \qquad\qquad x = -2$$
The x-intercepts are at $(0, 0)$ and $(-2, 0)$.

23. Refer to the graph in the back of the textbook. In $y = -2x^2 + 5x$, $a = -2$, $b = 5$, and $c = 0$. The vertex is where $x = \frac{-5}{2(-2)} = \frac{-5}{-4} = \frac{5}{4}$.

When $x = \frac{5}{4}$,
$$y = -2\left(\frac{5}{4}\right)^2 + 5\left(\frac{5}{4}\right)$$
$$= -2\left(\frac{25}{16}\right) + \frac{25}{4}$$
$$= \frac{-25}{8} + \frac{25}{4}$$
$$= \frac{25}{8}$$

so the vertex is at $\left(\frac{5}{4}, \frac{25}{8}\right)$.

For the x-intercepts, solve
$$0 = -2x^2 + 5x$$
$$0 = x(-2x + 5).$$
$$x = 0 \quad \text{or} \quad -2x + 5 = 0$$
$$-2x = -5$$
$$x = \frac{5}{2}$$
The x-intercepts are at $(0, 0)$ and $\left(\frac{5}{2}, 0\right)$.

25. The constants affect the width of the parabola and whether or not it is reflected about the x-axis. If the constant is between -1 and 1, the graph is wider than the standard parabola. If the constant is greater than 1 or less than -1, the graph is narrower than the standard parabola. If the constant is negative, the graph is reflected about the x-axis. Refer to the graph in the back of the textbook.

27. The constants affect the location of the x-intercepts and whether or not the parabola opens upward or downward. For each graph, one x-intercept is $(0, 0)$ but the other one depends on the values of the constants. If the squared term has a negative coefficient, the parabola opens downward. Refer to the graph in the back of the textbook.

29. a. II

   b. IV

   c. I

   d. III

   e. VI

   f. V

31. a. Let $y = 0$:

$$0 = 0.4x - 0.0001x^2$$
$$0 = x(0.4 - 0.0001x)$$

$x = 0$ or

$$0.4 - 0.0001x = 0$$
$$-0.0001x = -0.4$$
$$x = 4000$$

The x-intercepts are $(0, 0)$ and $(4000, 0)$. When the fish population has biomass 0 tons or biomass 4000 tons, there is no increase in biomass.

   b. $a = -0.0001$ and $b = 0.4$, so the x-coordinate of the vertex is

$$x = -\frac{b}{2a} = -\frac{0.4}{2(-0.0001)} = 2000.$$

The y-coordinate is

$$y = 0.4(2000) - 0.0001(2000)^2$$
$$= 400.$$

Hence the vertex is $(2000, 400)$. The largest annual increase in biomass, 400 tons, occurs when the biomass is 2000 tons.

   c. Refer to the graph in the back of the textbook.

   d. The biomass decreases for $4000 < x \le 5000$. If the population becomes too large, its supply of food may be inadequate.

Homework 3.5

Note: Simplifying radical expressions is not a required skill in this chapter. However, for the benefit of students who have already learned this skill, the simplified answers are noted, where relevant, after the required answer. See, for example, exercise 13. For information on simplifying radicals, see Chapter 6.

1. In $x^2 + 8x$, since one-half of 8 is 4, add $4^2 = 16$ to get
$$x^2 + 8x + 16 = (x+4)^2$$

3. In $x^2 - 7x$, since one-half of -7 is $-\frac{7}{2}$, add $\left(\frac{-7}{2}\right)^2 = \frac{49}{4}$ to get
$$x^2 - 7x + \frac{49}{4} = \left(x - \frac{7}{2}\right)^2$$

5. In $x^2 + \frac{3}{2}x$, since one-half of $\frac{3}{2}$ is $\frac{3}{4}$, add $\left(\frac{3}{4}\right)^2 = \frac{9}{16}$ to get
$$x^2 + \frac{3}{2}x + \frac{9}{16} = \left(x + \frac{3}{4}\right)^2$$

7. In $x^2 - \frac{4}{5}x$, since one-half of $-\frac{4}{5}$ is $-\frac{2}{5}$, add $\left(-\frac{2}{5}\right)^2 = \frac{4}{25}$ to get
$$x^2 - \frac{4}{5}x + \frac{4}{25} = \left(x - \frac{2}{5}\right)^2$$

9. Note that $\left[\frac{1}{2}(-2)\right]^2 = 1$, so this is already a perfect square.
$$x^2 - 2x + 1 = 0$$
$$(x-1)^2 = 0$$
$$x - 1 = 0$$
$$x = 1$$

11. $$x^2 + 9x + 20 = 0$$
$$x^2 + 9x = -20$$
One-half of 9 is $\frac{9}{2}$, so add
$$\left(\frac{9}{2}\right)^2 = \frac{81}{4} \text{ to both sides.}$$
$$x^2 + 9x + \frac{81}{4} = -20 + \frac{81}{4}$$
$$\left(x + \frac{9}{2}\right)^2 = \frac{1}{4}$$
$$x + \frac{9}{2} = \pm\sqrt{\frac{1}{4}}$$
$$x + \frac{9}{2} = \pm\frac{1}{2}$$

$$x + \frac{9}{2} = \frac{1}{2} \quad \text{or} \quad x + \frac{9}{2} = -\frac{1}{2}$$
$$x = -\frac{8}{2} \qquad\qquad x = -\frac{10}{2}$$
$$x = -4 \qquad\qquad x = -5$$

13. $$x^2 = 3 - 3x$$
$$x^2 + 3x = 3$$
Since one-half of 3 is $\frac{3}{2}$, add
$$\left(\frac{3}{2}\right)^2 = \frac{9}{4} \text{ to both sides.}$$
$$x^2 + 3x + \frac{9}{4} = 3 + \frac{9}{4}$$
$$\left(x + \frac{3}{2}\right)^2 = \frac{21}{4}$$
$$x + \frac{3}{2} = \pm\sqrt{\frac{21}{4}}$$
$$x = -\frac{3}{2} \pm \sqrt{\frac{21}{4}}$$

Note: This answer can be simplified to $x = -\frac{3}{2} \pm \frac{\sqrt{21}}{2}$.

15.
$$2x^2 + 4x - 3 = 0$$
$$\tfrac{1}{2}\left(2x^2 + 4x - 3\right) = \tfrac{1}{2}(0)$$
$$x^2 + 2x - \tfrac{3}{2} = 0$$
$$x^2 + 2x = \tfrac{3}{2}$$

Since one-half of 2 is 1, add $1^2 = 1$ to both sides.
$$x^2 + 2x + 1 = \tfrac{3}{2} + 1$$
$$(x+1)^2 = \tfrac{5}{2}$$
$$x + 1 = \pm\sqrt{\tfrac{5}{2}}$$
$$x = -1 \pm \sqrt{\tfrac{5}{2}}$$

Note: This answer can be simplified to $x = -1 \pm \dfrac{\sqrt{10}}{2}$.

17.
$$3x^2 + x = 4$$
$$\tfrac{1}{3}\left(3x^2 + x\right) = \tfrac{1}{3}(4)$$
$$x^2 + \tfrac{1}{3}x = \tfrac{4}{3}$$

Since one-half of $\tfrac{1}{3}$ is $\tfrac{1}{6}$, add
$$\left(\tfrac{1}{6}\right)^2 = \tfrac{1}{36}$$ to both sides.
$$x^2 + \tfrac{1}{3}x + \tfrac{1}{36} = \tfrac{4}{3} + \tfrac{1}{36}$$
$$\left(x + \tfrac{1}{6}\right)^2 = \tfrac{49}{36}$$
$$x + \tfrac{1}{6} = \pm\sqrt{\tfrac{49}{36}}$$
$$x = -\tfrac{1}{6} \pm \tfrac{7}{6}$$
$$x = -\tfrac{1}{6} + \tfrac{7}{6} \quad \text{or} \quad x = -\tfrac{1}{6} - \tfrac{7}{6}$$
$$x = 1 \qquad\qquad x = -\tfrac{4}{3}$$

19.
$$4x^2 - 3 = 2x$$
$$4x^2 - 2x - 3 = 0$$
$$\tfrac{1}{4}\left(4x^2 - 2x - 3\right) = \tfrac{1}{4}(0)$$
$$x^2 - \tfrac{1}{2}x - \tfrac{3}{4} = 0$$
$$x^2 - \tfrac{1}{2}x = \tfrac{3}{4}$$

Since one-half of $-\tfrac{1}{2}$ is $-\tfrac{1}{4}$, add
$$\left(-\tfrac{1}{4}\right)^2 = \tfrac{1}{16}$$ to both sides.
$$x^2 - \tfrac{1}{2}x + \tfrac{1}{16} = \tfrac{3}{4} + \tfrac{1}{16}$$
$$\left(x - \tfrac{1}{4}\right)^2 = \tfrac{13}{16}$$
$$x - \tfrac{1}{4} = \pm\sqrt{\tfrac{13}{16}}$$
$$x = \tfrac{1}{4} \pm \sqrt{\tfrac{13}{16}}$$

Note: This answer can be simplified to $x = \tfrac{1}{4} \pm \dfrac{\sqrt{13}}{4}$.

21.
$$3x^2 - x - 4 = 0$$
$$\tfrac{1}{3}\left(3x^2 - x - 4\right) = \tfrac{1}{3}(0)$$
$$x^2 - \tfrac{1}{3}x - \tfrac{4}{3} = 0$$
$$x^2 - \tfrac{1}{3}x = \tfrac{4}{3}$$

Since one-half of $-\tfrac{1}{3}$ is $-\tfrac{1}{6}$, add

$\left(-\tfrac{1}{6}\right)^2 = \tfrac{1}{36}$ to both sides.
$$x^2 - \tfrac{1}{3}x + \tfrac{1}{36} = \tfrac{4}{3} + \tfrac{1}{36}$$
$$\left(x - \tfrac{1}{6}\right)^2 = \tfrac{49}{36}$$
$$x - \tfrac{1}{6} = \pm\sqrt{\tfrac{49}{36}}$$
$$x = \tfrac{1}{6} \pm \tfrac{7}{6}$$
$$x = \tfrac{1}{6} + \tfrac{7}{6} \quad \text{or} \quad x = \tfrac{1}{6} - \tfrac{7}{6}$$
$$x = \tfrac{4}{3} \qquad\qquad x = -1$$

23.
$$5x^2 + 8x = 4$$
$$\tfrac{1}{5}\left(5x^2 + 8x\right) = \tfrac{1}{5}(4)$$
$$x^2 + \tfrac{8}{5}x = \tfrac{4}{5}$$

Since one-half of $\tfrac{8}{5}$ is $\tfrac{4}{5}$, add

$\left(\tfrac{4}{5}\right)^2 = \tfrac{16}{25}$ to both sides.
$$x^2 + \tfrac{8}{5}x + \tfrac{16}{25} = \tfrac{4}{5} + \tfrac{16}{25}$$
$$\left(x + \tfrac{4}{5}\right)^2 = \tfrac{36}{25}$$
$$x + \tfrac{4}{5} = \pm\sqrt{\tfrac{36}{25}}$$
$$x = -\tfrac{4}{5} \pm \tfrac{6}{5}$$
$$x = -\tfrac{4}{5} + \tfrac{6}{5} \quad \text{or} \quad x = -\tfrac{4}{5} - \tfrac{6}{5}$$
$$x = \tfrac{2}{5} \qquad\qquad x = -2$$

25.
$$x^2 + 2x + c = 0$$
$$x^2 + 2x = -c$$
$$x^2 + 2x + \left[\tfrac{1}{2}(2)\right]^2 = -c + \left[\tfrac{1}{2}(2)\right]^2$$
$$x^2 + 2x + 1 = -c + 1$$
$$(x + 1)^2 = 1 - c$$
$$x + 1 = \pm\sqrt{1 - c}$$
$$x = -1 \pm \sqrt{1 - c}$$

27.
$$x^2 + bx + 1 = 0$$
$$x^2 + bx = -1$$
$$x^2 + bx + \left[\tfrac{1}{2}(b)\right]^2 = -1 + \left[\tfrac{1}{2}(b)\right]^2$$
$$x^2 + bx + \tfrac{b^2}{4} = -1 + \tfrac{b^2}{4}$$
$$\left(x + \tfrac{b}{2}\right)^2 = \tfrac{b^2}{4} - 1$$
$$x + \tfrac{b}{2} = \pm\sqrt{\tfrac{b^2}{4} - 1}$$
$$x = -\tfrac{b}{2} \pm \sqrt{\tfrac{b^2}{4} - 1}$$
$$x = -\tfrac{b}{2} \pm \sqrt{\tfrac{b^2 - 4}{4}}$$

Note: This answer can be simplified

to $x = -\tfrac{b}{2} \pm \dfrac{\sqrt{b^2 - 4}}{2}$.

29.
$$ax^2 + 2x - 4 = 0$$
$$ax^2 + 2x = 4$$
$$\frac{1}{a}\left(ax^2 + 2x\right) = \frac{1}{a}(4)$$
$$x^2 + \frac{2}{a}x = \frac{4}{a}$$
$$x^2 + \frac{2}{a}x + \left[\frac{1}{2}\left(\frac{2}{a}\right)\right]^2 = \frac{4}{a} + \left[\frac{1}{2}\left(\frac{2}{a}\right)\right]^2$$
$$x^2 + \frac{2}{a}x + \frac{1}{a^2} = \frac{4}{a} + \frac{1}{a^2}$$
$$\left(x + \frac{1}{a}\right)^2 = \frac{4}{a} + \frac{1}{a^2}$$
$$x + \frac{1}{a} = \pm\sqrt{\frac{4}{a} + \frac{1}{a^2}}$$
$$x = -\frac{1}{a} \pm \sqrt{\frac{4}{a} + \frac{1}{a^2}}$$
$$x = -\frac{1}{a} \pm \sqrt{\frac{4a+1}{a^2}}$$

Note: This answer can be simplified
to $x = -\dfrac{1}{a} \pm \dfrac{\sqrt{4a+1}}{a}$.

31. $V = \pi(r-3)^2 h$, for r
$$\frac{V}{\pi h} = (r-3)^2$$
$$\pm\sqrt{\frac{V}{\pi h}} = r - 3$$
$$3 \pm \sqrt{\frac{V}{\pi h}} = r$$

33. $E = \frac{1}{2}mv^2 + mgh$, for v
$$E = \frac{1}{2}mv^2 + mgh$$
$$E - mgh = \frac{1}{2}mv^2$$
$$\frac{2(E - mgh)}{m} = v^2$$
$$\pm\sqrt{\frac{2(E - mgh)}{m}} = v$$

35. $V = 2\left(s^2 + t^2\right)w$, for t
$$V = 2\left(s^2 + t^2\right)w$$
$$\frac{V}{2w} = s^2 + t^2$$
$$\frac{V}{2w} - s^2 = t^2$$
$$\pm\sqrt{\frac{V}{2w} - s^2} = t$$

37. $x^2 y - y^2 = 0$
$$y\left(x^2 - y\right) = 0$$
$$y = 0 \quad \text{or} \quad x^2 - y = 0$$
$$x^2 = y$$

39. $(2y + 3x)^2 = 9$
$$2y + 3x = \pm\sqrt{9}$$
$$2y + 3x = \pm 3$$
$$2y = -3x \pm 3$$
$$y = \frac{-3x \pm 3}{2}$$

41. $4x^2 - 9y^2 = 36$
$$-9y^2 = -4x^2 + 36$$
$$9y^2 = 4x^2 - 36$$
$$y^2 = \frac{4x^2 - 36}{9}$$
$$y = \pm\sqrt{\frac{4x^2 - 36}{9}}$$

Note: This answer can be simplified
to $y = \pm\dfrac{\sqrt{4x^2 - 36}}{3} = \pm\dfrac{2\sqrt{x^2 - 9}}{3}$.

43. $4x^2 - 25y^2 = 0$

$$-25y^2 = -4x^2$$
$$25y^2 = 4x^2$$
$$y^2 = \frac{4x^2}{25}$$
$$y = \pm\sqrt{\frac{4x^2}{25}}$$

Note: This answer can be simplified to $y = \pm\frac{2x}{5}$.

45. $x^2 + bx + c = 0$

$$x^2 + bx = -c$$
$$x^2 + bx + \left[\frac{1}{2}(b)\right]^2 = -c + \left[\frac{1}{2}(b)\right]^2$$
$$x^2 + bx + \frac{b^2}{4} = -c + \frac{b^2}{4}$$
$$\left(x + \frac{b}{2}\right)^2 = \frac{-4c + b^2}{4}$$
$$x + \frac{b}{2} = \pm\sqrt{\frac{b^2 - 4c}{4}}$$
$$x = -\frac{b}{2} \pm \sqrt{\frac{b^2 - 4c}{4}}$$

Note: This answer can be simplified to $x = \frac{-b \pm \sqrt{b^2 - 4c}}{2}$.

47. a. Each side of the square has length $x + y$ so the area of the square is $A = (x + y)^2$.

b.
$$A = (x + y)^2$$
$$= (x + y)(x + y)$$
$$= x^2 + xy + xy + y^2$$
$$= x^2 + 2xy + y^2$$

c.

49. a. The area of the colored stripe is equal to the area of the large triangle minus the area of the small triangle: $A = \frac{1}{2}x^2 - \frac{1}{2}y^2$.

b.
$$A = \frac{1}{2}x^2 - \frac{1}{2}y^2$$
$$= \frac{1}{2}(x^2 - y^2)$$
$$= \frac{1}{2}(x - y)(x + y)$$

c. We are given that $x = 7\frac{1}{2}$ and $y = 4\frac{1}{2}$. So the area is

$$A = \frac{1}{2}\left(7\frac{1}{2} - 4\frac{1}{2}\right)\left(7\frac{1}{2} + 4\frac{1}{2}\right)$$
$$= \frac{1}{2}(3)(12)$$
$$= 18 \text{ ft}^2.$$

Homework 3.6

1. In $x^2 - x - 1 = 0$, $a = 1$, $b = -1$, and $c = -1$, so by the quadratic formula,

$$x = \frac{-(-1) \pm \sqrt{(-1)^2 - 4(1)(-1)}}{2(1)}$$

$$= \frac{1 \pm \sqrt{1 + 4}}{2}$$

$$= \frac{1 \pm \sqrt{5}}{2}$$

The solutions are

$$x = \frac{1 + \sqrt{5}}{2} \approx 1.618 \text{ and}$$

$$x = \frac{1 - \sqrt{5}}{2} \approx -0.618$$

3. $\qquad y^2 + 2y = 5$

$y^2 + 2y - 5 = 0$

Thus, $a = 1$, $b = 2$, and $c = -5$, so by the quadratic formula,

$$y = \frac{-2 \pm \sqrt{(2)^2 - 4(1)(-5)}}{2(1)}$$

$$= \frac{-2 \pm \sqrt{4 + 20}}{2}$$

$$= \frac{-2 \pm \sqrt{24}}{2}$$

The solutions are

$$y = \frac{-2 + \sqrt{24}}{2} \approx 1.449 \text{ and}$$

$$y = \frac{-2 - \sqrt{24}}{2} \approx -3.449$$

5. $3z^2 = 4.2z + 1.5$

$3z^2 - 4.2z - 1.5 = 0$

Thus, $a = 3$, $b = -4.2$, and $c = -1.5$. By the quadratic formula,

$$z = \frac{-(-4.2) \pm \sqrt{(-4.2)^2 - 4(3)(-1.5)}}{2(3)}$$

$$= \frac{4.2 \pm \sqrt{17.64 + 18}}{6}$$

$$= \frac{4.2 \pm \sqrt{35.64}}{6}$$

The solutions are

$$z = \frac{4.2 + \sqrt{35.64}}{6} \approx 1.695 \text{ and}$$

$$z = \frac{4.2 - \sqrt{35.64}}{6} \approx -0.295.$$

7. $0 = x^2 - \frac{5}{3}x + \frac{1}{3}$. Multiply both sides by 3 to clear fractions to get $0 = 3x^2 - 5x + 1$. Thus $a = 3$, $b = -5$, and $c = 1$. By the quadratic formula,

$$x = \frac{-(-5) \pm \sqrt{(-5)^2 - 4(3)(1)}}{2(3)}$$

$$= \frac{5 \pm \sqrt{25 - 12}}{6}$$

$$= \frac{5 \pm \sqrt{13}}{6}$$

The solutions are

$$x = \frac{5 + \sqrt{13}}{6} \approx 1.434 \text{ and}$$

$$x = \frac{5 - \sqrt{13}}{6} \approx 0.232$$

9. $-5.2z^2 + 176z + 1218 = 0$. Thus, $a = -5.2$, $b = 176$, and $c = 1218$ and by the quadratic formula,

$$z = \frac{-176 \pm \sqrt{(176)^2 - 4(-5.2)(1218)}}{2(-5.2)}$$

$$= \frac{-176 \pm \sqrt{30,976 + 25,334.4}}{-10.4}$$

$$= \frac{-176 \pm \sqrt{56,310.4}}{-10.4}$$

The solutions are

$$z = \frac{-176 + \sqrt{56,310.4}}{-10.4} \approx -5.894$$

and

$$z = \frac{-176 - \sqrt{56,310.4}}{-10.4} \approx 39.740.$$

11. a. In your graphing calculator, enter $Y_1 = \frac{x^2}{24} + \frac{x}{2}$. In your *table set* menu, enter Table Start as 10 and increment size (ΔTbl) as 10. This gives the following table, with the d values rounded off to the nearest integers.

s	10	20	30	40	50
d	9	27	53	87	129

s	60	70	80	90	100
d	180	239	307	383	467

b. Refer to the graph in the back of the textbook.

c. For the car to stop in 50 feet, solve for s.

$$50 = \frac{s^2}{24} + \frac{s}{2}$$

$$24(50) = 24\left(\frac{s^2}{24} + \frac{s}{2}\right)$$

$$1200 = s^2 + 12s$$

$$0 = s^2 + 12s - 1200$$

So $a = 1$, $b = 12$, and $c = -1200$.

$$s = \frac{-12 \pm \sqrt{(12)^2 - 4(1)(-1200)}}{2(1)}$$

$$= \frac{-12 \pm \sqrt{144 + 4800}}{2}$$

$$= \frac{-12 \pm \sqrt{4944}}{2}$$

The negative square root gives a negative answer, which for this problem, is not applicable. Thus, the speed of the car is

$$s = \frac{-12 + \sqrt{4944}}{2} \approx 29.157 \text{ mph.}$$

This answer is verified by tracing along the graph to the approximate point (29, 50).

13. a. Use Table Start = 0 and ΔTbl=5 as the increment size.

t	0	5	10	15	20	25
h	11,000	10,520	9240	7160	4280	600

b. Refer to the graph in the back of the textbook.

c. To find when her altitude is 1000 feet, solve

$1000 = -16t^2 - 16t + 11,000$ or

$16t^2 + 16t - 10,000 = 0$. So
$a = 16$, $b = 16$, and $c = -10,000$.

$$t = \frac{-16 \pm \sqrt{16^2 - 4(16)(-10000)}}{2(16)}$$

$$= \frac{-16 \pm \sqrt{640,256}}{32}$$

The answer must be positive, so

$$t = \frac{-16 + \sqrt{640,256}}{32} \approx 24.5 \text{ sec.}$$

d. The altitude of the marker is still given by the equation in the textbook, so solve

$0 = -16t^2 - 16t + 11,000$. Here $a = -16$, $b = -16$, and $c = 11,000$.

$$t = \frac{-(-16) \pm \sqrt{(-16)^2 - 4(-16)(11000)}}{2(-16)}$$

$$= \frac{16 \pm \sqrt{704,256}}{-32}$$

Again, t must be positive, so

$$t = \frac{16 - \sqrt{704,256}}{-32} \approx 25.7 \text{ sec.}$$

This is the time from when the marker and the skydiver leave the plane, so the time it takes the marker to reach the ground from when the skydiver releases it is $25.7 - 24.5 = 1.2$ seconds.

e. (24.5, 1000) and (25.7, 0).

15. a. Let w represent the width of a pen and l the length of the enclosure in feet:

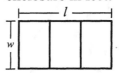

Then the amount of chain link fence is given by $4w + 2l = 100$.

b. $4w + 2l = 100$
$2l = 100 - 4w$
$l = 50 - 2w$

c. The area enclosed is
$A = wl = w(50 - 2w) = 50w - 2w^2$.
The area is 250 feet, so
$50w - 2w^2 = 250$
$0 = 2w^2 - 50w + 250$
$0 = w^2 - 25w + 125$
Thus $a = 1$, $b = -25$, and $c = 125$.

$$w = \frac{-(-25) \pm \sqrt{(-25)^2 - 4(1)(125)}}{2(1)}$$

$$= \frac{25 \pm \sqrt{625 - 500}}{2}$$

$$= \frac{25 \pm \sqrt{125}}{2}$$

The solutions are

$$w = \frac{25 + \sqrt{125}}{2} \approx 18.09 \text{ feet and}$$

$$w = \frac{25 - \sqrt{125}}{2} \approx 6.91 \text{ feet.}$$

d. When $w = 18.09$ feet,
$l = 50 - 2(18.09) = 13.82$ feet.
When $w = 6.91$ feet,
$l = 50 - 2(6.91) = 36.18$ feet.
The length of each pen is one-third the length of the whole enclosure, so dimensions of each pen are 18.09 feet by 4.61 feet or 6.91 feet by 12.06 feet.

17. a. Let x represent the distance that you can see from the top of a building that is h miles high, as shown below.

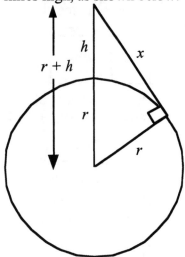

By the Pythagorean theorem, $r^2 + x^2 = (r+h)^2$. Since there are 5280 feet in a mile, the Petronas Tower is $\frac{1483}{5280} \approx 0.2809$ miles tall. The radius of the earth is $r = 3960$ miles. Substitute these values into the equation and solve for x:

$$3960^2 + x^2 = (3960 + 0.2809)^2$$
$$15,681,600 + x^2 = 15,683,824.5789$$
$$x^2 = 2024.5789$$
$$x = \pm\sqrt{2024.5789}$$
$$x = \pm 47.2$$

So you can see for 47.2 miles on a clear day. (Since distance cannot be negative, use only the positive square root.

b. To see for 100 miles, let $x = 100$ and solve for h:
$$3960^2 + 100^2 = (h + 3960)^2$$
$$15,691,600 = (h + 3960)^2$$
$$\pm\sqrt{15,691,600} = h + 3960$$
$$\pm 3961.2624 = h + 3960$$
$$-3960 \pm 3961.2624 = h$$

Since the height of the building must be positive, use the positive square root to find $h = -3960 + 3961.2624 = 1.2624$. Hence the building must be 1.26 miles tall.

19. a. Since solutions of quadratic equations come in conjugate pairs, if $2 + \sqrt{5}$ is a solution, then $2 - \sqrt{5}$ must also be a solution.

b. A quadratic equation that has these solutions is

$$0 = \left(x - \left(2 + \sqrt{5}\right)\right)\left(x - \left(2 - \sqrt{5}\right)\right)$$
$$0 = \left((x - 2) - \sqrt{5}\right)\left((x - 2) + \sqrt{5}\right)$$
$$0 = (x - 2)^2 - \sqrt{25}$$
$$0 = (x - 2)^2 - 5$$
$$0 = x^2 - 4x + 4 - 5$$
$$0 = x^2 - 4x - 1$$

21. a. Since solutions of quadratic equations come in conjugate pairs, if $4 - 3i$ is a solution, then $4 + 3i$ must also be a solution.

b. A quadratic equation that has these solutions is

$$0 = \left(x - (4 - 3i)\right)\left(x - (4 + 3i)\right)$$
$$0 = \left((x - 4) + 3i\right)\left((x - 4) - 3i\right)$$
$$0 = (x - 4)^2 - 9i^2$$
$$0 = (x - 4)^2 - 9(-1)$$
$$0 = x^2 - 8x + 16 + 9$$
$$0 = x^2 - 8x + 25$$

23. a. Refer to the graph in the back of the textbook. Using TRACE we find that the x-intercepts of $y = x^2 - 6x + 5$ are $(5, 0)$ and $(1, 0)$ and that the x-intercept of $y = x^2 - 6x + 9$ is $(3, 0)$. The graph of $y = x^2 - 6x + 12$ does not pass through the x-axis and therefore does not have any x-intercepts.

b. Solution for $x^2 - 6x + 5 = 0$:

$$x = \frac{-(-6) \pm \sqrt{(-6)^2 - 4(1)(5)}}{2(1)}$$
$$= \frac{6 \pm 4}{2}$$

Hence there are two distinct real solutions, $x = 1$ and $x = 5$. This means that the graph of $y = x^2 - 6x + 5$ has two x-intercepts.

Solution for $x^2 - 6x + 9 = 0$:

$$x = \frac{-(-6) \pm \sqrt{(-6)^2 - 4(1)(9)}}{2(1)}$$
$$= \frac{6 \pm 0}{2} = 3$$

Hence there is one real solution, $x = 3$, and so the graph of $y = x^2 - 6x + 9$ has one x-intercept.

Solution for $x^2 - 6x + 12 = 0$:

$$x = \frac{-(-6) \pm \sqrt{(-6)^2 - 4(1)(12)}}{2(1)}$$
$$= \frac{6 \pm \sqrt{-12}}{2} = \frac{6 \pm 2i\sqrt{3}}{2}$$

Hence there are two complex solutions, $x = 3 \pm i\sqrt{3}$, and so the graph of $y = x^2 - 6x + 12$ has no x-intercepts.

25. Solve $A = 2w^2 + 4lw$, for w. Write the equation in standard form: $0 = 2w^2 + 4lw - A$. So $a = 2$, $b = 4l$, and $c = -A$.

$$w = \frac{-4l \pm \sqrt{(4l)^2 - 4(2)(-A)}}{2(2)}$$

$$= \frac{-4l \pm \sqrt{16l^2 + 8A}}{4}$$

27. Solve $h = 4t - 16t^2$, for t. Write the equation in standard form: $16t^2 - 4t + h = 0$ So $a = 16$, $b = -4$, and $c = h$.

$$t = \frac{-(-4) \pm \sqrt{(-4)^2 - 4(16)(h)}}{2(16)}$$

$$= \frac{4 \pm \sqrt{16 - 64h}}{32}$$

29. Solve $s = vt - \frac{1}{2}at^2$, for t. Multiply both sides by 2 to clear fractions: $2s = 2vt - at^2$. Next, write the equation in standard form: $at^2 - 2vt + 2s = 0$. So $a = a$, $b = -2v$, and $c = 2s$.

$$t = \frac{-(-2v) \pm \sqrt{(-2v)^2 - 4(a)(2s)}}{2(a)}$$

$$= \frac{2v \pm \sqrt{4v^2 - 8as}}{2a}$$

31. Solve $3x^2 + xy + y^2 = 2$, for y. Write the equation in standard form: $y^2 + xy + 3x^2 - 2 = 0$. So $a = 1$, $b = x$, and $c = 3x^2 - 2$.

$$y = \frac{-x \pm \sqrt{x^2 - 4(1)(3x^2 - 2)}}{2(1)}$$

$$= \frac{-x \pm \sqrt{x^2 - 12x^2 + 8}}{2}$$

$$= \frac{-x \pm \sqrt{8 - 11x^2}}{2}$$

33. Let S_1 and S_2 be the solutions to the quadratic equation $ax^2 + bx + c = 0$ given by the quadratic formula. So

$$S_1 = \frac{-b + \sqrt{b^2 - 4ac}}{2a} \text{ and}$$

$$S_2 = \frac{-b - \sqrt{b^2 - 4ac}}{2a}.$$

Summing these two solutions, we have:

$S_1 + S_2$

$$= \frac{-b + \sqrt{b^2 - 4ac}}{2a} + \frac{-b - \sqrt{b^2 - 4ac}}{2a}$$

$$= \frac{-b + \sqrt{b^2 - 4ac} - b - \sqrt{b^2 - 4ac}}{2a} .$$

$$= \frac{-2b}{2a}$$

$$= -\frac{b}{a}$$

Chapter 3 Review

1. $(2x-5)^2 = 9$

$$2x - 5 = \pm\sqrt{9}$$
$$2x - 5 = \pm 3$$
$$2x = 5 \pm 3$$
$$x = \frac{5 \pm 3}{2}$$

$x = \frac{5+3}{2}$ or $x = \frac{5-3}{2}$

$x = 4$ $x = 1$

3.

$$x(3x + 2) = (x + 2)^2$$
$$3x^2 + 2x = x^2 + 4x + 4$$
$$2x^2 - 2x - 4 = 0$$
$$2\left(x^2 - x - 2\right) = 0$$
$$2(x - 2)(x + 1) = 0$$

$x - 2 = 0$ or $x + 1 = 0$

$x = 2$ $x = -1$

5. $4x - (x + 1)(x + 2) = -8$

$$4x - \left(x^2 + 3x + 2\right) = -8$$
$$4x - x^2 - 3x - 2 = -8$$
$$-x^2 + x - 2 = -8$$
$$0 = x^2 - x - 6$$
$$0 = (x - 3)(x + 2)$$

$x - 3 = 0$ or $x + 2 = 0$

$x = 3$ $x = -2$

7. $\left(x - \left(\frac{-3}{4}\right)\right)(x - 8) = 0$

$$\left(x + \frac{3}{4}\right)(x - 8) = 0$$
$$x^2 - \frac{29}{4}x - 6 = 0$$
$$4\left(x^2 - \frac{29}{4}x - 6\right) = 4(0)$$
$$4x^2 - 29x - 24 = 0$$

9. Refer to the graph in the back of the textbook. The x-intercepts are at $x = 3$ and $x = -2.4$, so the equation in factored form is $y = (x - 3)(x + 2.4)$.

11. a. In $y = \frac{1}{2}x^2$, $a = \frac{1}{2}$, $b = 0$, and $c = 0$. The vertex is where

$$x = \frac{-0}{2\left(\frac{1}{2}\right)} = \frac{0}{1} = 0.$$

When $x = 0$,

$y = \frac{1}{2}(0)^2 = \frac{1}{2}(0) = 0$, so the vertex is at $(0, 0)$. This point is also the x-intercept and y-intercept.

b. Refer to the graph in the back of the textbook.

13. a. In $y = x^2 - 9x$, $a = 1$, $b = -9$, and $c = 0$. The vertex is where

$$x = \frac{-(-9)}{2(1)} = \frac{9}{2}.$$

When $x = \frac{9}{2}$, $y = \left(\frac{9}{2}\right)^2 - 9\left(\frac{9}{2}\right)$

$= \frac{81}{4} - \frac{81}{2} = -\frac{81}{4}$, so the vertex is

at $\left(\frac{9}{2}, -\frac{81}{4}\right)$. The y-intercept is at $(0, 0)$. For the x-intercept, solve:

$$x^2 - 9x = 0$$
$$x(x - 9) = 0$$

The solutions are $x = 0$ and $x = 9$, so the x-intercepts are $(0, 0)$ and $(9, 0)$.

b. Refer to the graph in the back of the textbook.

15. $x^2 - 4x = 6$. Since one-half of -4 is -2, add $(-2)^2 = 4$ to both sides.

$$x^2 - 4x + 4 = 6 + 4$$
$$(x - 2)^2 = 10$$
$$x - 2 = \pm\sqrt{10}$$
$$x = 2 \pm \sqrt{10}$$

17.
$$2x^2 + 3 = 6x$$
$$2x^2 - 6x = -3$$
$$\frac{1}{2}\left(2x^2 - 6x\right) = \frac{1}{2}(-3)$$
$$x^2 - 3x = -\frac{3}{2}$$

Since one-half of -3 is $-\frac{3}{2}$, add

$\left(-\frac{3}{2}\right)^2 = \frac{9}{4}$ to both sides.

$$x^2 - 3x + \frac{9}{4} = -\frac{3}{2} + \frac{9}{4}$$
$$\left(x - \frac{3}{2}\right)^2 = \frac{3}{4}$$
$$x - \frac{3}{2} = \pm\sqrt{\frac{3}{4}}$$
$$x = \frac{3}{2} \pm \sqrt{\frac{3}{4}}$$

Note: This answer can be simplified

to $x = \frac{3}{2} \pm \frac{\sqrt{3}}{2}$.

19.
$$\frac{1}{2}x^2 + 1 = \frac{3}{2}x$$
$$\frac{1}{2}x^2 - \frac{3}{2}x + 1 = 0$$
$$2\left(\frac{1}{2}x^2 - \frac{3}{2}x + 1\right) = 2(0)$$
$$x^2 - 3x + 2 = 0$$

So $a = 1$, $b = -3$, and $c = 2$.

$$x = \frac{-(-3) \pm \sqrt{(-3)^2 - 4(1)(2)}}{2(1)}$$
$$= \frac{3 \pm \sqrt{9 - 8}}{2}$$
$$= \frac{3 \pm \sqrt{1}}{2}$$
$$= \frac{3 \pm 1}{2}$$

$x = \frac{3 + 1}{2}$ or $x = \frac{3 - 1}{2}$
$x = 2$ $x = 1$

21. In $x^2 - 4x + 2 = 0$, we have $a = 1$, $b = -4$, and $c = 2$.

$$x = \frac{-(-4) \pm \sqrt{(-4)^2 - 4(1)(2)}}{2(1)}$$
$$= \frac{4 \pm \sqrt{16 - 8}}{2}$$
$$= \frac{4 \pm \sqrt{8}}{2}$$

Note: This answer can be simplified

to $\frac{4 \pm 2\sqrt{2}}{2} = 2 \pm \sqrt{2}$.

23. Solve $K = \frac{1}{2}mv^2$, for v.

$$2K = mv^2$$
$$\frac{2K}{m} = v^2$$
$$\pm\sqrt{\frac{2K}{m}} = v$$

25. To solve $h = 6t - 3t^2$ for t, first write the equation in standard form: $3t^2 - 6t + h = 0$. Then use the quadratic formula with $a = 3$, $b = -6$, and $c = h$:

$$t = \frac{-(-6) \pm \sqrt{(-6)^2 - 4(3)(h)}}{2(3)}$$

$$= \frac{6 \pm \sqrt{36 - 12h}}{6}$$

27.
$$\frac{n(n-1)}{2} = 36$$
$$n(n-1) = 72$$
$$n^2 - n = 72$$
$$n^2 - n - 72 = 0$$
$$(n-9)(n+8) = 0$$

The solutions are $n = 9$ and $n = -8$, so the organizers should invite 9 players.

29.
$$2464.20 = 2000(1+r)^2$$
$$1.2321 = (1+r)^2$$
$$\pm\sqrt{1.2321} = 1+r$$
$$-1 \pm 1.11 = r$$

The solutions are $r = -1 + 1.11 = 0.11$ and $r = -1 - 1.11 = -2.11$. A negative interest rate makes no sense, so the solution is $r = 0.11$. The account had an interest rate of 11%.

31. Let x represent the width of the coops and let l represent the length of the coops along the henhouse, as shown below.

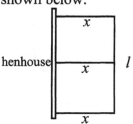

So $3x + l = 66$ and $l = 66 - 3x$. Since area $= lw$, and Irene wants the area of the coops to be 360 ft²,

$$x(66 - 3x) = 360$$
$$66x - 3x^2 = 360$$
$$0 = 3x^2 - 66x + 360$$
$$0 = 3\left(x^2 - 22x + 120\right)$$
$$0 = 3(x-12)(x-10)$$

The solutions are $x = 12$ and $x = 10$. The dimensions of each coop are x feet by $\frac{1}{2}l$ feet.

When $x = 12$, $l = 66 - 3(12) = 66 - 36 = 30$ and the coops are 12 feet by 15 feet.

When $x = 10$, $l = 66 - 3(10) = 66 - 30 = 36$ and the coops are 10 feet by 18 feet.

33. a. Substitute $v_0 = 100$ and $g = 5.6$ to get $h = 100t - 2.8t^2$.

b.

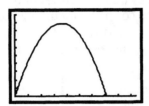

c. Trace along the curve until you get to the vertex, which has a y-value of approximately 893. Hence the golf ball reaches a maximum height of 893 feet.

d. Let $h = 880$ so that $880 = 100t - 2.8t^2$ or $2.8t^2 - 100t + 880 = 0$.
Use the quadratic formula with $a = 2.8$, $b = -100$, and $c = 880$.

$$t = \frac{-(-100) \pm \sqrt{(-100)^2 - 4(2.8)(880)}}{2(2.8)} = \frac{100 \pm \sqrt{144}}{5.6} = \frac{100 \pm 12}{5.6}$$

So $t = \dfrac{100 + 12}{5.6} = 20$ or $t = \dfrac{100 - 12}{5.6} \approx 15.7$.

The golf ball will reach a height of 880 feet after 15.7 seconds. (After 20 seconds, the golf ball is again at 880 feet, but this time it is on its way down.)

35. The area of each of the triangles in the first picture is $A_T = \dfrac{1}{2}y^2$. The area of the

square is $A_S = x^2$. The area of the shaded strip within the square is equal to the area of the square minus twice the area of one of the triangles:

$$A_1 = A_S - 2A_T = x^2 - 2\left(\frac{1}{2}y^2\right) = x^2 - y^2.$$

The area of the rectangle is $A_2 = lw = (x + y)(x - y) = x^2 - y^2$. Thus $A_1 = A_2$.

Chapter Four: Applications of Quadratic Models

Homework 4.1

1. In $y = 2 + 3x - x^2$, $a = -1$, $b = 3$, and $c = 2$. The vertex is where
$x = \dfrac{-3}{2(-1)} = \dfrac{-3}{-2} = \dfrac{3}{2}$. When $x = \dfrac{3}{2}$,

$$y = 2 + 3\left(\dfrac{3}{2}\right) - \left(\dfrac{3}{2}\right)^2$$
$$= 2 + \dfrac{9}{2} - \dfrac{9}{4}$$
$$= \dfrac{8}{4} + \dfrac{18}{4} - \dfrac{9}{4}$$
$$= \dfrac{17}{4}$$

so the vertex is $\left(\dfrac{3}{2},\ \dfrac{17}{4}\right)$.

Since $a < 0$, the parabola opens downward, so the vertex is the maximum point on the graph.

3. In $y = \dfrac{1}{2}x^2 - \dfrac{2}{3}x + \dfrac{1}{3}$, $a = \dfrac{1}{2}$, $b = \dfrac{-2}{3}$, and $c = \dfrac{1}{3}$. The vertex is

where $x = \dfrac{-\left(\frac{-2}{3}\right)}{2\left(\frac{1}{2}\right)} = \dfrac{\frac{2}{3}}{1} = \dfrac{2}{3}$.

When $x = \dfrac{2}{3}$,

$$y = \dfrac{1}{2}\left(\dfrac{2}{3}\right)^2 - \dfrac{2}{3}\left(\dfrac{2}{3}\right) + \dfrac{1}{3}$$
$$= \dfrac{1}{2}\left(\dfrac{4}{9}\right) - \dfrac{4}{9} + \dfrac{1}{3}$$
$$= \dfrac{2}{9} - \dfrac{4}{9} + \dfrac{3}{9}$$
$$= \dfrac{1}{9}$$

so the vertex is $\left(\dfrac{2}{3},\ \dfrac{1}{9}\right)$.

Since $a > 0$, the parabola opens upward, so the vertex is the minimum point on the graph.

5. In $y = 2.3 - 7.2x - 0.8x^2$, $a = -0.8$, $b = -7.2$, and $c = 2.3$.
The vertex is where
$x = \dfrac{-(-7.2)}{2(-0.8)} = \dfrac{7.2}{-1.6} = -4.5$.
When $x = -4.5$,

$$y = 2.3 - 7.2(-4.5) - 0.8(4.5)^2$$
$$= 2.3 + 32.4 - 16.2$$
$$= 18.5$$

so the vertex is at $(-4.5, 18.5)$.
Since $a < 0$, the parabola opens downward, so the vertex is the maximum point on the graph.

7. a. In $y = -2x^2 + 7x + 4$, $a = -2$, $b = 7$, and $c = 4$. The vertex is where $x = \dfrac{-7}{2(-2)} = \dfrac{-7}{-4} = \dfrac{7}{4}$.

When $x = \dfrac{7}{4}$,

$$y = -2\left(\frac{7}{4}\right)^2 + 7\left(\frac{7}{4}\right) + 4$$
$$= -2\left(\frac{49}{16}\right) + \frac{49}{4} + 4$$
$$= \frac{-49}{8} + \frac{49}{4} + 4$$
$$= \frac{-49}{8} + \frac{98}{8} + \frac{32}{8} = \frac{81}{8}$$

so the vertex is at $\left(\dfrac{7}{4}, \dfrac{81}{8}\right)$. The y-intercept is at $(0, 4)$. Note that the y-intercept is always at $(0, c)$. For the x-intercepts, solve

$0 = -2x^2 + 7x + 4$ or

$2x^2 - 7x - 4 = 0$, which is $(2x + 1)(x - 4) = 0$.

$\begin{array}{ll} 2x + 1 = 0 & \text{or} \quad x - 4 = 0 \\ 2x = -1 & \qquad\quad x = 4 \\ x = \dfrac{-1}{2} & \end{array}$

x-intercepts: $\left(-\dfrac{1}{2}, 0\right)$ and $(4, 0)$.

b, c. Refer to the graph in the back of the textbook.

9. a. In $y = 0.6x^2 + 0.6x - 1.2$, $a = 0.6$, $b = 0.6$, and $c = -1.2$. The vertex is where

$$x = \frac{-0.6}{2(0.6)} = \frac{-1}{2} = -0.5.$$

When $x = -0.5$,

$$y = 0.6(-0.5)^2 + 0.6(-0.5) - 1.2$$
$$= 0.6(0.25) - 0.3 - 1.2$$
$$= 0.15 - 1.5$$
$$= -1.35$$

so the vertex is at $(-0.5, -1.35)$. The y-intercept is at $(0, -1.2)$, while to find the x-intercepts, solve

$$0 = 0.6x^2 + 0.6x - 1.2$$
$$0 = 0.6\left(x^2 + x - 2\right)$$
$$0 = 0.6(x - 1)(x + 2)$$

$\begin{array}{ll} x - 1 = 0 & \text{or} \quad x + 2 = 0 \\ x = 1 & \qquad\quad x = -2 \end{array}$

The x-intercepts are at $(1, 0)$ and $(-2, 0)$.

b, c. Refer to the graph in the back of the textbook.

11. a. In $y = x^2 + 4x + 7$, $a = 1$, $b = 4$, and $c = 7$. The vertex is where

$x = \dfrac{-4}{2(1)} = -2$. When $x = -2$,

$$y = (-2)^2 + 4(-2) + 7$$
$$= 4 - 8 + 7$$
$$= 3$$

so the vertex is at $(-2, 3)$. The y-intercept is at $(0, 7)$. Note that the parabola opens upward since $a > 0$ and that the vertex is above the x-axis. Therefore, there are no x-intercepts.

b, c. Refer to the graph in the back of the textbook.

13. a. In $y = x^2 + 2x - 1$, $a = 1$, $b = 2$, and $c = -1$. The vertex is where $x = \frac{-2}{2(1)} = -1$. When $x = -1$,

$$y = (-1)^2 + 2(-1) - 1$$
$$= 1 - 2 - 1$$
$$= -2$$

so the vertex is at $(-1, -2)$. The y-intercept is at $(0, -1)$. For the x-intercepts, solve $0 = x^2 + 2x - 1$ by using the quadratic formula.

$$x = \frac{-2 \pm \sqrt{(2)^2 - 4(1)(-1)}}{2(1)}$$
$$= \frac{-2 \pm \sqrt{4 + 4}}{2}$$
$$= \frac{-2 \pm \sqrt{8}}{2}$$

$$x = \frac{-2 + \sqrt{8}}{2} \approx 0.41 \text{ and }$$

$$x = \frac{-2 - \sqrt{8}}{2} \approx -2.41 \text{ so the }$$

x-intercepts are at approximately $(0.41, 0)$ and $(-2.41, 0)$.

b, c. Refer to the graph in the back of the textbook.

15. a. In $y = -2x^2 + 6x - 3$, $a = -2$, $b = 6$, and $c = -3$. The vertex is where $x = \frac{-6}{2(-2)} = \frac{-6}{-4} = \frac{3}{2}$.

When $x = \frac{3}{2}$,

$$y = -2\left(\frac{3}{2}\right)^2 + 6\left(\frac{3}{2}\right) - 3$$
$$= -2\left(\frac{9}{4}\right) + 9 - 3$$
$$= \frac{-9}{2} + 6$$
$$= \frac{3}{2}$$

so the vertex is at $\left(\frac{3}{2}, \frac{3}{2}\right)$. The y-intercept is at $(0, -3)$. For the x-intercepts, solve $0 = -2x^2 + 6x - 3$ by using the quadratic formula.

$$x = \frac{-6 \pm \sqrt{(6)^2 - 4(-2)(-3)}}{2(-2)}$$
$$= \frac{-6 \pm \sqrt{36 - 24}}{-4}$$
$$= \frac{-6 \pm \sqrt{12}}{-4}$$

$$x = \frac{-6 + \sqrt{12}}{-4} \approx 0.63 \text{ and }$$

$$x = \frac{-6 - \sqrt{12}}{-4} \approx 2.37 \text{ so the }$$

x-intercepts are at approximately $(0.63, 0)$ and $(2.37, 0)$.

b, c. Refer to the graph in the back of the textbook.

17. a. IV, since the parabola opens downward and has vertex at (0, 1).

b. V, since the parabola opens upward from (–2, 0).

c. I, since the parabola opens upward from the origin.

d. VII, since the parabola has x-intercepts (4, 0) and (–2, 0).

19. a. If the x-intercepts are at $x = 2$, and $x = -3$, the parabola can have the equation $y = (x - 2)(x + 3)$ or

$$y = x^2 + x - 6.$$

b. Another parabola with the same x-intercepts is $y = 2x^2 + 2x - 12$, or, in general $y = kx^2 + kx - 6k$ where k is any real number.

21. a. The vertex is (3, 4).

b. To write the equation in standard form, multiply and combine like terms:

$$y = 2(x - 3)^2 + 4$$
$$y = 2\left(x^2 - 6x + 9\right) + 4$$
$$y = 2x^2 - 12x + 18 + 4$$
$$y = 2x^2 - 12x + 22$$

23. a. The vertex is (–4, –3).

b. To write the equation in standard form, multiply and combine like terms:

$$y = -\frac{1}{2}(x + 4)^2 - 3$$
$$y = -\frac{1}{2}\left(x^2 + 8x + 16\right) - 3$$
$$y = -\frac{1}{2}x^2 - 4x - 8 - 3$$
$$y = -\frac{1}{2}x^2 - 4x - 11$$

25. a. The equation of a parabola with vertex (–2, 6) is

$$y = a(x - (-2))^2 + 6, \text{ or}$$
$$y = a(x + 2)^2 + 6, \text{ where } a \text{ can be}$$
any real number.

b. If the y-intercept is 18, then (0, 18) is a point on the graph:

$$18 = a(0 + 2)^2 + 6$$
$$18 = 4a + 6$$
$$12 = 4a$$
$$3 = a$$

27. a. The equation of a parabola with vertex (0, –3) is $y = a(x - 0)^2 - 3$, or $y = ax^2 - 3$. To find the value of a, use the point on the graph (2, 0):

$$0 = a(2)^2 - 3$$
$$0 = 4a - 3$$
$$3 = 4a$$
$$\frac{3}{4} = a$$

The equation is $y = \frac{3}{4}x^2 - 3$.

b. From part (a), use the equation $y = ax^2 - 3$. Since the vertex is below the x-axis, the parabola will have no x-intercepts if it opens downward. Therefore, choose any negative value of a. For example, if $a = -1$, then the equation is $y = -x^2 - 3$.

29. The *x*-intercepts are at $x = -3$ and $x = 3$. Using the factored form, the equation is $y = a(x-3)(x+3)$. To find the value of a, note that $(0, -9)$ is also a point on the graph. This gives

$-9 = a(0-3)(0+3)$

$-9 = a(-3)(3)$

$-9 = -9a$

$1 = a$

So the equation is:

$y = (1)(x-3)(x+3)$ or $y = x^2 - 9$.

Note: This same answer could have been obtained by first using the vertex form with the vertex $(0, -9)$ so that $y = a(x-0)^2 - 9$ or

$y = ax^2 - 9$. Then use any point on the graph to find a. For example, using the *x*-intercept $(-3, 0)$,

$0 = a(-3)^2 - 9$

$9 = 9a$

$1 = a$

So, as above, we have $y = x^2 - 9$.

31. The vertex is at $(0, 0)$. Using the vertex form, $y = a(x-0)^2 + 0$ or

$y = ax^2$. To find the value of a, note that $(1, -2)$ is also a point on the graph so

$-2 = a(1)^2$

$-2 = a$

Thus the equation is $y = -2x^2$.

33. The *x*-intercepts are at $x = -3$ and $x = 5$. Using the factored form, the equation is $y = a(x-5)(x+3)$. To find the value of a, note that $(1, -16)$ is also a point on the graph. This gives

$-16 = a(1-5)(1+3)$

$-16 = a(-4)(4)$

$-16 = -16a$

$1 = a$

So the equation is:

$y = (1)(x-5)(x+3)$ or

$y = x^2 - 2x - 15$.

Note: This same answer could have been obtained by first using the vertex form with the vertex $(1, -16)$ so that $y = a(x-1)^2 - 16$. Then use any point on the graph to find a. For example, using the *x*-intercept $(-3, 0)$,

$0 = a(-3-1)^2 - 16$

$0 = 16a - 16$

$16 = 16a$

$1 = a$

This gives the equation

$y = (1)(x-1)^2 - 16$

$y = (x^2 - 2x + 1) - 16$

$y = x^2 - 2x - 15$

which is the same as the answer obtained using the factored form.

35. The vertex is at (2, 1). Using the vertex form, $y = a(x-2)^2 + 1$. To find the value of a, note that (0, 5) is also a point on the graph so

$$5 = a(0-2)^2 + 1$$
$$5 = 4a + 1$$
$$4 = 4a$$
$$1 = a$$

This gives the equation

$$y = 1(x-2)^2 + 1$$
$$y = \left(x^2 - 4x + 4\right) + 1.$$
$$y = x^2 - 4x + 5$$

37. a.

# of price increases	Price of room	# of rooms rented	Total revenue
0	20	60	1200
1	22	57	1254
2	24	54	1296
3	26	51	1326
4	28	48	1344
5	30	45	1350
6	32	42	1344
7	34	39	1326
8	36	36	1296
10	40	30	1200
12	44	24	1056
16	52	12	624
20	60	0	0

b. The price of a room is $20 + 2x$, the number of rooms rented is $60 - 3x$, and the total revenue earned at that price is $(20 + 2x)(60 - 3x)$.

c. Enter $Y_1 = 20 + 2x$, $Y_2 = 60 - 3x$, and $Y_3 = (20 + 2x)(60 - 3x)$ in your calculator. In your *table setup* menu, use Table Start $= 0$ and increment ΔTbl $= 1$. The values in the calculator's table should agree with the table drawn in part (a).

d. If $x = 20$, the total revenue is 0.

e. Refer to the graph in the back of the textbook.

f. The owner must charge at least $24 but no more than $36 per room to make her revenue at least $1296 per night.

g. The maximum revenue from one night is $1350, which is obtained by charging $30 per room and renting 45 rooms at this price.

Homework 4.2

1. To fit the points on a parabola, use the equations:

$$0 = a(-1)^2 + b(-1) + c$$
$$12 = a(2)^2 + b(2) + c$$
$$8 = a(-2)^2 + b(-2) + c$$

which simplify to

$$a - b + c = 0 \quad (1)$$
$$4a + 2b + c = 12 \quad (2)$$
$$4a - 2b + c = 8 \quad (3)$$

Subtract (3) from (2):

$$4a + 2b + c = 12 \quad (2)$$
$$\underline{-4a + 2b - c = -8} \quad (3a)$$
$$4b = 4$$
$$b = 1 \quad (4)$$

Subtract (1) from (2):

$$4a + 2b + c = 12 \quad (2)$$
$$\underline{-a + b - c = 0} \quad (1a)$$
$$3a + 3b = 12 \quad (5)$$

Put the value for b from (4) into (5):

$$3a + 3(1) = 12$$
$$3a + 3 = 12$$
$$3a = 9$$
$$a = 3$$

Put the values for a and b into (1):

$$3 - 1 + c = 0$$
$$2 + c = 0$$
$$c = -2$$

Therefore $a = 3$, $b = 1$, $c = -2$ and the equation of the parabola is

$$y = 3x^2 + x - 2.$$

Note: This system of equations may also be solved using matrices. See Section 2.4.

3. a. To fit the points $(15, 4)$, $(20, 13)$, and $(30, 7)$, onto a parabola we have the equations:

$$(15)^2 a + 15b + c = 4$$
$$(20)^2 a + 20b + c = 13$$
$$(30)^2 a + 30b + c = 7$$

Which simplify to

$$225a + 15b + c = 4 \quad (1)$$
$$400a + 20b + c = 13 \quad (2)$$
$$900a + 30b + c = 7 \quad (3)$$

Subtract (2) from (3):

$$900a + 30b + c = 7 \quad (3)$$
$$\underline{-400a - 20b - c = -13} \quad (2a)$$
$$500a + 10b = -6 \quad (4)$$

Subtract (1) from (2):

$$400a + 20b + c = 13 \quad (2)$$
$$\underline{-225a - 15b - c = -4} \quad (1a)$$
$$175a + 5b = 9 \quad (5)$$

Subtract twice (5) from (4):

$$500a + 10b = -6 \quad (4)$$
$$\underline{-350a - 10b = -18} \quad (5a)$$
$$150a = -24$$
$$a = -0.16 \quad (6)$$

Put this value for a into (4):

$$500(-0.16) + 10b = -6$$
$$-80 + 10b = -6$$
$$10b = 74$$
$$b = 7.4$$

Substitute the values for a and b into (1):

$$225(-0.16) + 15(7.4) + c = 4$$
$$-36 + 111 + c = 4$$
$$c = -71$$

The parabola is

$$P = -0.16x^2 + 7.4x - 71.$$

b. Using $x = 25$:
$$P = -0.16(25)^2 + 7.4(25) - 71$$
$$= -0.16(625) + 185 - 71$$
$$= -100 + 114 = 14$$
The parabola predicts that 14% of 25-year-olds use marijuana regularly.

c. Refer to the graph in the back of the textbook.

5. a. Since t is the number of years since 1970, the year 1970 corresponds to $t = 0$, 1985 to $t = 15$, and 1990 to $t = 20$. Therefore use the points $(0, 27.7)$, $(15, 36.1)$, and $(20, 42.1)$ to form the equations:
$$27.7 = a(0)^2 + b(0) + c$$
$$36.1 = a(15)^2 + b(15) + c$$
$$42.1 = a(20)^2 + b(20) + c$$

which simplify to
$$c = 27.7 \quad (1)$$
$$225a + 15b + c = 36.1 \quad (2)$$
$$400a + 20b + c = 42.1 \quad (3)$$

Substitute the value of c from (1) into equation (2) and simplify the result:
$$225a + 15b + 27.7 = 36.1$$
$$225a + 15b = 8.4 \quad (4)$$

Substitute the value of c from (1) into equation (3) and simplify the result:
$$400a + 20b + 27.7 = 42.1$$
$$400a + 20b = 14.4 \quad (5)$$

Multiply (4) by -4 and (5) by 3 and add the results:

$$-900a - 60b = -33.6 \quad (4a)$$
$$\underline{1200a + 60b = 43.2} \quad (5a)$$
$$300a = 9.6$$
$$a = 0.032$$

Substitute this value of a into (4):
$$225(0.032) + 15b = 8.4$$
$$7.2 + 15b = 8.4$$
$$15b = 1.2$$
$$b = 0.08$$

Thus the equation is
$$C = 0.032t^2 + 0.08t + 27.7.$$

b. In the year 1992, t = 22, so
$$C = 0.032(22)^2 + 0.08(22) + 27.7$$
$$= 15.488 + 1.76 + 27.7$$
$$= 44.948$$

According to the equation, Americans' per capita consumption of chicken in 1992 was 44.948 pounds. (Compare this to 45.9 in the table.)

c. Refer to the graph in the back of the textbook.

7. Let D represent the number of diagonals in a polygon of n sides. We are looking for a quadratic of the form $D = an^2 + bn + c$. Using the data given for polygons with 4, 5, and 6 sides:

$16a + 4b + c = 2$ (1)
$25a + 5b + c = 5$ (2)
$36a + 6b + c = 9$ (3)

Subtract (1) from (2):

$$\begin{array}{rl} 25a + 5b + c = & 5 \quad (2) \\ \underline{-16a - 4b - c = -2} & \quad (1a) \\ 9a + b = 3 & \quad (4) \end{array}$$

Subtract (2) from (3):

$$\begin{array}{rl} 36a + 6b + c = & 9 \quad (3) \\ \underline{-25a - 5b - c = -5} & \quad (2a) \\ 11a + b = 4 & \quad (5) \end{array}$$

Subtract (4) from (5):

$$\begin{array}{rl} 11a + b = & 4 \quad (5) \\ \underline{-9a - b = -3} & \quad (4a) \\ 2a = 1 & \\ a = \tfrac{1}{2} & \quad (6) \end{array}$$

Put the value from (6) into (4):

$$9\left(\tfrac{1}{2}\right) + b = 3$$
$$\tfrac{9}{2} + b = 3$$
$$b = -\tfrac{3}{2}$$

Substitute these values into (1):

$$16\left(\tfrac{1}{2}\right) + 4\left(-\tfrac{3}{2}\right) + c = 2$$
$$8 - 6 + c = 2$$
$$c = 0$$

The parabola is $D = \tfrac{1}{2}n^2 - \tfrac{3}{2}n$. (You should have this same parabola even if you choose a different set of 3 points from the data.)

9. Since the vertex is given, use the vertex form of the equation:

$$y = a(x - 30)^2 + 280.$$

To find a, use the point (20, 80):

$$80 = a(20 - 30)^2 + 280$$
$$80 = 100a + 280$$
$$-200 = 100a$$
$$-2 = a$$

Thus the equation is

$$y = -2(x - 30)^2 + 280.$$

11. a. Let x represent the horizontal distance and y the vertical distance that the clay pigeon travels. From the data given, form the ordered pairs $(0, 4)$ and $(80, 164)$. Since 164 is the maximum height, $(80, 164)$ must be the vertex of the parabola. Use vertex form:
$y = a(x - 80)^2 + 164$. To find a, use the point $(0, 4)$:

$$4 = a(0 - 80)^2 + 164$$
$$4 = 6400a + 164$$
$$-160 = 6400a$$
$$-0.025 = a$$

Thus the equation is
$y = -0.025(x - 80)^2 + 164$.

b. Let $y = 0$ and solve for x:

$$0 = -0.025(x - 80)^2 + 164$$
$$-164 = -0.025(x - 80)^2$$
$$6560 = (x - 80)^2$$
$$\pm\sqrt{6560} = x - 80$$
$$x = 80 \pm \sqrt{6560}$$

The negative square root gives a negative value of x, which doesn't make sense in this problem. Therefore
$x = 80 + \sqrt{6560} \approx 160.99$ and so the clay pigeon will it the ground 160.99 feet from the launch site.

13. The vertex is $(2000, 20)$ and another point on the cable is $(0, 500)$. Using vertex form, $y = a(x - 2000)^2 + 20$. To find the value of a, use the point $(0, 500)$:

$$500 = a(0 - 2000)^2 + 20$$
$$480 = 4,000,000a$$
$$0.00012 = a$$

The shape of the cable is given by the equation
$y = 0.00012(x - 2000)^2 + 20$.

15. a. Enter the data in the statistical menu of your calculator. Let x represent the time since the projectile was fired and let y represent the projectile's height. The regression line given by the calculator is $y = 8.24x + 38.89$.

b. Using $x = 15$,
$y = 8.24(15) + 38.89 \approx 162.5$.
According to the regression line, the projectile will be 162.5 meters high after 15 seconds.

c. Refer to the graph in the back of the textbook.

d. Using the quadratic regression feature of the statistical mode, we find the equation
$y = -0.81x^2 + 21.2x$.

e. Using $x = 15$,
$y = -0.81(15)^2 + 21.2(15) \approx 136$.
According to the quadratic equation, the projectile will be 136 meters high after 15 seconds.

f. Refer to the graph in the back of the textbook.

g. The quadratic model appears to best fit the data, and therefore is a more appropriate model.

17. a. Enter the data in the statistical menu of your calculator. Let x represent the number of days since January 1 and let y represent the number of hours of daylight. The regression line given by the calculator is $y = 0.0051x + 11.325$.

b. Using $x = 365$,
$y = 0.0051(365) + 11.325 \approx 13.2$.
According to the regression line, there will be 13.2 hours of daylight 365 days after January 1.

c. Refer to the graph in the back of the textbook.

d. Using the quadratic regression feature of the statistical mode, we find the equation
$y = -0.00016x^2 + 0.053x + 9.32$.

e. Using $x = 365$, the y-value is $y = -0.00016(365)^2 + 0.053(365) + 9.32$
≈ 7.35
According to the quadratic equation, there will be 7.35 hours of daylight 365 days after January 1.

f. Refer to the graph in the back of the textbook.

g. 365 days after January 1 is again January 1. According to the data given in the problem, there are 9.8 hours of daylight on January 1. The quadratic equation gave much too low of an estimate and the linear equation gave much too high of an estimate so neither of the models is appropriate for estimating number of daylight hours a full year in advance.

Homework 4.3

1. The equation $d = 96t - 16t^2$ is a parabola which opens downward so the maximum value occurs at the vertex. Here, $a = -16$ and $b = 96$, so the vertex is where

$$t = \frac{-96}{2(-16)} = \frac{-96}{-32} = 3.$$

The rocket reaches its greatest height 3 seconds after it is launched.

When $t = 3$,

$$d = 96(3) - 16(3)^2$$
$$= 288 - 16(9)$$
$$= 288 - 144$$
$$= 144$$

so the rocket reaches a maximum height of 144 feet.

The vertex is (3, 144) and the vertical intercept is (0, 0). Graph $y = 96x - 16x^2$ using Xmin = 0, Xmax = 10, Ymin = –200, and Ymax = 200.

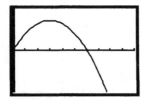

Tracing along the curve, the vertex appears to be at (3, 144) which confirms our answer found algebraically.

3. a. The perimeter of the rectangle must be 100 inches. If l represents the length of the rectangle and w represents the width, $2w + 2l = 100$, so $w + l = 50$ and $l = 50 - w$. The area is length times width or

$$A = lw = (50 - w)w = 50w - w^2.$$

b. The graph of the area is a parabola which opens downward, so the maximum value will occur at the vertex. Here $a = -1$ and $b = 50$ so the vertex is where

$$w = \frac{-50}{2(-1)} = \frac{-50}{-2} = 25.$$

The maximum area is

$$A = 50(25) - (25)^2$$
$$= 1250 - 625$$
$$= 625.$$

So Sheila can enclose a maximum area of 625 square inches.

Graph $y = 50x - x^2$ using Xmin = 0, Xmax = 50, Ymin = 0, and Ymax = 800.

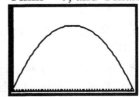

Tracing along the curve, the vertex appears to be at (25, 625) which confirms our answer found algebraically.

5. a. Let w represent the width of the rectangle (the lengths of the fences perpendicular to the river). Then there are two fenced sides of length w and one fenced side of length l. Thus $2w + l = 300$, since there are 300 yards of fence available, and $l = 300 - 2w$. The area is

$$A = lw = (300 - 2w)w$$

$$= 300w - 2w^2.$$

b. The graph of the area is a parabola which opens downward, so the maximum value occurs at the vertex. Here $a = -2$ and $b = 300$, so the vertex is at

$$w = \frac{-300}{2(-2)} = \frac{-300}{-4} = 75.$$

When $w = 75$, the area is

$$A = 300(75) - 2(75)^2$$

$$= 22,500 - 11,250$$

$$= 11,250.$$

The maximum area that can be enclosed is $11,250$ yd^2.

Graph $y = 300x - 2x^2$ using Xmin = 0, Xmax = 200, Ymin = 0, and Ymax = 15,000.

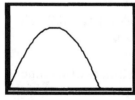

Tracing along the curve, the vertex appears to be at $(75, 11,250)$ which confirms our answer found algebraically.

7. a. The total number of people signed up is $16 + x$, and the price per person is $2400 - 100x$, so the total revenue is

$$(16 + x)(2400 - 100x)$$

$$= 38,400 + 800x - 100x^2.$$

b. The graph of the revenue is a parabola which opens downward, so the maximum revenue occurs at the vertex. Here $a = -100$ and $b = 800$, so the vertex is where

$$x = \frac{-800}{2(-100)} = \frac{-800}{-200} = 4 \text{ and } 4$$

additional people (20 people total) must sign up for the travel agent to maximize her revenue.

Graph

$y = 38,400 + 800x - 100x^2$ using Xmin = 0, Xmax = 25, Ymin = 20,000, and Ymax = 50,000.

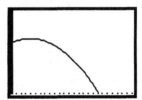

Tracing along the curve, the vertex appears to be where $x = 4$, which confirms our answer found algebraically.

9. The equation $C = 0.01x^2 - 2x + 120$ is a parabola which opens upward so the minimum value occurs at the vertex. Here, $a = 0.01$ and $b = -2$, so the vertex is where

$$x = \frac{-(-2)}{2(0.01)} = 100.$$

When $x = 100$,

$$C = 0.01(100)^2 - 2(100) + 120 = 20.$$

The minimum cost per basket occurs when they produce 100 baskets at $20 per basket. The total production cost is the number of baskets times the price per basket. So the total production cost is $(100)(20) = \$2000$.

Graph $y = 0.01x^2 - 2x + 120$ using Xmin = 0, Xmax = 200, Ymin = 0, and Ymax = 100.

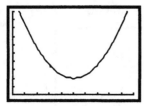

Tracing along the curve, the vertex appears to be at (100, 20) which confirms our answer found algebraically.

11. We want to minimize
$$
\begin{aligned}
V &= 576a^2 + 5184(1-a)^2 \\
&= 576a^2 + 5184\left(1 - 2a + a^2\right) \\
&= 576a^2 + 5184 - 10{,}368a + 5184a^2 \\
&= 5760a^2 - 10{,}368a + 5184
\end{aligned}
$$
The graph of V is a parabola which opens upward so the minimum value occurs at the vertex. The vertex is where
$$a = \frac{-(-10{,}368)}{2(5760)} = \frac{10{,}368}{11{,}520} = 0.9.$$
Graph

$y = 5760x^2 - 10{,}368x + 5184$ using Xmin = 0, Xmax = 2, Ymin = 0, and Ymax = 10,000.

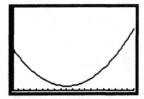

Tracing along the curve, the vertex appears to be where $x = 0.9$ which confirms our answer found algebraically.

The refined estimate for total income is
$$
\begin{aligned}
I &= (0.9)860 + (1 - 0.9)918 \\
&= 774 + 91.8 \\
&= 865.8
\end{aligned}
$$
so the estimate of income is $865.80 per month.

13. Given $y = x^2 - 4x + 7$ and $y = 11 - x$, equate the expressions for y:

$$x^2 - 4x + 7 = 11 - x$$
$$x^2 - 3x - 4 = 0$$
$$(x - 4)(x + 1) = 0$$

So $x = 4$, and $x = -1$. When $x = 4$, $y = 11 - 4 = 7$ and when $x = -1$, $y = 11 - (-1) = 12$. The solution points are $(4, 7)$ and $(-1, 12)$. Graph the equations using Xmin $= -10$, Xmax $= 10$, Ymin $= -20$, and Ymax $= 20$.

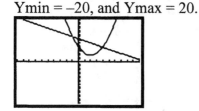

15. Given $y = -x^2 - 2x + 7$ and $y = 2x + 11$, equate the expressions for y:

$$-x^2 - 2x + 7 = 2x + 11$$
$$0 = x^2 + 4x + 4$$
$$0 = (x + 2)^2$$

So $x = -2$. When $x = -2$, $y = 2(-2) + 11 = 7$. The solution point is $(-2, 7)$. Graph the equations using Xmin $= -10$, Xmax $= 10$, Ymin $= -20$, and Ymax $= 20$.

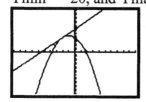

17. Given $y = x^2 + 8x + 8$ and $3y + 2x = -36$, we solve the second equation for y

$$3y + 2x = -36$$
$$3y = -2x - 36$$
$$y = -\frac{2}{3}x - 12$$

and then equate the two expressions for y:

$$x^2 + 8x + 8 = -\frac{2}{3}x - 12$$
$$x^2 + \frac{26}{3}x + 20 = 0$$
$$3\left(x^2 + \frac{26}{3}x + 20\right) = 3(0)$$
$$3x^2 + 26x + 60 = 0$$

Solve using the quadratic formula with $a = 3$, $b = 26$, and $c = 60$:

$$x = \frac{-26 \pm \sqrt{26^2 - 4(3)(60)}}{2(3)}$$
$$= \frac{-26 \pm \sqrt{-44}}{6}$$

Since there is a negative under the radical, there are no real solutions to the system. (The graphs do not intersect.) Graph the equations using Xmin $= -10$, Xmax $= 10$, Ymin $= -20$, and Ymax $= 20$.

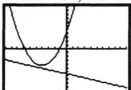

19. Given $y = x^2 - 9$ and

$y = -2x^2 + 9x + 21$, equate the
expressions for y:

$$x^2 - 9 = -2x^2 + 9x + 21$$
$$3x^2 - 9x - 30 = 0$$
$$3(x^2 - 3x - 10) = 0$$
$$3(x - 5)(x + 2) = 0$$

So $x = 5$ and $x = -2$. When $x = 5$,

$y = (5)^2 - 9 = 16$ and when $x = -2$,

$y = (-2)^2 - 9 = -5$. The solution
points are (5, 16) and (–2, –5).
Graph the equations using
Xmin = –10, Xmax = 10,
Ymin = –50, and Ymax = 50.

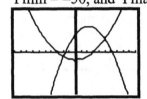

21. Given $y = x^2 - 0.5x + 3.5$ and

$y = -x^2 + 3.5x + 1.5$, equate the
expressions for y:

$$x^2 - 0.5x + 3.5 = -x^2 + 3.5x + 1.5$$
$$2x^2 - 4x + 2 = 0$$
$$2(x^2 - 2x + 1) = 0$$
$$2(x - 1)^2 = 0$$

So $x = 1$. Using the first equation,

$y = (1)^2 - 0.5(1) + 3.5 = 4$. The
solution point is (1, 4).
Graph the equations using
Xmin = –10, Xmax = 10,
Ymin = –10, and Ymax = 10.

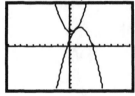

23. Given $y = x^2 - 4x + 4$ and

$y = x^2 - 8x + 16$, equate the
expressions for y:

$$x^2 - 4x + 4 = x^2 - 8x + 16$$
$$4x - 12 = 0$$
$$4x = 12$$
$$x = 3$$

The solution is when $x = 3$.
Substituting this value of x into the
first equation, $y = (3)^2 - 4(3) + 4 = 1$.
The solution point is (3, 1).

Graph the equations using
Xmin = –10, Xmax = 10,
Ymin = –10, and Ymax = 10.

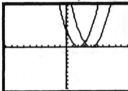

25. a. Refer to the graph in the back of the textbook.

  b. If the biomass is currently 2500 tons, the rate of growth is
  $$y = 0.4(2500) - 0.0001(2500)^2$$
  $$= 375 \text{ fish per year.}$$
  Thus if 300 tons of fish are harvested, the population will still be larger next year by 75 tons.

  If the biomass is currently 3500 tons, the rate of growth is
  $$y = 0.4(3500) - 0.0001(3500)^2$$
  $$= 175 \text{ fish per year.}$$
  So if 300 tons of fish are harvested, the population will be smaller next year by 125 tons.

  c. For the population to remain stable while harvesting 300 tons annually, we need
  $y = 300$ or $0.4x - 0.0001x^2 = 300$. Write the equation in standard form:
  $0 = 0.0001x^2 - 0.4x + 300$ and use the quadratic formula with $a = 0.0001$,
  $b = -0.4$, and $c = 300$:

  $$x = \frac{-(-0.4) \pm \sqrt{(-0.4)^2 - 4(0.0001)(300)}}{2(0.0001)} = \frac{0.4 \pm \sqrt{0.04}}{0.0002} = \frac{0.4 \pm 0.2}{0.0002}.$$

  This gives the values $x = \dfrac{0.4 - 0.2}{0.0002} = 1000$ and $x = \dfrac{0.4 + 0.2}{0.0002} = 3000$.
  (Note: Since these answers are rational, we see that we could have solved the quadratic equation by factoring.)

  Hence the biomass must be 1000 tons or 3000 tons for the population to remain stable while harvesting 300 tons of fish. These answers are confirmed by the intersection points $(1000, 300)$ and $(3000, 300)$ on the graph in part (a).

  d. From the graph in part (a), we see that if $x < 1000$, then $y < 300$, which means that if the biomass falls below 1000 tons, the rate of growth is less than 300 tons. So if 300 tons are harvested, the population will decrease each year and eventually die out.

27. a. Refer to the graph in the back of the textbook.

   b. When the bear population is 1200, $N = 0.0002(1200)(2000 - 1200) = 192$ and
   $K = 0.2(1200) = 240$, so the value of K is greater. Since the number of bears
   killed is greater than the annual increase by $240 - 192 = 48$, the population will
   decrease next year by 48 bears.

   c. When the bear population is 900, $N = 0.0002(900)(2000 - 900) = 198$ and
   $K = 0.2(900) = 180$. Since the annual increase is greater than the number of
   bears killed by $198 - 180 = 18$, the population will increase next year by 18
   bears.

   d. To find out what sizes of bear population will remain stable after hunting, let
   $N = K$ and solve the resulting equation:
   $$0.0002x(2000 - x) = 0.2x$$
   $$0.4x - 0.0002x^2 = 0.2x$$
   $$0.2x - 0.0002x^2 = 0$$
   $$x(0.2 - 0.0002x) = 0$$
   $$x = 0 \quad \text{or} \quad 0.2 - 0.0002x = 0$$
   $$-0.0002x = -0.2$$
   $$x = 1000$$
   Disregard the answer of 0 since this doesn't allow for a bear population at all.
   Thus, the bear population will remain stable at 1000 bears. This is confirmed by
   the intersection point at $x = 1000$ on the graph in part (a).

   e. From the graph in part (a), we see that $N > K$ for $x < 1000$ and $N < K$ for
   $x > 1000$. Therefore, despite hunting, a population between 0 and 1000 will
   increase and a population over 1000 will decrease.

   f. From the results of part (e), we see that the population will tend toward 1000
   bears, unless the population is zero in which case it will remain at zero.

   g. As in part (d), let $N = K$ and solve the resulting equation, but now use $K = 0.3x$.
   $$0.0002x(2000 - x) = 0.3x$$
   $$0.4x - 0.0002x^2 = 0.3x$$
   $$0.1x - 0.0002x^2 = 0$$
   $$x(0.1 - 0.0002x) = 0$$
   $$x = 0 \quad \text{or} \quad 0.1 - 0.0002x = 0$$
   $$-0.0002x = -0.1$$
   $$x = 500$$
   Thus, the bear population in this case will tend toward 500, where it will remain
   stable. (Unless of course the population is zero in which case it will remain at
   zero.)

29. a. Let $C = R$ to form the equation $0.0075x^2 + x + 2100 = 13x$. Write this in standard form $0.0075x^2 - 12x + 2100 = 0$ and use the quadratic formula with $a = 0.0075$, $b = -12$, and $c = 2100$ to solve for x:

$$x = \frac{-(-12) \pm \sqrt{(-12)^2 - 4(0.0075)(2100)}}{2(0.0075)} = \frac{12 \pm \sqrt{81}}{0.015} = \frac{12 \pm 9}{0.015}.$$

This gives the values $x = \dfrac{12 - 9}{0.015} = \dfrac{3}{0.015} = 200$ and $x = \dfrac{12 + 9}{0.015} = \dfrac{21}{0.015} = 1400$.

When $x = 200$, $R = 13(200) = 2600$ and when $x = 1400$, $R = 13(1400) = 18{,}200$. The break-even points are $(200, 2600)$ and $(1400, 18{,}200)$.

b. Refer to the graph in the back of the textbook.

c. Profit = Revenue − Cost

$$= 13x - \left(0.0075x^2 + x + 2100\right)$$

$$= 13x - 0.0075x^2 - x - 2100$$

$$= -0.0075x^2 + 12x - 2100$$

which is a parabola opening downward so the maximum occurs at the vertex. Using $a = -0.0075$ and $b = 12$, the x-value of the vertex is

$$x = \frac{-12}{2(-0.0075)} = 800.$$ Maximum profit occurs from producing 800 pens.

31. a. Let $C = R$ to form the equation $1.625x^2 + 33{,}150 = 650x$. Write this in standard form $1.625x^2 - 650x + 33{,}150 = 0$ and use the quadratic formula with $a = 1.625$, $b = -650$, and $c = 33{,}150$ to solve for x:

$$x = \frac{-(-650) \pm \sqrt{(-650)^2 - 4(1.625)(33{,}150)}}{2(1.625)} = \frac{650 \pm \sqrt{207{,}025}}{3.25} = \frac{650 \pm 455}{3.25}.$$

This gives the values $x = \dfrac{650 - 455}{3.25} = 60$ and $x = \dfrac{650 + 455}{3.25} = 340$.

When $x = 60$, $R = 650(60) = 39{,}000$ and when $x = 340$, $R = 650(340) = 221{,}000$. The break-even points are $(60, 39{,}000)$ and $(340, 221{,}000)$.

b. Refer to the graph in the back of the textbook.

c. Profit = Revenue − Cost

$$= 650x - \left(1.625x^2 + 33{,}150\right)$$

$$= 650x - 1.625x^2 - 33{,}150$$

$$= -1.625x^2 + 650x - 33{,}150$$

which is a parabola opening downward so the maximum occurs at the vertex. Using $a = -1.625$ and $b = 650$, the x-value of the vertex is $x = \dfrac{-650}{2(-1.625)} = 200$.

Maximum profit occurs from producing 200 washing machines.

Midchapter 4 Review

1. In $y = -2x^2 + 5x - 1$, $a = -2$ and $b = 5$, so the parabola has vertex where $x = \dfrac{-5}{2(-2)} = \dfrac{5}{4}$. For $x = \dfrac{5}{4}$,

$$y = -2\left(\frac{5}{4}\right)^2 + 5\left(\frac{5}{4}\right) - 1$$
$$= -2\left(\frac{25}{16}\right) + \frac{25}{4} - 1$$
$$= -\frac{25}{8} + \frac{25}{4} - 1$$
$$= -\frac{25}{8} + \frac{50}{8} - \frac{8}{8}$$
$$= \frac{17}{8}$$

The vertex is $\left(\dfrac{5}{4}, \dfrac{17}{8}\right)$. The parabola opens downward so the vertex is a maximum point.

3. **a.** In $y = x^2 + x - 6$, $a = 1$ and $b = 1$ so the vertex is where $x = \dfrac{-1}{2(1)} = -\dfrac{1}{2}$. When $x = -\dfrac{1}{2}$,

$$y = \left(-\frac{1}{2}\right)^2 + \left(-\frac{1}{2}\right) - 6$$
$$= \frac{1}{4} - \frac{1}{2} - 6$$
$$= \frac{1}{4} - \frac{2}{4} - \frac{24}{4}$$
$$= -\frac{25}{4}$$

The vertex is $\left(-\dfrac{1}{2}, -\dfrac{25}{4}\right)$.

For the x-intercepts, let $y = 0$:
$$0 = x^2 + x - 6$$
$$0 = (x + 3)(x - 2)$$
which has solutions $x = -3$ and $x = 2$. Hence the x-intercepts are $(-3, 0)$ and $(2, 0)$. The y-intercept is $(0, -6)$. (Note that the y-intercept is always $(0, c)$).

b,c. Refer to the graph in the back of the textbook.

5. **a.** The vertex is $(-3, -2)$.

b. To write the equation in standard form, multiply and combine terms:
$$y = \frac{1}{3}(x + 3)^2 - 2$$
$$y = \frac{1}{3}\left(x^2 + 6x + 9\right) - 2$$
$$y = \frac{1}{3}x^2 + 2x + 3 - 2$$
$$y = \frac{1}{3}x^2 + 2x + 1$$

c. Refer to the graph in the back of the textbook.

x	−6	0	1
y	1	1	10/3

132

7. a. To fit the points on a parabola, use the equations:

$$-2 = a(0)^2 + b(0) + c$$
$$1 = a(-6)^2 + b(-6) + c$$
$$6 = a(4)^2 + b(4) + c$$

which simplify to

$$c = -2 \quad (1)$$
$$36a - 6b + c = 1 \quad (2)$$
$$16a + 4b + c = 6 \quad (3)$$

Substitute $c = -2$ into (2):

$$36a - 6b + (-2) = 1$$
$$36a - 6b = 3 \quad (4)$$

Substitute $c = -2$ into (3):

$$16a + 4b + (-2) = 6$$
$$16a + 4b = 8 \quad (5)$$

Multiply (4) by 2 and (5) by 3 and add the results:

$$72a - 12b = 6 \quad (4a)$$
$$\underline{48a + 12b = 24} \quad (5a)$$
$$120a = 30$$
$$a = \frac{1}{4}$$

Substitute this value of a into (5):

$$16\left(\frac{1}{4}\right) + 4b = 8$$
$$4 + 4b = 8$$
$$4b = 4$$
$$b = 1$$

Substitute a and b into (2):

$$36\left(\frac{1}{4}\right) - 6(1) + c = 1$$
$$9 - 6 + c = 1$$
$$c = -2$$

The equation is $y = \frac{1}{4}x^2 + x - 2$.

Note: This system of equations may also be solved using matrices. See Section 2.4.

b. Refer to the graph in the back of the textbook.

9. a. Billy is standing 4 feet from the railing, which means at $x = -4$. The bridge is 42 feet high, and so the ball is at the height $h = 45$ when it leaves Billy's hand. This gives the ordered pair $(-4, 45)$. The vertex of the trajectory is 3 ft above the railing, which is 46 feet high, so the vertex is $(0, 49)$.

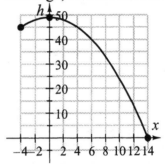

b. Since the vertex is $(0, 49)$, the equation in vertex form is

$$h = a(x - 0)^2 + 49 \text{ or}$$

$h = ax^2 + 49$. To find the value of a, use the point $(-4, 45)$:

$$45 = a(-4)^2 + 49$$
$$45 = 16a + 49$$
$$-4 = 16a$$
$$-\frac{1}{4} = a$$

The equation is $h = -\frac{1}{4}x^2 + 49$.

c. Let $h = 0$ and solve for x:

$$0 = -\frac{1}{4}x^2 + 49$$
$$\frac{1}{4}x^2 = 49$$
$$x^2 = 196$$
$$x = \pm\sqrt{196}$$
$$x = \pm 14$$

The object is 14 feet from the origin when it hits the water. (Disregard the negative answer.)

11. a. The quadratic regression equation given by the calculator is $y = -0.05x^2 - 0.003x + 234.2$. Refer to the graph in the back of the textbook.

b. Since $a = -0.05$ and $b = -0.003$, the vertex is where
$$x = \frac{-(-0.003)}{2(-0.05)} = -0.03.$$
When $x = -0.3$, $y =$
$$-0.05(-0.03)^2 - 0.003(-0.03) + 234.2$$
$$\approx 234.2$$
The vertex is at $(-0.03, 234.2)$. According to the regression equation, the velocity is -0.03 meters per second at the maximum height of 234.2 meters. At the maximum height, the velocity of the debris should be zero.

13. a. In $P = -0.4x^2 + 36x$, $a = -0.4$ and $b = 36$, so the vertex is where
$$x = \frac{-36}{2(-0.4)} = 45. \text{ Kiyoshi}$$
should produce and sell 45 floral arrangements to maximize profit.

b. When $x = 45$,
$$P = -0.4(45)^2 + 36(45) = 810.$$
His maximum profit is $810.

c. Graph $Y_1 = -0.4x^2 + 36x$ using Xmin = 0, Xmax = 100, Ymin = 0, and Ymax = 1000.

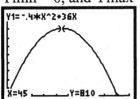

15. a. Given $y = \frac{1}{2}x^2 - \frac{3}{2}x$ and $y = -\frac{1}{2}x^2 + \frac{1}{2}x + 3$, set the y-values equal to each other:
$$\frac{1}{2}x^2 - \frac{3}{2}x = -\frac{1}{2}x^2 + \frac{1}{2}x + 3$$
$$x^2 - 2x - 3 = 0$$
$$(x - 3)(x + 1) = 0$$
So $x = 3$ or $x = -1$.
When $x = 3$,
$$y = \frac{1}{2}(3)^2 - \frac{3}{2}(3) = 0.$$
When $x = -1$,
$$y = \frac{1}{2}(-1)^2 - \frac{3}{2}(-1) = 2.$$
The solution points are $(3, 0)$ and $(-1, 2)$.

b. Refer to the graph in the back of the textbook.

Homework 4.4

1.a,b. Refer to the graph in the back of the textbook.

   c. The parts of the graph where $y > 9$, and therefore the solutions to $x^2 > 9$, are $x > 3$ or $x < -3$. The inequality $x > 3$ is an incomplete answer because it does not include the values where $x < -3$.

3. a. To graph $y = x^2 - 2x - 3$, first find the x-intercepts by solving:
$$x^2 - 2x - 3 = 0$$
$$(x - 3)(x + 1) = 0$$
So the parabola crosses the x-axis at $x = 3$ and $x = -1$. The x-value of the vertex is directly between the x-intercepts, at $x = 1$. When $x = 1$, $y = 1^2 - 2(1) - 3 = -4$, so the vertex is $(1, -4)$. The y-intercept is $(0, -3)$. Refer to the graph in the back of the textbook.

   b. Refer to the graph in the back of the textbook.

   c. The parts of the graph where $y > 0$, and therefore the solutions to $x^2 - 2x - 3 > 0$, are $x < -1$ or $x > 3$.

5. a. The solutions to the equation are the x-values of the x-intercepts of the graph, which appear to be $x = -12$ and $x = 15$.

   b. The parts of the graph with positive y-coordinate appear to be where $x < -12$ or $x > 15$.

7. a. The solutions to the equation are the x-values of the x-intercepts of the graph, which appear to be $x = 0.3$ and $x = 0.5$.

   b. The part of the graph with zero or positive y-coordinates appears to be where $0.3 \le x \le 0.5$.

9. $(-5, 3]$

11. $[-4, 0]$

13. $(-6, \infty)$

15. $(-\infty, -3) \cup [-1, \infty)$

17. $[-6, -4) \cup (-2, 0]$

135

19. Use the *x*-intercepts at $x = -3$ and $x = 6$ from the first graph for parts (a) and (b). Use the intersections at $x = -2$ and $x = 5$ from the second graph for parts (c) and (d).

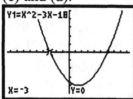

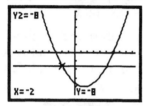

a. $y > 0$ for $x < -3$ or $x > 6$. Using interval notation:
$$(-\infty, -3) \cup (6, \infty)$$

b. $y < 0$ for $-3 < x < 6$. Using interval notation: $(-3, 6)$

c. $y \leq -8$ for $-2 \leq x \leq 5$. Using interval notation: $[-2, 5]$

d. $y \geq -8$ for $x \leq -2$ or $x \geq 5$. Using interval notation:
$$(-\infty, -2] \cup [5, \infty)$$

21. Use the *x*-intercepts at $x = -4$ and $x = 4$ from the first graph for parts (a) and (b). Use the intersections at $x = -3$ and $x = 3$ from the second graph for parts (c) and (d).

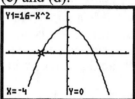

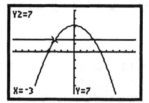

a. $y > 0$ for $-4 < x < 4$. Using interval notation: $(-4, 4)$

b. $y < 0$ for $x < -4$ or $x > 4$. Using interval notation:
$$(-\infty, -4) \cup (4, \infty)$$

c. $y \leq 7$ for $x \leq -3$ or $x \geq 3$. Using interval notation:
$$(-\infty, -3] \cup [3, \infty)$$

d. $y \geq 7$ for $-3 \leq x \leq 3$. Using interval notation: $[-3, 3]$.

23. Graph $y = (x - 3)(x + 2)$. Then $y > 0$ for $x < -2$ or $x > 3$.

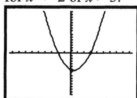

25. Graph $y = k(4 - k)$. Then $y \geq 0$ for $0 \leq k \leq 4$.

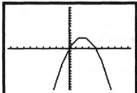

27. Graph $y = 6 + 5p - p^2$. Then $y < 0$ for $p < -1$ or $p > 6$.

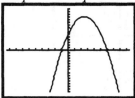

29. Graph the equations
$Y_1 = x^2 - 1.4x - 20$ and $Y_2 = 9.76$.

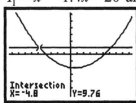

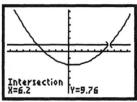

The solutions are where $Y_1 < Y_2$.
Hence $-4.8 < x < 6.2$.

31. Graph the equations
$Y_1 = 5x^2 + 39x + 27$ and $Y_2 = 5.4$.

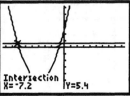

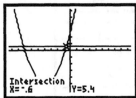

The solutions are where $Y_1 \geq Y_2$.
Hence $x \geq -0.6$ or $x \leq -7.2$.

33. Graph the equations
$Y_1 = -8x^2 + 112x - 360$ and
$Y_2 = 6.08$.

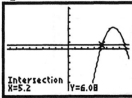

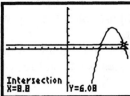

The solutions are where $Y_1 < Y_2$.
Hence $x < 5.2$ or $x > 8.8$.

35. Graph $Y_1 = x^2$ and $Y_2 = 12.2$ using Xmin $= -10$, Xmax $= 10$, Ymin $= -5$, and Ymax $= 15$.

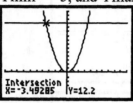

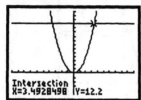

The solutions are where $Y_1 > Y_2$.
Hence $x < -3.5$ or $x > 3.5$.

37. Graph $Y_1 = -3x^2 + 7x - 25$ using Xmin $= -10$, Xmax $= 10$, Ymin $= -100$, and Ymax $= 20$.

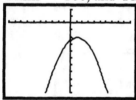

The solutions are where $Y_1 \leq 0$.
Since the parabola opens downward with vertex below the x-axis, all points on the parabola have negative y-coordinates, so all values of x are solutions to the inequality.

39. Graph $Y_1 = 0.4x^2 - 54x$ and $Y_2 = 620$ using Xmin $= -200$, Xmax $= 200$, Ymin $= -2000$, and Ymax $= 2000$.

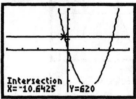

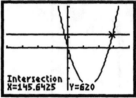

The solutions are where $Y_1 < Y_2$.
Hence $-10.6 < x < 145.6$.

Homework 4.5

1. The solutions of $(x+3)(x-4)=0$ are $x=-3$ and $x=4$. These are also the x-intercepts of the parabola $y=(x+3)(x-4)$. Since this parabola opens upward, the solution set to $(x+3)(x-4)<0$ is $(-3, 4)$.

3. To solve $28-3x-x^2=0$, factor so that $(7+x)(4-x)=0$. The solutions to the equation are $x=-7$ and $x=4$. (Alternatively, one could solve the equation using the quadratic formula.) $x=-7$ and $x=4$ are also the x-intercepts of the parabola $y=28-3x-x^2$. Since this parabola opens downward, the solution set to $28-3x-x^2 \geq 0$ is $[-7, 4]$.

5. $\quad 2z^2-7z>4$
$2z^2-7z-4>0$
Solve $2z^2-7z-4=0$ by factoring:
$2z^2-7z-4=0$
$(2z+1)(z-4)=0$
$2z+1=0 \quad$ or $\quad z-4=0$
$z=-\dfrac{1}{2} \qquad\qquad z=4$
(Alternatively, one could solve the equation using the quadratic formula.) $z=-\dfrac{1}{2}$ and $z=4$ are the z-intercepts of the parabola $y=2z^2-7z-4$. Since this parabola opens upward, the solution set to $2z^2-7z-4>0$ is
$\left(-\infty, -\dfrac{1}{2}\right) \cup (4, \infty)$.

7. Solve $64-t^2=0$:
$64-t^2=0$
$64=t^2$
$\pm\sqrt{64}=t$
$\pm 8=t$
Thus $t=-8$ and $t=8$ are the t-intercepts of the parabola $y=64-t^2$. Since this parabola opens downward, the solution set to $64-t^2>0$ is $(-8, 8)$.

9. $\quad v^2<5$
$v^2-5<0$
Solve $v^2-5=0$:
$v^2-5=0$
$v^2=5$
$v=\pm\sqrt{5}$
Thus, $v=-\sqrt{5}$ and $v=\sqrt{5}$ are the v-intercepts of the parabola $y=v^2-5$. Since this parabola opens upward, the solution set to $v^2-5<0$ is $(-\sqrt{5}, \sqrt{5})$ or, approximately, $(-2.24, 2.24)$.

11. $\quad 5a^2-32a+12=0$
$(5a-2)(a-6)=0$
$5a-2=0 \quad$ or $\quad a-6=0$
$a=\dfrac{2}{5} \qquad\qquad a=6$
Thus $a=\dfrac{2}{5}$ and $a=6$ are the a-intercepts of $y=5a^2-32a+12$. Since this parabola opens upward, the solution set to $5a^2-32a+12 \geq 0$ is
$\left(-\infty, -\dfrac{2}{5}\right] \cup [6, \infty)$.

13.
$$4x^2 + x \geq -2x^2 + 2$$
$$6x^2 + x - 2 \geq 0$$
Solve $6x^2 + x - 2 = 0$:
$$6x^2 + x - 2 = 0$$
$$(3x + 2)(2x - 1) = 0$$
$$3x + 2 = 0 \quad \text{or} \quad 2x - 1 = 0$$
$$x = -\frac{2}{3} \qquad\qquad x = \frac{1}{2}$$
Thus $x = -\frac{2}{3}$ and $x = \frac{1}{2}$ are the x-intercepts of $y = 6x^2 + x - 2$.. Since this parabola opens upward, the solution set to $6x^2 + x - 2 \geq 0$ is
$$\left(-\infty, -\frac{2}{3}\right] \cup \left[\frac{1}{2}, \infty\right).$$

15. Solve $x^2 - 4x + 1 = 0$ using the quadratic formula:
$$x = \frac{-(-4) \pm \sqrt{(-4)^2 - 4(1)(1)}}{2(1)}$$
$$= \frac{4 \pm \sqrt{12}}{2}$$
Thus $x = \frac{4 - \sqrt{12}}{2} \approx 0.27$ and $x = \frac{4 + \sqrt{12}}{2} \approx 3.73$ are the x-intercepts of $y = x^2 - 4x + 1$. Since this parabola opens upward, the solution set to $x^2 - 4x + 1 \geq 0$ is $(-\infty, 0.27] \cup [3.73, \infty)$

Note: The solution set in exact form is $\left(-\infty, \frac{4 - 2\sqrt{3}}{2}\right] \cup \left[\frac{4 + 2\sqrt{3}}{2}, \infty\right)$ which simplifies to
$$\left(-\infty, 2 - \sqrt{3}\right] \cup \left[2 + \sqrt{3}, \infty\right).$$

17. Solve $-3 - m^2 = 0$:
$$-3 - m^2 = 0$$
$$-3 = m^2$$
$$\pm\sqrt{-3} = m$$
so the equation $-3 - m^2 = 0$ has no real solutions, which means that the parabola $y = -3 - m^2$ has no m-intercepts. Since this parabola opens downward and never passes through the m-axis, all of the y-coordinates are negative. Therefore, the solution to $-3 - m^2 < 0$ is all values of m, or using interval notation, $(-\infty, \infty)$.

Note: The inequality $-3 - m^2 < 0$ can also be solved by noting that since m^2 is positive for all m, $-3 - m^2$ is negative for all m.

19. Solve $w^2 - w + 4 = 0$ using the quadratic formula:
$$w = \frac{-(-1) \pm \sqrt{(-1)^2 - 4(1)(4)}}{2(1)}$$
$$= \frac{1 \pm \sqrt{-15}}{2}$$
so the equation $w^2 - w + 4 = 0$ has no real solutions, which means that the parabola $y = w^2 - w + 4$ has no w-intercepts. Since this parabola opens upward and never passes through the w-axis, all of the y-coordinates are positive. Therefore, the inequality $w^2 - w + 4 \leq 0$ has no solutions.

21. a. To solve $320t - 16t^2 > 1024$ or $-16t^2 + 320t - 1024 > 0$, solve

$$320t - 16t^2 = 1024$$

$$-16t^2 + 320t - 1024 = 0$$

$$-16\left(t^2 - 20t + 64\right) = 0$$

$$-16(t - 4)(t - 16) = 0$$

So $t = 4$ and $t = 16$ are the *t*-intercepts of the parabola $y = -16t^2 + 320t - 1024$. Since the parabola opens downward and we are looking for points with positive *y*-coordinates, $4 < t < 16$. The rocket is above 1024 feet from 4 to 16 seconds into its flight.

b. Graph $Y_1 = 320x - 16x^2$ and $Y_2 = 1024$ using Xmin = 0, Xmax = 20, Ymin = 0, and Ymax = 2000. The solutions are where $Y_1 > Y_2$. Hence $4 < x < 16$, which agrees with our answer in part (a).

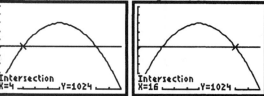

23. a. To solve $-0.02x^2 + 14x + 1600 < 2800$, or $-0.02x^2 + 14x - 1200 < 0$ with $0 \le x \le 700$, solve $-0.02x^2 + 14x - 1200 = 0$ by using the quadratic formula.

$$x = \frac{-14 \pm \sqrt{14^2 - 4(-0.02)(-1200)}}{2(-0.02)} = \frac{-14 \pm \sqrt{100}}{-0.04} = \frac{-14 \pm 10}{-0.04}$$

Thus, $x = \dfrac{-14 + 10}{-0.04} = 100$ and $x = \dfrac{-14 - 10}{-0.04} = 600$ are the *x*-intercepts of the parabola. Since the parabola opens downward and we are looking for points with negative *y*-coordinates, the solution is when $0 \le x < 100$ or $600 < x \le 700$. (The given equation is valid only for $0 \le x \le 700$.) The company must produce fewer than 100 pairs of shears or more than 600 but at most 700 pairs of shears.

b. Graph $Y_1 = -0.02x^2 + 14x + 1600$ and $Y_2 = 2800$ using Xmin = 0, Xmax = 1000, Ymin = 0, and Ymax = 5000. The solutions are where $Y_1 < Y_2$. Hence $x < 100$ or $x > 600$, and when we consider the requirement that $0 \le x \le 700$, we have the same answer as in part (a).

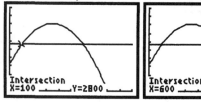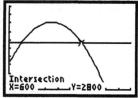

25. a. Revenue $= (1200 - 30p)p = 1200p - 30p^2$. To solve $1200p - 30p^2 > 9000$ or $-30p^2 + 1200p - 9000 > 0$, first solve $-30p^2 + 1200p - 9000 = 0$ using the quadratic formula.

$$p = \frac{-1200 \pm \sqrt{(1200)^2 - 4(-30)(-9000)}}{2(-30)} = \frac{-1200 \pm \sqrt{360,000}}{-60} = \frac{-1200 \pm 600}{-60}$$

Thus, $p = \dfrac{-1200 + 600}{-60} = 10$ and $p = \dfrac{-1200 - 600}{-60} = 30$ are the p-intercepts of the parabola. Since the parabola opens downward and we are looking for points with positive y-coordinates, the solution set is $10 < p < 30$, so the price of the sweatshirts should be between \$10 and \$30.

b. Graph $Y_1 = 1200x - 30x^2$ and $Y_2 = 9000$ using Xmin = 0, Xmax = 40, Ymin = 0, and Ymax = 15,000. The solutions are where $Y_1 > Y_2$, so $10 < x < 30$, which agrees with our answer in part (a)

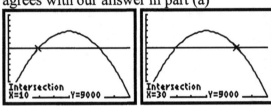

27. a. The volume of a cylinder is $V = \pi r^2 h$. If $h = 20$, then $V = 20\pi r^2$. So we need to solve the inequality $500\pi < 20\pi r^2 < 2880\pi$. Divide each expression by 20π so that $25 < r^2 < 144$. Now $r^2 = 25$ for both $r = -5$ and $r = 5$ and $r^2 = 144$ for both $r = -12$ and $r = 12$, but since the radius must be positive, we can disregard the negative values. Therefore, $25 < r^2 < 144$ for $5 < r < 12$. The radius should be between 5 and 12 feet.

b. Graph $Y_1 = 20\pi x^2$, $Y_2 = 500\pi$, and $Y_3 = 2880\pi$ using Xmin = 0, Xmax = 15, Ymin = 0, and Ymax = 10,000. The solution is the set of nonnegative values of x where $Y_2 < Y_1 < Y_3$ and so $5 < x < 12$ which agrees with our results in part (a).

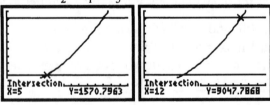

29. a.

Additional people	Size of group	Price per person	Total income
0	20	600	12,000
5	25	550	13,750
10	30	500	15,000
15	35	450	15,750
20	40	400	16,000
25	45	350	15,750
30	50	300	15,000
35	55	250	13,750
40	60	200	12,000
45	65	150	9,750
50	70	100	7,000
55	75	50	3,750
60	80	0	0

b. If x additional people sign up, the size of the group is $20 + x$ and the price per person is $600 - 10x$.

c. Let $y = $ the travel agency's total income. Then y is equal to the size of the group times the price per person: $y = (20 + x)(600 - 10x) = 12,000 + 400x - 10x^2$.

d. Enter the equation from part (c) into your calculator, set Table Start $= 0$, and set ΔTbl $= 5$. Your calculator's table values should agree with the table in part (a).

e. The parabola opens downward, so the maximum occurs at the vertex. With $a = -10$ and $b = 400$, the vertex is where $x = \dfrac{-400}{2(-10)} = 20$. When $x = 20$,

$$y = 12,000 + 400(20) - 10(20)^2 = 16,000.$$

The travel agency can earn a maximum of \$16,000, which they obtain if 20 additional people sign up (and therefore 40 people total attend).

f. We need to solve the inequality $-10x^2 + 400x + 12,000 \geq 15,750$, or $-10x^2 + 400x - 3750 \geq 0$. First, solve the equation $-10x^2 + 400x - 3750 = 0$ by dividing both sides by -10 to get $x^2 - 40x + 375 = 0$. By the quadratic formula,

$$x = \frac{-(-40) \pm \sqrt{(-40)^2 - 4(1)(375)}}{2(1)} = \frac{40 \pm \sqrt{100}}{2} = \frac{40 \pm 10}{2}.$$

Thus $x = \dfrac{40 - 10}{2} = 15$ and $x = \dfrac{40 + 10}{2} = 25$ are the x-intercepts of the parabola $y = -10x^2 + 400x - 3750$. This parabola opens downward, so the solution set to $-10x^2 + 400x - 3750 \geq 0$ is $15 \leq x \leq 25$. Since x is the number of additional people, a total of at least 35 but no more than 45 must sign up.

g. Refer to the graph in the back of the textbook.

Chapter 4 Review

1. a. In $y = x^2 - x - 12$, $a = 1$, $b = -1$, and $c = -12$. The vertex is where $x = \frac{-(-1)}{2(1)} = \frac{1}{2}$. When $x = \frac{1}{2}$.

$y = \left(\frac{1}{2}\right)^2 - \frac{1}{2} - 12 = \frac{-49}{4}$, so the vertex is at $\left(\frac{1}{2}, \frac{-49}{4}\right)$.

The y-intercept is at $(0, -12)$. For the x-intercepts, solve:

$0 = x^2 - x - 12$
$0 = (x - 4)(x + 3)$

The solutions are $x = 4$ and $x = -3$, so the x-intercepts are $(4, 0)$ and $(-3, 0)$.

b. Refer to the graph in the back of the textbook.

3. a. In $y = -x^2 + 2x + 4$, $a = -1$, $b = 2$, and $c = 4$. The vertex is where $x = \frac{-2}{2(-1)} = \frac{-2}{-2} = 1$. When $x = 1$, $y = -(1)^2 + 2(1) + 4 = 5$ so the vertex is at $(1, 5)$. The y-intercept is at $(0, 4)$. For the x-intercepts, use the quadratic formula.

$x = \frac{-2 \pm \sqrt{(2)^2 - 4(-1)(4)}}{2(-1)}$

$= \frac{-2 \pm \sqrt{20}}{-2}$

The solutions are

$x = \frac{-2 - \sqrt{20}}{-2} \approx 3.24$ and

$x = \frac{-2 + \sqrt{20}}{-2} \approx -1.24$

so the x-intercepts are $(3.24, 0)$ and $(-1.24, 0)$.

Note: The expression $\frac{-2 \pm \sqrt{20}}{-2}$ can be simplified to $\frac{-2 \pm 2\sqrt{5}}{-2} = 1 \pm \sqrt{5}$.

b. Refer to the graph in the back of the textbook.

5. To solve the inequality, first solve $(x - 3)(x + 2) = 0$. The solutions are $x = 3$ and $x = -2$. Since we are looking for where the y-coordinate is positive and since the parabola $y = (x - 3)(x + 2)$ opens upward, the solution to $(x - 3)(x + 2) > 0$ is where $x < -2$ or $x > 3$. In interval notation, $(-\infty, -2) \cup (3, \infty)$.
Graph using the standard window.

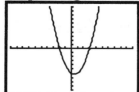

7. An equivalent inequality is $2y^2 - y - 3 \leq 0$. To solve this inequality, first solve:
$$2y^2 - y - 3 = 0$$
$$(2y - 3)(y + 1) = 0$$
$$2y - 3 = 0 \quad \text{or} \quad y + 1 = 0$$
$$y = \frac{3}{2} \qquad\qquad y = -1$$
Let y correspond to the horizontal axis and z the vertical axis. Since the parabola $z = 2y^2 - y - 3$ opens upward and we are looking for where the z-coordinate is zero or negative, the solution set is $\left[-1, \frac{3}{2}\right]$.
Graph using the standard window.

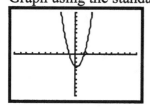

9. An equivalent inequality is $s^2 - 4 \leq 0$. To solve this inequality, first solve
$$s^2 - 4 = 0$$
$$(s + 2)(s - 2) = 0$$
$$s + 2 = 0 \quad \text{or} \quad s - 2 = 0$$
$$s = -2 \qquad\qquad s = 2$$
Since we are looking for where the graph of $y = s^2 - 4$ is on or below the s-axis and since the parabola opens upward, the solution set is $[-2, 2]$.

Graph using the standard window.

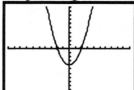

11. a. The revenue from selling $220 - \frac{1}{4}p$ sandwiches at p cents each is

$$R = \left(220 - \frac{1}{4}p\right)p = 220p - \frac{1}{4}p^2.$$

b. Since $\$480 = 48{,}000$ cents, we need to solve the inequality

$220p - \frac{1}{4}p^2 > 48{,}000$ or $-\frac{1}{4}p^2 + 220p - 48{,}000 > 0$. Solve the equation

$-\frac{1}{4}p^2 + 220p - 48{,}000 = 0$ by multiplying both sides by -4 to get

$p^2 - 880p + 192{,}000 = 0$. By the quadratic formula,

$$p = \frac{-(-880) \pm \sqrt{(-880)^2 - 4(1)(192{,}000)}}{2(1)} = \frac{880 \pm \sqrt{6400}}{2} = \frac{880 \pm 80}{2}.$$

So $p = \frac{880 - 80}{2} = 400$ and $p = \frac{880 + 80}{2} = 480$ are the p-intercepts of the

graph of $y = -\frac{1}{4}p^2 + 220p - 48{,}000$. Since this parabola opens downward and

we are looking for the parts of the graph above the p-axis, the solution to

$-\frac{1}{4}p^2 + 220p - 48{,}000 > 0$ is $400 < p < 480$. The sandwiches can sell for

between $\$4.00$ and $\$4.80$.

13. a. Let x represent the number of additional hives per square mile that should be installed. The total number of hives per square mile is $4 + x$ and the production amount per hive is $32 - 4x$. So the total amount of honey produced on the 60 square miles of pastureland is

$$y = 60(4 + x)(32 - 4x) = 60\left(128 + 16x - 4x^2\right) = 7680 + 960x - 240x^2.$$

b. The graph of y is a parabola which opens downward, so the maximum value

occurs at the vertex. The vertex is where $x = \frac{-960}{2(-240)} = 2$. Thus, to maximize

honey production, the beekeeper should install 2 additional hives per square mile.

15. Refer to the graph in the back of the textbook. Substitute the value $y = 3$ from the second equation into the first equation:

$$3 + x^2 = 4$$
$$x^2 = 1$$
$$x = \pm\sqrt{1}$$
$$x = \pm 1$$

If $x = 1$, then $y = 3$ and if $x = -1$, then $y = 3$ so the solutions are $(1, 3)$ and $(-1, 3)$.

17. Refer to the graph in the back of the textbook. Equate the expressions for y:

$$x^2 - 5 = 4x$$
$$x^2 - 4x - 5 = 0$$
$$(x + 1)(x - 5) = 0$$

So $x = -1$ and $x = 5$. When $x = -1$, $y = 4(-1) = -4$ and when $x = 5$, $y = 4(5) = 20$. The solutions are $(-1, -4)$ and $(5, 20)$.

19. Refer to the graph in the back of the textbook. Equate the expressions for y:

$$x^2 - 6x + 20 = 2x^2 - 2x - 25$$
$$0 = x^2 + 4x - 45$$
$$0 = (x + 9)(x - 5)$$

The solutions are when $x = -9$ and $x = 5$. When $x = -9$,

$$y = 2(-9)^2 - 2(-9) - 25$$
$$= 162 + 18 - 25$$
$$= 155$$

When $x = 5$,

$$y = 2(5)^2 - 2(5) - 25$$
$$= 50 - 10 - 25$$
$$= 15$$

The solutions are $(-9, 155)$ and $(5, 15)$.

21. From the points given, the equations are

$$-4 = a(-1)^2 + b(-1) + c$$
$$-6 = a(0)^2 + b(0) + c$$
$$6 = a(4)^2 + b(4) + c$$

which simplify to

$$a - b + c = -4 \quad (1)$$
$$c = -6 \quad (2)$$
$$16a + 4b + c = 6 \quad (3)$$

From (2), $c = -6$. Put this value for c into (1):

$$a - b - 6 = -4$$
$$a - b = 2 \quad (4)$$

Substitute $c = -6$ into (3):

$$16a + 4b - 6 = 6$$
$$16a + 4b = 12 \quad (5)$$

Add 4 times (4) to (5):

$$4a - 4b = 8 \quad (4a)$$
$$\underline{16a + 4b = 12 \quad (5)}$$
$$20a = 20$$
$$a = 1$$

Put this value into (4):

$$1 - b = 2$$
$$-b = 1$$
$$b = -1$$

Hence $a = 1$, $b = -1$, and $c = -6$.

23. Using vertex form, the equation is $y = a(x - 15)^2 - 6$. To find the value of a, use the point $(3, 22.8)$:

$$22.8 = a(3 - 15)^2 - 6$$
$$22.8 = 144a - 6$$
$$28.8 = 144a$$
$$0.2 = a$$

Hence $y = 0.2(x - 15)^2 - 6$.

Chapter Five: Functions and Their Graphs

Homework 5.1

1. Function; the sales tax is uniquely determined from the price

3. Not a function; the annual income may differ between people with the same number of years of education.

5. Function; the weight is uniquely determined from the volume.

7. Input variable: items purchased; output variable: price of an item. The relationship is a function since each item has only one price.

9. Input variable: topic of the indexed item; output variable: page or pages on which the topic occurs. The relationship is not a function since a topic can be found on more than one page.

11. Input variable: student's name; output variable: grade(s) on various tests. This is not a function since the same student may have different grades on different tests.

13. Input variable: person stepping on scale; output variable: person's weight. The relationship is a function since a person cannot have two different weights at the same time.

15. Not a function: Some values of x have more than one value of t. For example, when $x = -1$, t can be either 2 or 5.

17. Function: Each value of x has a unique value of y.

19. Function: Each value of r has a unique value of v.

21. Function: Each value of p has a unique value of v.

23. Not a function: Some values of T have more than one value of h. For example, on both Jan. 1 and Jan. 7 the temperature is 34 degrees, but there are two different humidity values: 42% and 48%.

25. Function: Each value of I has a unique value of T.

27. a. When $p = 25$, $v = 60.0$, so $g(25) = 60.0$.

b. When $p = 40$, $v = 37.5$, so $g(40) = 37.5$.

c. $x = 30$ (when $p = 30$, $v = 50.0$).

29. a. Note that $7010 < 8750 < 9169$. So when $I = 8750$, $T = 15\%$. Hence $T(8750) = 15\%$.

b. Note that $4750 < 6249 < 7009$. So when $I = 6249$, $T = 14\%$. Hence $T(6249) = 14\%$.

c. $x =$ any number between \$7010 and \$9169 inclusive (when $7010 \leq I \leq 9169$, $T = 15\%$).

31. a. When $C = 2$, $t = 1$. So 2000 students considered themselves computer literate in 1991.

  b. When $C = 4$, $t = 2$. So the value of C doubled from 2 to 4 in one year.

  c. When $C = 8$, $t = 3$. So the value of C doubled from 4 to 8 in one year.

  d. Assume that t starts in the beginning of January 1990. In the beginning of January 1992, $t = 2$ and $C = 4$, so 4000 students were computer literate. In the beginning of June 1993,
$t = 3\dfrac{5}{12} \approx 3.4$ and C ≈ 11, so 11,000 students were computer literate. Thus $11{,}000 - 4000 = 7000$ students became computer literate between January 1992 and June 1993.

33. a. When $d = 6.5$, $R \approx 360$. If the theatre charges $6.50 per ticket, the revenue is about $360.

  b. When $R = 250$, $d = 2$ or $d = 8$. The theatre should charge $2 or $8 per ticket.

  c. R > 250 for $3.30 < d <$ $6.70.

35. a. The temperature does not rise at a constant rate. Initially, the temperature rises rapidly, then it levels off at 55°, and finally it slowly increases.

  b. Refer to the graph in the back of the textbook.

  c. At the melting point, extra energy is expended so the temperature levels off for a time. Thus the melting point of stearic acid is 55°, which occurs between 3 and 8 minutes.

37. a.

Initial clutch size laid	Experimental clutch size				
	4	5	6	7	8
5	0.3	0.7	0.5	0.3	0
6	1.7	1.9	2.8	0.8	1.2
7	3.5	2.3	3.1	3.6	2.4
8	2.5	3.5	3.5	4.3	4.5

  b. The optimal experimental clutch size is the experimental clutch size that results in the highest number of young fledglings.

Initial clutch size	5	6	7	8
Optimum clutch size	5	6	7	8

  c. In each category, the natural clutch size produced the largest number of fledglings. Thus, optimal clutch size seems to depend upon the individual bird.

39. a. $f(3) = 6 - 2(3) = 6 - 6 = 0$

  b. $f(-2) = 6 - 2(-2) = 6 + 4 = 10$

  c. $f(12.7) = 6 - 2(12.7) = 6 - 25.4 = -19.4$

  d. $f\left(\dfrac{2}{3}\right) = 6 - 2\left(\dfrac{2}{3}\right) = 6 - \dfrac{4}{3} = 4\dfrac{2}{3}$

41. a. $h(0) = 2(0)^2 - 3(0) + 1 =$
$0 + 0 + 1 = 1$

b. $h(-1) = 2(-1)^2 - 3(-1) + 1$
$= 2 + 3 + 1 = 6$

c. $h\left(\frac{1}{4}\right) = 2\left(\frac{1}{4}\right)^2 - 3\left(\frac{1}{4}\right) + 1$
$= \frac{1}{8} - \frac{3}{4} + 1 = \frac{3}{8}$

d. $h(-6.2) = 2(-6.2)^2 - 3(-6.2) + 1$
$= 76.88 + 18.6 + 1 = 96.48$

43. a. $H(4) = \frac{2(4) - 3}{(4) + 2} = \frac{5}{6}$

b. $H(-3) = \frac{2(-3) - 3}{(-3) + 2} = \frac{-9}{-1} = 9$

c. $H\left(\frac{4}{3}\right) = \frac{2\left(\frac{4}{3}\right) - 3}{\left(\frac{4}{3}\right) + 2} = \frac{-\frac{1}{3}}{\frac{10}{3}} = -\frac{1}{10}$

d. $H(4.5) = \frac{2(4.5) - 3}{(4.5) + 2} = \frac{6}{6.5} = \frac{12}{13}$
≈ 0.923

45. a. $E(16) = \sqrt{(16) - 4} = \sqrt{12} = 2\sqrt{3}$
$\approx 3.464.$

b. $E(4) = \sqrt{(4) - 4} = \sqrt{0} = 0$

c. $E(7) = \sqrt{(7) - 4} = \sqrt{3}$

d. $E(4.2) = \sqrt{(4.2) - 4} = \sqrt{0.2}$
≈ 0.447

47. a.

t	2	5	8	10
V	23,520	16,800	10,080	5,600

b. $V(12) = 28,000\left(1 - 0.08(12)\right)$
$= 28,000(1 - 0.96) = 1120$
After 12 years, the value of the sports car is $1120.

49. a.

p	5000	8000	10,000	12,000
N	2400	1500	1200	1000

b. $N(6000) = \frac{12,000,000}{6000} = 2000$
This gives the number of cars that will be sold at a price of $6000.

51. a.

d	20	50	80	100
v	15.5	24.5	31.0	34.6

b. $v(250) = \sqrt{12(250)} = \sqrt{3000}$
$\approx 54.8.$ If the skid mark is 250 feet long, the car was traveling 54.8 miles per hour.

Homework 5.2

1. a. The points $(-3, -2)$, $(1, 0)$, and $(3, 5)$ lie on the graph, so $h(-3) = -2$, $h(1) = 0$, and $h(3) = 5$.

  b. The point $(2, 3)$ lies on the graph, so $h(z) = 3$ when $z = 2$.

  c. The z-intercepts are $(-2, 0)$ and $(1, 0)$. The h-intercept is $(0, -2)$. $h(-2) = 0$, $h(1) = 0$, and $h(0) = -2$.

  d. The highest point is $(3, 5)$ so the maximum value of $h(z)$ is 5.

  e. The maximum occurs for $z = 3$.

  f. The function is increasing on the intervals $(-4, -2)$ and $(0, 3)$. The function is decreasing on the interval $(-2, 0)$.

3. a. The points $(1, -1)$ and $(3, 2)$ lie on the graph, so $R(1) = -1$ and $R(3) = 2$.

  b. The points $(3, 2)$ and $(\approx -1.5, 2)$ lie on the graph, so $R(p) = 2$ when $p = 3$ and $p \approx -1.5$.

  c. The p-intercepts are $(-2, 0)$, $(2, 0)$, and $(4, 0)$ The R-intercept is $(0, 4)$. $R(-2) = 0$, $R(0) = 4$, $R(2) = 0$, and $R(4) = 0$.

  d. The highest point is $(0, 4)$ and the lowest is $(5, -5)$, so $R(p)$ has a maximum value of 4 and a minimum value of -5.

  e. The maximum occurs for $p = 0$. The minimum occurs for $p = 5$.

  f. The function is increasing on the intervals $(-3, 0)$ and $(1, 3)$. The function is decreasing on the intervals $(0, 1)$ and $(3, 5)$.

5. a. The points $(0, 0)$, $\left(\frac{1}{6}, \frac{1}{2}\right)$, and $(-1, 0)$ lie on the graph, so $S(0) = 0$, $S\left(\frac{1}{6}\right) = \frac{1}{2}$, and $S(-1) = 0$.

  b. The point $\left(\frac{1}{3}, \approx \frac{5}{6}\right)$ lies on the graph, so $S\left(\frac{1}{3}\right) \approx \frac{5}{6}$.

  c. The points $\left(-\frac{5}{6}, -\frac{1}{2}\right)$, $\left(-\frac{1}{6}, -\frac{1}{2}\right)$, $\left(\frac{7}{6}, -\frac{1}{2}\right)$, and $\left(\frac{11}{6}, -\frac{1}{2}\right)$ lie on the graph, so $S(x) = -\frac{1}{2}$ when $x = -\frac{5}{6}$, $-\frac{1}{6}$, $\frac{7}{6}$, and $\frac{11}{6}$.

  d. The maximum value for $S(x)$ is 1. The minimum value is -1.

  e. The maximum occurs when $x = -\frac{3}{2}$ and $x = \frac{1}{2}$. The minimum value occurs when $x = -\frac{1}{2}$ and $x = \frac{3}{2}$.

7. a. The points (–3, 2), (–2, 2), and (2, 1) lie on the graph, so $F(-3) = 2$, $F(-2) = 2$, and $F(2) = 1$.

b. $F(s) = -1$ for $-6 \leq s < -4$ and for $0 \leq s < 2$.

c. The maximum value of $F(s)$ is 2. The minimum is value is –1.

d. The maximum occurs when $-3 \leq s < -1$ or $3 \leq s < 5$. The minimum occurs when $-6 \leq s < -4$ or $0 \leq s < 2$.

9. a. Function, since no vertical line intersects the graph in more than one point.

b. Not a function, since there are vertical lines which intersect the graph in more than one point.

c. Not a function, since there are vertical lines which intersect the graph in more than one point.

d. Function, since no vertical line intersects the graph in more than one point.

e. Not a function, since there are vertical lines which intersect the graph in more than one point.

11. a. Refer to the graph in the back of the textbook.

x	–2	–1	0	1	2
$g(x)$	–4	3	4	5	12

b. Graph $g(x) = x^3 + 4$ using Xmin = –2, Xmax = 2, Ymin = –4, and Ymax = 12. In your table menu, choose Table Start = –2 and ΔTbl = 1. Your calculator's graph and table should agree with those in part (a).

13. a. Refer to the graph in the back of the textbook.

x	–5	–4	–3	–2	–1	0	1	2	3	4
$G(x)$	3	2.8	2.6	2.4	2.2	2	1.7	1.4	1	0

b. Graph $G(x) = \sqrt{4-x}$ using Xmin = –5, Xmax = 4, Ymin = 0, and Ymax = 3. In your table menu, choose Table Start = –5 and ΔTbl = 1. Your calculator's graph and table should agree with those in part (a).

15. a. Refer to the graph in the back of the textbook.

x	–3	–2	–1	0	1	2	3
$v(x)$	10	–3	–4	1	6	5	–8

b. Graph $v(x) = 1 + 6x - x^3$ using Xmin = –3, Xmax = 3, Ymin = –8, and Ymax = 10. In your table menu, choose Table Start = –3 and ΔTbl = 1. Your calculator's graph and table should agree with those in part (a).

17. a. The points (–1, 4) and (1, 4) lie on the graph so the solution to the equation is $x = -1$ or $x = 1$.

b. We are looking for where the y-values on the graph are between 1 and 2. This occurs for $-3 < x < -2$ or $2 < x < 3$.

19. a. $g(q) = 0$ for $q = -2$ or $q = 2$.

b. $g(q) = 16$ for $q \approx -2.8$, $q = 0$, or $q \approx 2.8$.

c. $g(q) < 6$ for $-2.5 < q < -1.25$ or $1.25 < q < 2.5$

d. $g(q)$ is increasing for $-2 < q < 0$ and $q > 2$.

21. (−4.8, 3.6) and (4.8, 3.6)

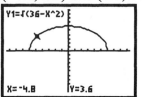

23. (−1.6, 4.352) and (1.6, −4.352)

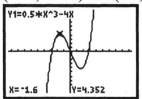

25. a. No. As the precipitation increases, the sediment yield doesn't always increase.

b. A maximum sediment yield of 800 tons per square mile occurs when precipitation is 12 inches per year.

c. Sediment yield decreases for $12 < p < 60$.

d. In desert shrub, increased precipitation causes increased runoff. In grassland, runoff decreases to 350 tons per square mile as precipitation increases to 30 inches annually, because vegetation increases. In forests, runoff decreases slightly with increased precipitation, but levels off beyond 50 inches at about 300 tons per square mile.

27. a. On the surface, the temperature is 20°C and at the deepest level shown, the temperature is 2°C. There is an 18° difference.

b. The temperature changes most rapidly between 0.2 km and 0.5 km.

c. The thermocline is between depths 0.2 km and 0.8 km. At 0.2 km, the temperature is 20°C. At 0.8 km, the temperature is about 4°C. The average rate of change is $\dfrac{20-4}{0.18-0.8} = \dfrac{16}{-0.62} \approx -26°$ per kilometer.

29. Refer to the graphs in the back of the textbook. In the standard window, the graph cannot even be seen since the y-values of the function are so small. In the second window, we can see a curve with one turning point.

31. Refer to the graphs in the back of the textbook. In the standard window, we cannot see the turning points or the y-intercept of the graph since the y-values of the function are so large. In the standard window, the graph appears to be two lines while in the second window we can see a curve with two turning points.

33, 35, 37, 39, 41. Refer to the graphs in the back of the textbook.

43. a. $f(0) = 2(0)^2 + 3(0) = 0 + 0 = 0$;
$g(0) = 5 - 6(0) = 5 - 0 = 5$

b. $f(x) = 0$
$2x^2 + 3x = 0$
$x(2x + 3) = 0$
$x = 0 \text{ or } 2x + 3 = 0$
$x = 0 \text{ or } x = -\dfrac{3}{2}$

c. $g(x) = 0$
$5 - 6x = 0$
$x = \dfrac{5}{6}$

d. $f(x) = g(x)$
$2x^2 + 3x = 5 - 6x$
$2x^2 + 9x - 5 = 0$
$(x + 5)(2x - 1) = 0$
$x + 5 = 0 \text{ or } 2x - 1 = 0$
$x = -5 \text{ or } x = \dfrac{1}{2}$

e. Refer to the graph in the back of the textbook.

45. a. $f(0) = 2(0)^2 - 2(0) = 0$;
$g(0) = 0^2 + 3 = 3$

b. $f(x) = 0$
$2x^2 - 2x = 0$
$2x(x - 1) = 0$
$2x = 0 \text{ or } x - 1 = 0$
$x = 0 \text{ or } x = 1$

c. $g(x) = 0$
$x^2 + 3 = 0$
$x^2 = -3$
$x = \pm\sqrt{-3}$
There are no real solutions.

d. $f(x) = g(x)$
$2x^2 - 2x = x^2 + 3$
$x^2 - 2x - 3 = 0$
$(x - 3)(x + 1) = 0$
$x = 3 \text{ or } x = -1$

e. Refer to the graph in the back of the textbook.

Homework 5.3

1. a. $\sqrt[3]{512} = 8; \; 8^3 = 512$

  b. $\sqrt[3]{-125} = -5; \; (-5)^3 = -125$

  c. $\sqrt[3]{-0.064} = -0.4;$
  $(-0.4)^3 = -0.064$

  d. $\sqrt[3]{1.728} = 1.2; \; 1.2^3 = 1.728$

3. a. $\dfrac{4 - 3\sqrt[3]{64}}{2} = \dfrac{4 - 3(4)}{2}$

$= \dfrac{4 - 12}{2}$

$= \dfrac{-8}{2}$

$= -4$

  b. $\dfrac{4 + \sqrt[3]{-216}}{8 - \sqrt[3]{8}} = \dfrac{4 + (-6)}{8 - 2}$

$= \dfrac{-2}{6}$

$= -\dfrac{1}{3}$

5. a. $-|-9| = -(+9) = -9$

  b. $-(-9) = 9$

7. a. $|-8| - |12| = 8 - 12 = -4$

  b. $|-8 - 12| = |-20| = 20$

9. $4 - 9|2 - 8| = 4 - 9|-6| = 4 - 9 \cdot 6$
$= 4 - 54 = -50$

11. $|-4 - 5||1 - 3(-5)| = |-9||1 + 15|$
$= 9|16| = 9 \cdot 16 = 144$

13. $||-5| - |-6|| = |5 - 6| = |-1| = 1$

15. Construct a table like the following, plot the points, and draw a smooth curve through the points. Refer to the graph in the back of the textbook.

x	$f(x) = x^3$
-3	-27
-2	-8
-1	-1
0	0
1	1
2	8
3	27

17. Construct a table like the following, plot the points, and draw a smooth curve through the points. Note that it isn't possible to plot points with negative values of x since $f(x)$ is only defined for $x \geq 0$. Refer to the graph in the back of the textbook.

x	$f(x) = \sqrt{x}$
0	0
1	1
2	≈ 1.41
4	2
9	3

19. Construct a table like the following, plot the points, and draw a smooth curve through the points. Refer to the graph in the back of the textbook.

x	$f(x) = \dfrac{1}{x}$
-5	$-\dfrac{1}{5}$
-2	$-\dfrac{1}{2}$
-1	-1
$-\dfrac{1}{2}$	-2
$-\dfrac{1}{5}$	-5
0	Undefined
$\dfrac{1}{5}$	5
$\dfrac{1}{2}$	2
1	1
2	$\dfrac{1}{2}$
5	$\dfrac{1}{5}$

21. a. $(1.4)^3 \approx 2.7$ since there is a point on the graph at about $(1.4, 2.7)$.

b. The answer is -2.7 since the point on the graph with y-value -20 is about $(-2.7, -20)$.

c. $\sqrt[3]{24} \approx 2.9$ since there is a point on the graph at about $(2.9, 24)$.

d. The answer is 1.8 since there is a point on the graph at about $(1.8, 6)$.

e. We are looking for where the y-values on the graph are between -12 and 15 inclusive. This occurs for $-2.3 \le x \le 2.5$.

23. a. $\dfrac{1}{3.4} \approx 0.3$ since there is a point on the graph at about $(3.4, 0.3)$.

b. The answer is -0.4 since there is a point on the graph at about $(-0.4, -2.5)$.

c. The answer is 0.2 since there is a point on the graph at about $(0.2, 4.8)$

d. We are looking for where the y-values on the graph are between 0.3 and 4.5 inclusive. This occurs for $0.2 \le x \le 3.3$.

25. Refer to the graph in the back of the textbook. g is shifted 2 units down and h is shifted 1 unit up.

27. Refer to the graph in the back of the textbook. g is shifted 1.5 units to the left and h is shifted 1 unit to the right.

29. Refer to the graph in the back of the textbook. g is reflected about the x-axis and h is reflected about the y-axis.

31. a. $y = \sqrt{x}$ (shifted right)

b. $y = \sqrt[3]{x}$ (shifted up)

c. $y = |x|$ (shifted down)

d. $y = \dfrac{1}{x}$ (reflected about x-axis)

e. $y = x^3$ (reflected about x-axis and shifted up)

f. $y = \dfrac{1}{x^2}$ (reflected about x-axis and shifted up)

33. a. $x \approx 12$
Check algebraically:
$\sqrt{12} - 2 \approx 3.5 - 2 \approx 1.5$

b. $x \approx 18$
Check algebraically:
$\sqrt{18} - 2 \approx 4.24 - 2 \approx 2.25$

c. $x < 9$
Check algebraically:
$\sqrt{9} - 2 = 3 - 2 = 1$
and so if $x < 9$, $\sqrt{x} - 2 < 1$.

d. Approximately $x > 3$
Check algebraically:
$\sqrt{3} - 2 \approx 1.73 - 2 \approx -0.25$ and so
if $x > 3$, $\sqrt{x} - 2 > -0.25$.

35. a. $t \approx -3.1$
Check algebraically:
$-10(-3.1+1)^3 + 10$
$= -10(-2.1)^3 + 10$
$= -10(-9.261) + 10$
$= 92.61 + 10$
$= 102.61 \approx 100$

b. $t \approx 1.5$
Check algebraically:
$-10(1.5+1)^3 + 10$
$= -10(2.5)^3 + 10$
$= -10(15.625) + 10$
$= -156.25 + 10$
$= -146.25 \approx -140$

c. Approximately $t < 0.8$
Check algebraically:
$-10(0.8+1)^3 + 10$
$= -10(1.8)^3 + 10$
$= -10(5.832) + 10$
$= -58.32 + 10$
$= -48.32 \approx -50$

d. Approximately $-2.4 < t < 0.4$
Check algebraically:
$-10(-2.4+1)^3 + 10$
$= -10(-1.4)^3 + 10$
$= -10(-2.744) + 10$
$= 27.44 + 10$
$= 37.44 \approx 40$
and
$-10(0.4+1)^3 + 10$
$= -10(1.4)^3 + 10$
$= -10(2.744) + 10$
$= -27.44 + 10$
$= -17.44 \approx -20$

37. a. (vi)

    b. (ii)

    c. (iv)

    d. (i)

    e. (v)

    f. (iii)

39. Graph in the specified window. Use TRACE to locate the x-values corresponding to the given y-values. For example, when $y = 16$, $x = 41$ as shown below.

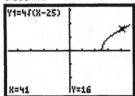

    a. $x = 41$

    Check algebraically:

$$4\sqrt{41 - 25} = 4\sqrt{16} = 4 \cdot 4 = 16$$

    b. $29 < x \le 61$

$$4\sqrt{29 - 25} = 4\sqrt{4} = 4 \cdot 2 = 8$$

and

$$4\sqrt{61 - 25} = 4\sqrt{36} = 4 \cdot 6 = 24$$

41. Graph in the specified window. Use TRACE to locate the x-values corresponding to the given y-values. For example, when $y = -6.25$, $x = -5$ or $x = 17$, as shown below.

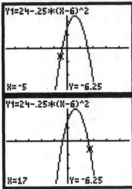

    a. $x = -5$ or $x = 17$

    Check algebraically:

$$24 - 0.25(-5 - 6)^2$$
$$= 24 - 0.25(-11)^2$$
$$= 24 - 0.25(121) = -6.25$$

and

$$24 - 0.25(17 - 6)^2$$
$$= 24 - 0.25(11)^2$$
$$= 24 - 0.25(121) = -6.25$$

    b. $-1 < x < 13$

    Check algebraically:

$$24 - 0.25(-1 - 6)^2$$
$$= 24 - 0.25(-7)^2$$
$$= 24 - 0.25(49) = 11.75$$

and

$$24 - 0.25(13 - 6)^2$$
$$= 24 - 0.25(7)^2$$
$$= 24 - 0.25(49) = 11.75$$

Midchapter 5 Review

1. Each social security number has a unique person assigned to it. And, each person has a unique age. (No two persons are born at *exactly* the same time.) Therefore, a person's age is a function of the social security number.

3. No, since there is an x value that has two different y values. For example, when $x = 1$, $y = 9$ or 6.

5. **a.** $s = h(t)$

   b. The height of a duck 3 seconds after it was flushed out of the bushes was 7 meters.

7. **a.** $f(300) = 315$

   b. If no money is spent on advertising, the revenue is $20,000.

9. **a.** Refer to the graph in the back of the textbook.

x	0	1	2	3	4	5	6	7	8	9
$f(x)$	2	1	.586	.268	0	-.236	-.449	-.646	-.828	-1

   b. Use Table Start = 0 and ΔTbl = 1. The calculator's table should agree with the one in (a). Graph $f(x) = 2 - \sqrt{x}$ using Xmin = 0, Xmax = 9, Ymin = -1, and Ymax = 1. The calculator's graph should agree with the one in (a).

11. Refer to the graph in the back of the textbook.

13. $-8 + 5\sqrt[3]{27} = -8 + 5(3) = -8 + 15 = 7$

15. $10 - 5|5 - 10| = 10 - 5|-5|$
    $= 10 - 5(5) = 10 - 25 = -15$

Homework 5.4

1. a. $\dfrac{\text{tax}}{\text{price}} = \dfrac{1.17}{18} = 0.065$

$\dfrac{\text{tax}}{\text{price}} = \dfrac{1.82}{28} = 0.065$

$\dfrac{\text{tax}}{\text{price}} = \dfrac{0.78}{12} = 0.065$

Yes, the tax is proportional to the price. The tax rate is 6.5%.

b. $T = 0.065p$

c. Refer to the graph in the back of the textbook.

3. a. length · width = $(12)(2) = 24$
length · width = $(9.6)(2.5) = 24$
length · width = $(8)(3) = 24$
The constant of inverse proportionality is $k = 24$.

b. $L = \dfrac{24}{w}$

c. Refer to the graph in the back of the textbook.

5. a.

L	W	P	A
10	8	36	80
12	8	40	96
15	8	46	120
20	8	56	160

b. Since the ratio $\dfrac{P}{L}$ is not constant, the perimeter does not vary directly with the length.

c. $P = 2L + 16$

d. The area does vary directly with the length.

e. $A = 8L$

7. Only (b) could describe direct variation. It starts from the origin, and as x increases, y also increases.

9. Only (c) has y getting larger without limit as x approaches 0, and y approaching 0 as x gets larger without limit, so only (c) could represent inverse variation.

11. a. $m = kw$ and $m = 24.75$ when $w = 150$, so $k = \dfrac{24.75}{150} = 0.165$.
Thus $m = 0.165w$.

w	100	150	200	400
m	16.5	24.75	33	66

Xmin = 0, Xmax = 400,
Ymin = 0, Ymax = 100

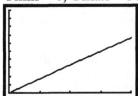

b. $m = 0.165(120) = 19.8$ lb.

c. $w = \dfrac{50}{0.165} \approx 303$ lb.

d. Locate the points on your graph by tracing to (120, 19.8) and (303, 50).

13. a. $L = kT^2$ and $L = 3.25$ ft. when $T = 2$ sec, so $3.25 = k(2)^2$. This gives $k = 0.8125$. Thus, $L = 0.8125T^2$.

T	1	5	10	20
L	0.8125	20.3	81.25	325

Xmin = 0, Xmax = 25,
Ymin = 0, Ymax = 400

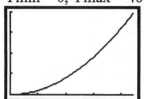

b. $L = 0.8125(17)^2 \approx 234.8$ ft.

c. 9 inches = 0.75 feet so solve:
$0.75 = 0.8125T^2$ or
$$T^2 = \frac{0.75}{0.8125}$$
$$T = \sqrt{\frac{0.75}{0.8125}} \approx 0.96 \text{ sec}$$

d. Locate the points on your graph by tracing to (17, 234.8) and (0.96, 0.75).

15. a. $F = \dfrac{k}{d}$ and $F = 22$ milligauss when $d = 4$ in., so $22 = \dfrac{k}{4}$
$k = 22(4) = 88$. Thus, $F = \dfrac{88}{d}$.

d	2	4	12	24
B	44	22	7.3	3.7

Xmin = 0, Xmax = 25,
Ymin = 0, Ymax = 50

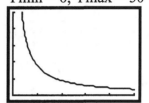

b. $L = \dfrac{88}{10} = 8.8$ milligauss.

c. $L = 4.3 = \dfrac{88}{d}$, so $d = \dfrac{88}{4.3} \approx 20.5$ inches. The computer user should sit at least 20.5 inches from the screen.

d. Locate the points on your graph by tracing to (10, 8.8) and (20.5, 4.3).

17. a. $P = kw^3$ and $P = 7300$ kilowatts when $w = 32$ mph.

So $7300 = k(32)^3$ and

$$k = \frac{7300}{32^3} \approx 0.2228.$$

Thus, $P \approx 0.2228w^3$.

w	10	20	40	80
P	223	1782	14,259	114,074

Xmin = 0, Xmax = 90,
Ymin = 0, Ymax = 130,000

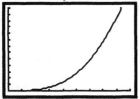

b. $P \approx 0.2228(15)^3 \approx 752$ kW.

c. $P = 10,000 \approx 0.2228w^3$, so

$$w^3 = \frac{10,000}{0.2228} \approx 44,883.3$$

$$w \approx \sqrt[3]{44,883.3} \approx 35.5 \text{ mph.}$$

d. Locate the points on your graph by tracing to (15,752) and (35.5, 10,000).

19. a. $y = kx$ and $y = 1.5$ when $x = 5$. So

$$1.5 = k(5) \text{ and } k = \frac{1.5}{5} = 0.3.$$

Thus, $y = 0.3x$.

b.

x	y
2	0.6
5	1.5
8	2.4
12	3.6
15	4.5

d. y doubles when x doubles.

21. a. $y = kx^2$ and $y = 24$ when $x = 6$. So

$24 = k(6)^2$ and $k = \frac{24}{36} = \frac{2}{3}$. Thus,

$$y = \frac{2}{3}x^2.$$

b.

x	y
3	6
6	24
9	54
12	96
15	150

c. y quadruples when x doubles.

23. a. $y = \frac{k}{x}$ and $y = 6$ when $x = 20$. So

$6 = \frac{k}{20}$ and $k = 6(20) = 120$.

Thus, $y = \frac{120}{x}$.

b.

x	y
4	30
8	15
20	6
30	4
40	3

c. y becomes one-half as big when x doubles.

25.

x	y	$\dfrac{y}{x}$	$\dfrac{y}{x^2}$
2	2	1	0.5
3	4.5	1.5	0.5
5	12.5	2.5	0.5
8	32	4	0.5

(b) y varies directly as x^2 with $k = 0.5$. $(y = 0.5x^2)$

27.

x	y	$\dfrac{y}{x}$	$\dfrac{y}{x^2}$	$\dfrac{y}{x^3}$	etc.
1.5	3	2	1.33	0.89	etc.
2.4	5.3	2.21	0.92	0.38	etc.

(c) y divided by powers of x does not yield constants, so y does not vary directly as a power of x.

29.

x	y	xy	x^2y
0.5	288	144	72
2	18	36	72
3	8	24	72
6	2	12	72

(b) y varies inversely as x^2, with

$k = 72.$ $\left(y = \dfrac{72}{x^2} \right).$

31.

x	y	$x + y$
1	4	5
1.3	3.7	5
3	2	5
4	1	5

(c) $y = 5 - x$, rather than $y = \dfrac{k}{x^n}$, so y does not vary inversely with a power of x.

33.

v	d	$\dfrac{d}{v^2}$
10	0.5	0.005
20	2	0.005
40	8	0.005

a. $d = 0.005v^2$

b. $d = 0.005(100)^2 = 50$ m.

35.

p	m	pm
1	8	8
2	4	8
4	2	8
8	1	8

a. $m = \dfrac{8}{p}$

b. $m = \dfrac{8}{10} = 0.8$ tons.

37.

d	t	dt
0.5	12	6
1	6	6
2	3	6
3	2	6

a. $t = \dfrac{6}{d}$

b. $t = \dfrac{6}{6} = 1°$ C.

39.

d	w	$\dfrac{w}{d^2}$
0.5	150	600
1.0	600	600
1.5	1350	600
2.0	2400	600

a. $w = 600d^2$

b. $w = 600(1.2)^2 = 864$ newtons.

Homework 5.5

1. (b) Your pulse rate rises during the class and falls as you rest after the class.

3. (a) Your income rises at a steady rate as the number of hours you work increases.

5. Refer to the graph in the back of the textbook. The height is always positive because your head is always above ground. The height increases as you move upward, then decreases as the wheel spins you back downward. This pattern repeats over and over again until the ride comes to an end.

7. Refer to the graph in the back of the textbook. Your distance from the English classroom increases as you leave class, decreases to zero as you return for the book, and then increases as you walk to math class.

9. Refer to the graph in the back of the textbook. The distance increases from 0 as she drives to the gym, stays constant while she is at the gym, increases as she drives to her friend's home, stays constant while she is at her friend's home, then decreases back to 0 as she returns home.

11. (b) Inflation is increasing, but at a slower and slower rate.

13. a. II. Since $V = \pi r^2 h$, as r increases, V increases but at a faster rate.

    b. IV. Since $t = \dfrac{d}{r}$, as r increases t decreases and for $r > 0$, t cannot equal zero.

    c. I, because $I = PR$ is a linear function starting from the origin.

    d. III. As the number present increases, the number absent decreases.

15. NOTE: In each of the tables, the difference between successive x's is always 1, so the successive slopes are the same as the successive Δy's.

    a. (4) & (c) since the concentration increases but at a slower and slower rate until it is nearly constant.

    b. (3) & (b) since the rate of increase is constant. (The slope between the points doesn't change.)

    c. (1) & (d) since the population is increasing, and at an increasing rate.

    d. (2) & (a) since production always increases, but at a decreasing rate.

17. a. III

    b. III

19. a. Let d represent the diameter and let s represent the speed. The ratio $\dfrac{s}{\sqrt{d}}$ remains constant at 50 for all of the data values in the table. Hence $k = 50$ and $s = 50\sqrt{d}$.

b. $s = 50\sqrt{0.36} = 30$ cm/sec.

21. a. The ratio $\dfrac{d}{\sqrt{h}}$ remains constant at 1.225 for all of the data values in the table. Hence $k = 1.225$ and $d = 1.225\sqrt{h}$.

b. $d = 1.225\sqrt{20,000} \approx 173.24$ mi.

23. The distance from x to 0 is six units. $|x| = 6$

25. $|p - (-3)| = 5$ so $|p + 3| = 5$.

27. The distance from t to 6 is less than or equal to 3, so $|t - 6| \le 3$.

29. The distance from b to -1 is greater than or equal to 0.5, so $|b - (-1)| \ge 0.5$ which simplifies to $|b + 1| \ge 0.5$.

31. a. The distance between x and -12 is $|x - (-12)| = |x + 12|$. The distance between x and -4 is $|x - (-4)| = |x + 4|$. The distance between x and 24 is $|x - 24|$.

b. $y = |x + 12| + |x + 4| + |x - 24|$

c. Graph the function from part (b) using Xmin $= -20$, Xmax $= 30$, Ymin $= -10$, and Ymax $= 90$.

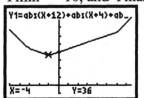

The potter should stand at x-coordinate -4.

33. Let the road be the x-axis and the point where the river crosses the road be the point 0. Then Richard works at $+10$, Marian works at -6, and the health club is at $+2$. Let x be the coordinate of their apartment. For minimum driving, the apartment should be between the two work places, that is, $-6 \le x \le 10$. Since Marian's distance to work and back is $2|x + 6|$, which is $2(x + 6)$ for $x \ge -6$, and Richard's is $2|x - 10|$ which is $2(10 - x)$ for $x \le 10$, their total driving distance to and from work is $2(x + 6) + 2(10 - x) = 32$ miles regardless of where they live. Thus, to minimize total driving distance, we need only minimize total driving distance to the health club. Each of them drives $2|x - 2|$ miles to the health club and back, for a total of $4|x - 2|$ miles, which would be a minimum of 0 for $x = 2$, i.e., they should live as close to the health club as possible.

35. Graph of $y = |x + 3|$:

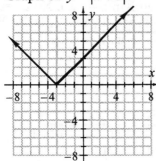

39. $-\dfrac{9}{2} < x < -\dfrac{3}{2}$

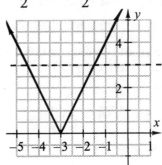

a. We are looking for the x-values where $y = 2$, so $x = -5$ or -1.

b. We are looking for the x-values on the graph at or below points $(-7, 4)$ and $(1, 4)$, so $-7 \le x \le 1$.

c. We are looking for the x-values on the graph above points $(-8, 5)$ and $(2, 5)$, so $x < -8$ or $x > 2$.

41. $x \le -2$ or $x \ge 5$

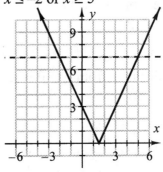

37. $x = -\dfrac{3}{2}$ or $x = \dfrac{5}{2}$

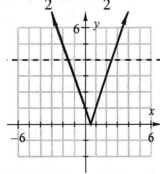

Chapter 5 Review

1. Function; each x value has a unique value of y.

3. Not a function; two students with IQ test scores of 98 had different SAT scores. (However, IQ test score is a function of SAT score, and each is a function of student's ID letter.)

5. $N(10) = 2000 + 500(10)$
   $= 2000 + 5000 = 7000$.
   Ten days after the well is opened, 7000 barrels of oil are pumped.

7. $F(0) = \sqrt{1 + 4(0)^2} = \sqrt{1 + 4 \cdot 0}$
   $= \sqrt{1 + 0} = \sqrt{1} = 1$
   $F(-3) = \sqrt{1 + 4(-3)^2} = \sqrt{1 + 4 \cdot 9}$
   $= \sqrt{1 + 36} = \sqrt{37} \approx 6.08$

9. $h(8) = 6 - |4 - 2(8)| = 6 - |4 - 16|$
   $= 6 - |-12| = 6 - 12 = -6$
   $h(-8) = 6 - |4 - 2(-8)| = 6 - |4 + 16|$
   $= 6 - |20| = 6 - 20 = -14$

11. a. $P(0) = (0)^2 - 6(0) + 5$
    $= 0 - 0 + 5 = 5$

    b. $x^2 - 6x + 5 = 0$
    $(x - 1)(x - 5) = 0$
    $x - 1 = 0$ or $x - 5 = 0$
    $x = 1$ or $x = 5$

13. a. Since $(-2, 3)$ is on the graph, $f(-2) = 3$. Since $(2, 5)$ is on the graph, $f(2) = 5$.

    b. Since $(1, 4)$ and $(3, 4)$ are on the graph, $f(1) = 4$ and $f(3) = 4$ so $t = 1$ or $t = 3$.

    c. The t-intercepts are $(-3, 0)$ and $(4, 0)$. The $f(t)$-intercept is $(0, 2)$.

    d. Since $(2, 5)$ is the highest point on the graph, the maximum value of f is 5, and this occurs at $t = 2$.

15. Function; each vertical line crosses the graph only once.

17. Not a function; a vertical line at the right end of the graph crosses it infinitely many times.

19. Refer to the graph in the back of the textbook.

21. Refer to the graph in the back of the textbook.

23. Refer to the graph in the back of the textbook.

    a. $x \approx 0.5$

    b. $x \approx 3.4$

    c. $x > 4.9$ (approximate)

    d. $x \leq 2.0$ (approximate)

25. Refer to the graph in the back of the textbook.

    a. $x \approx -5.8$ or $x \approx 5.8$

    b. $x = -0.4$ or $x = 0.4$

    c. $-2.5 < x < 0$ or $0 < x < 2.5$

    d. $x \leq -0.5$ or $x \geq 0.5$

27.

x	y	$\dfrac{y}{x^2}$
2	4.8	1.2
5	30	1.2
8	76.8	1.2
11	145.2	1.2

$y = 1.2x^2$

29.

x	y	xy
0.5	40	20
2	10	20
4	5	20
8	2.5	20

$y = \dfrac{20}{x}$

31. a. $s = kt^2$ and $s = 28$ cm when $t = 4$ sec. So $28 = k(4)^2 = 16k$, and $k = \dfrac{28}{16} = 1.75$. Thus, $s = 1.75t^2$.

  b. $s = 1.75(6)^2 = 1.75(36) = 63$ cm

33. Refer to the graph in the back of the textbook. $d = \dfrac{k}{p}$ and $d = 600$ bottles when $p = \$8$. So $(600) = \dfrac{k}{(8)}$ which gives $k = (600)(8) = 4800$ and the equation $d = \dfrac{4800}{p}$. Thus, $d(10) = \dfrac{4800}{(10)} = 480$. The company can sell 480 bottles at $10 each.

35. a. $w(r) = \dfrac{k}{r^2}$

  b. Refer to the graph in the back of the textbook.

  c. For w to be $\dfrac{1}{3}$ as large, r^2 must be 3 times as large, so r must be $\sqrt{3}$ times as large, or $3960\sqrt{3} \approx 6858.92$ mi.

37. Refer to the graph in the back of the textbook.

39. $|x| = 4$

41. $|p - 7| < 4$

43. $-\dfrac{2}{3} < x < 2$

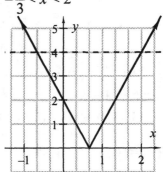

45. $y \le -0.9$ or $y \ge 0.1$

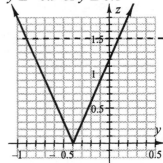

47. $y = 0.25x^2$. Refer to the graph in the back of the textbook.

49. $y = \dfrac{2}{x}$. Refer to the graph in the back of the textbook.

51. a. Refer to the graph in the back of the textbook.

b.

x	$g(x)$	$xg(x)$
2	12	24
3	8	24
4	6	24
6	4	24
8	3	24
12	2	24

$$g(x) = \frac{24}{x}$$

53. a.

x	0	4	8	14	16	22
y	24	20	16	10	8	2

b. The graph is a line with y-intercept (0, 24) and slope $m = -1$. Therefore, using $y = mx + b$, the equation is $y = -x + 24$.

55. a.

x	0	1	4	9	16	25
y	0	1	2	3	4	5

b. The y values are the square root of the x-values, so $y = \sqrt{x}$.

57. a.

x	-3	-2	-1	0	1	2
y	5	0	-3	-4	-3	0

b. The graph looks like a parabola that opens upward from the vertex (0, -4). Then, in vertex form, $y = a(x - 0)^2 - 4$ or $y = ax^2 - 4$. To find a, use one of the points from the table. For example, using (-3, 5):
$$5 = a(-3)^2 - 4$$
$$9 = 9a$$
$$1 = a$$
Thus, $y = x^2 - 4$. The points in the table satisfy this equation.

Chapter Six: Powers and Roots

Homework 6.1

1. a. $2^{-1} = \dfrac{1}{2^1} = \dfrac{1}{2}$

b. $(-5)^{-2} = \dfrac{1}{(-5)^2} = \dfrac{1}{25}$

c. $\left(\dfrac{1}{3}\right)^{-3} = \left(\dfrac{3}{1}\right)^3 = 3^3 = 27$

d. $\dfrac{1}{(-2)^{-4}} = (-2)^4 = 16$

3. a. $\dfrac{5}{4^{-3}} = 5 \cdot 4^3 = 5 \cdot 64 = 320$

b. $(2q)^{-5} = \dfrac{1}{(2q)^5} = \dfrac{1}{2^5 q^5} = \dfrac{1}{32q^5}$

c. $-4x^{-2} = \dfrac{-4}{x^2} = -\dfrac{4}{x^2}$

d. $\dfrac{8}{b^{-3}} = \dfrac{8b^3}{1} = 8b^3$

5. a. $(m-n)^{-2} = \dfrac{1}{(m-n)^2}$; Note: This answer can also be written as

$\dfrac{1}{(m-n)(m-n)} = \dfrac{1}{m^2 - 2mn + n^2}$

b. $y^{-2} + y^{-3} = \dfrac{1}{y^2} + \dfrac{1}{y^3}$

Note: The answer can be written in a single fraction as $\dfrac{y+1}{y^3}$.

c. $2pq^{-4} = \dfrac{2p}{q^4}$

d. $\dfrac{-5y^{-2}}{x^{-5}} = \dfrac{-5x^5}{y^2} = -\dfrac{5x^5}{y^2}$

7. a. $2^3 = 8$

b. $(-2)^3 = -8$

c. $2^{-3} = \dfrac{1}{2^3} = \dfrac{1}{8}$

d. $(-2)^{-3} = \dfrac{1}{(-2)^3} = \dfrac{1}{-8} = -\dfrac{1}{8}$

9. a. $\left(\dfrac{1}{2}\right)^3 = \dfrac{1^3}{2^3} = \dfrac{1}{8}$

b. $\left(-\dfrac{1}{2}\right)^3 = \dfrac{(-1)^3}{2^3} = \dfrac{-1}{8} = -\dfrac{1}{8}$

c. $\left(\dfrac{1}{2}\right)^{-3} = \left(\dfrac{2}{1}\right)^3 = 2^3 = 8$

d. $\left(-\dfrac{1}{2}\right)^{-3} = \left(-\dfrac{2}{1}\right)^3 = (-2)^3 = -8$

11. a. *Values rounded to two decimals:*

x	1	2	4	8	16
x^{-2}	1	0.25	0.06	0.02	0.00

b. As x increases, the values of x^2 increase, and since x^{-2} is the reciprocal of x^2, the values of $f(x) = x^{-2}$ decrease.

c. *Values rounded to two decimals:*

x	1	0.5	0.25	0.125	0.0625
x^{-2}	1	4	16	64	256

d. As x decreases towards 0, the values of x^2 decrease, and since x^{-2} is the reciprocal of x^2, the values of $f(x) = x^{-2}$ increase.

13. a. $f(x) = x^2$

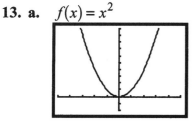

b. $f(x) = x^{-2}$

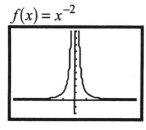

c. $f(x) = \dfrac{1}{x^2}$

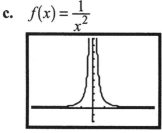

d. $f(x) = \left(\dfrac{1}{x}\right)^2$

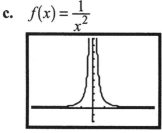

e. The graphs in parts (b), (c), and (d) are the same because

$$x^{-2} = \dfrac{1}{x^2} = \left(\dfrac{1}{x}\right)^2.$$

15. a. $F(r) = \dfrac{3}{r^4} = 3r^{-4}$

b. $G(w) = \dfrac{2}{5w^3} = \dfrac{2}{5}w^{-3}$

c. $H(z) = \dfrac{1}{(3z)^2} = \dfrac{1}{9z^2} = \dfrac{1}{9}z^{-2}$

17. a. $T = kr^2 d^4$

b. When $d = 2$ feet and $r = 100$ rotations per minute, $T = 1000$ pounds. Substitute into part (a):

$1000 = k(100)^2 2^4$

$1000 = k(10,000)(16)$

$1000 = 160,000k$

$k = \dfrac{1000}{160,000} = \dfrac{1}{160} = 0.00625$

19. a. $E = \dfrac{k}{x^3}$ or $F = kx^{-3}$

b. Substitute $x = 1$ and $E = \dfrac{2qL}{4\pi\in}$ into the equation from part (a):

$\dfrac{2qL}{4\pi\in} = \dfrac{k}{1^3}$ so $k = \dfrac{2qL}{4\pi\in}$.

21. a. $a^{-3} \cdot a^8 = a^{-3+8} = a^5$

b. $5^{-4} \cdot 5^{-3} = 5^{-4+(-3)} = 5^{-7} = \dfrac{1}{5^7} = \dfrac{1}{78,125}$

c. $\dfrac{p^{-7}}{p^{-4}} = p^{-7-(-4)} = p^{-3} = \dfrac{1}{p^3}$

d. $\left(7^{-2}\right)^5 = 7^{(-2)(5)} = 7^{-10} = \dfrac{1}{7^{10}} = \dfrac{1}{282,475,249}$

23. a. $\left(4x^{-5}\right)\left(5x^2\right) = 4 \cdot 5 \cdot x^{-5} \cdot x^2 = 20x^{-5+2} = 20x^{-3} = \dfrac{20}{x^3}$

b. $\dfrac{3u^{-3}}{9u^9} = \dfrac{u^{-3}}{3u^9} = \dfrac{1}{3u^3u^9} = \dfrac{1}{3u^{3+9}} = \dfrac{1}{3u^{12}}$

c. $\dfrac{5^6 t^0}{5^{-2} t^{-1}} = 5^{6-(-2)} \cdot t^{0-(-1)} = 5^8 t^1 = 390,625t$

25. a. $\left(3x^{-2}y^3\right)^{-2} = 3^{-2}\left(x^{-2}\right)^{-2}\left(y^3\right)^{-2} = 3^{-2}x^4 y^{-6} = \dfrac{x^4}{3^2 y^6} = \dfrac{x^4}{9y^6}$

b. $\left(\dfrac{6a^{-3}}{b^2}\right)^{-2} = \dfrac{6^{-2}\left(a^{-3}\right)^{-2}}{\left(b^2\right)^{-2}} = \dfrac{6^{-2}a^6}{b^{-4}} = \dfrac{a^6 b^4}{6^2} = \dfrac{a^6 b^4}{36}$

c. $\dfrac{5h^{-3}\left(h^4\right)^{-2}}{6h^{-5}} = \dfrac{5h^{-3}h^{-8}}{6h^{-5}} = \dfrac{5h^{-3+(-8)}}{6h^{-5}} = \dfrac{5h^{-11}}{6h^{-5}} = \dfrac{5h^{-11-(-5)}}{6} = \dfrac{5h^{-6}}{6} = \dfrac{5}{6h^6}$

27. a. No, since $(x+y)^{-2} = \dfrac{1}{(x+y)^2}$, $x^{-2} + y^{-2} = \dfrac{1}{x^2} + \dfrac{1}{y^2}$ and $\dfrac{1}{(x+y)^2} \neq \dfrac{1}{x^2} + \dfrac{1}{y^2}$.

b. For example, let $x = 2$ and $y = 3$. Then $(2+3)^{-2} = 5^{-2} = \dfrac{1}{5^2} = \dfrac{1}{25}$ while

$2^{-2} + 3^{-2} = \dfrac{1}{2^2} + \dfrac{1}{3^2} = \dfrac{1}{4} + \dfrac{1}{9} = \dfrac{9}{36} + \dfrac{4}{36} = \dfrac{13}{36}$. So $(2+3)^{-2} \neq 2^{-2} + 3^{-2}$.

29. a. $\dfrac{x}{3} + \dfrac{3}{x} = \dfrac{1}{3}x + 3x^{-1}$

b. $\dfrac{x - 6x^2}{4x^3} = \dfrac{x}{4x^3} - \dfrac{6x^2}{4x^3} = \dfrac{1}{4x^2} - \dfrac{3}{2x} = \dfrac{1}{4}x^{-2} - \dfrac{3}{2}x^{-1}$

31. a. $\dfrac{2}{x^4}\left(\dfrac{x^2}{4}+\dfrac{x}{2}-\dfrac{1}{4}\right)=\dfrac{2}{x^4}\cdot\dfrac{x^2}{4}+\dfrac{2}{x^4}\cdot\dfrac{x}{2}-\dfrac{2}{x^4}\cdot\dfrac{1}{4}=\dfrac{1}{2x^2}+\dfrac{1}{x^3}-\dfrac{1}{2x^4}$

$\qquad=\dfrac{1}{2}x^{-2}+x^{-3}-\dfrac{1}{2}x^{-4}$

b. $\dfrac{x^2}{3}\left(\dfrac{2}{x^4}-\dfrac{1}{3x^2}+\dfrac{1}{2}\right)=\dfrac{x^2}{3}\cdot\dfrac{2}{x^4}-\dfrac{x^2}{3}\cdot\dfrac{1}{3x^2}+\dfrac{x^2}{3}\cdot\dfrac{1}{2}=\dfrac{2}{3x^2}-\dfrac{1}{9}+\dfrac{x^2}{6}$

$\qquad=\dfrac{2}{3}x^{-2}-\dfrac{1}{9}+\dfrac{1}{6}x^2$

33. $x^{-1}\left(x^2-3x+2\right)=x^2\cdot x^{-1}-3x\cdot x^{-1}+2\cdot x^{-1}=x^1-3x^0-2x^{-1}=x-3-2x^{-1}$

35. $-3t^{-2}\left(t^2-2-4t^{-2}\right)=-3t^0+6t^{-2}+12t^{-4}=-3+6t^{-2}+12t^{-4}$

37. $2u^{-3}\left(-2u^3-u^2+3u\right)=-4u^0-2u^{-1}+6u^{-2}=-4-2u^{-1}+6u^{-2}$

39. $4x^2+16x^{-2}=4x^{-2}\left(x^4+4\right)$

41. $3a^{-3}-3a+a^3=a^{-3}\left(3-3a^4+a^6\right)$

43. a. $285=2.85\times10^2$ (the decimal moves 2 places to the left)

b. $8,372,000=8.372\times10^6$ (the decimal moves 6 places to the left)

c. $0.024=2.4\times10^{-2}$ (the decimal moves 2 places to the right)

d. $0.000523=5.23\times10^{-4}$ (the decimal moves 4 places to the right)

45. a. $2.4\times10^2=240$ (the decimal moves 2 places to the right)

b. $6.87\times10^{15}=6,870,000,000,000,000$ (the decimal moves 15 places to the right)

c. $5.0\times10^{-3}=0.005$ (the decimal moves 3 places to the left)

d. $2.02\times10^{-4}=0.000202$ (the decimal moves 4 places to the left)

47. You can compute these expressions entirely using your calculator. The following screen shows the input and answers to both parts (a) and (b). Note: The "E" on the screen comes from the "EE" button on your calculator.

```
2.4E-8*6.5E32/5.
2E18
          3000000
7.5E-13*3.6E-9/(
1.5E-15*1.6E-11)
          112500
```

For comparison, the worked-out answers are below.

a. $\dfrac{\left(2.4\times10^{-8}\right)\left(6.5\times10^{32}\right)}{5.2\times10^{18}} = \dfrac{(2.4)(6.5)}{5.2}\cdot\dfrac{10^{-8}10^{32}}{10^{18}} = 3\cdot\dfrac{10^{-8+32}}{10^{18}} = 3\cdot\dfrac{10^{24}}{10^{18}}$

$= 3\cdot10^{24-18} = 3\cdot10^{6} = 3,000,000$

b. $\dfrac{\left(7.5\times10^{-13}\right)\left(3.6\times10^{-9}\right)}{\left(1.5\times10^{-15}\right)\left(1.6\times10^{-11}\right)} = \dfrac{(7.5)(3.6)}{(1.5)(1.6)}\cdot\dfrac{10^{-13}10^{-9}}{10^{-15}10^{-11}} = \dfrac{27}{2.4}\cdot\dfrac{10^{-13+(-9)}}{10^{-15+(-11)}}$

$= 11.25\cdot\dfrac{10^{-22}}{10^{-26}} = 11.25\cdot10^{-22-(-26)} = 11.25\cdot10^{4} = 112,500$

49. a. $5,605,304,000,000 = 5.605304\times10^{12}$.

b. We first write 272,712,000 in scientific notation: $272,712,000 = 2.72712\times10^{8}$. To find the per capita debt, we divide the total debt by the population:

$\dfrac{5.605304\times10^{12}}{2.72712\times10^{8}} = 20,553.93$

(See problem 53 for an example on how to enter this expression into your calculator.) So the per capita debt in 1999 was $20,553.93 per person!

51. a. The sun travels one revolution about the galactic center in 240,000,000 years. In scientific notation, this number is 2.4×10^{8}. Since the sun is the midpoint between the center and the edge of the disk, its distance away from the center is one-fourth the diameter of the disk. If r equals the distance between the sun and the center, then $r = \dfrac{1.2\times10^{18}}{4} = 3\times10^{17}$. The distance the sun travels in one revolution is given by the circumference of the circle it forms in orbit, which is $C = 2\pi r = 2\pi\cdot3\times10^{17} = 6\pi\times10^{17}$ kilometers. The speed of the sun is the distance it travels in one revolution divided by the time it takes to make that revolution: $\dfrac{6\pi\times10^{17}\text{ kilometers}}{2.4\times10^{8}\text{ years}} \approx 7.85398\times10^{9}$ kilometers per year.

b. There are 10^3 meters in a kilometer, so to find the speed in meters per year, multiply the answer to part (a) by 10^3. The speed is then $\left(7.85398\times10^9\right)\cdot\left(10^3\right) = 7.85398\times10^{12}$ meters per year. There are $365(24)(3600) = 31,536,000$ seconds in one year, so the speed is $(31,536,000)\left(7.85398\times10^{12}\right) \approx 248,000$ meters per second.

53. a. Convert 5×10^{-10} watts to picowatts:

$$5\times10^{-10} \text{ watts} = 5\times10^{-10} \text{ watts}\cdot\frac{1 \text{ picowatt}}{10^{-12} \text{ watts}} = \frac{5\times10^{-10}}{10^{-12}} \text{ picowatts}$$

$$= 5\times10^2 \text{ picowatts} = 500 \text{ picowatts}.$$

b. $P = \dfrac{k}{d^4}$ and $P = 500$ picowatts when $d = 2$ nautical miles. So $500 = \dfrac{k}{2^4}$ and $k = 500\cdot2^4 = 8000$. The function is $P = \dfrac{8000}{d^4}$, or, using a negative exponent as directed, $P = 8000d^{-4}$.

c.

d (nautical miles)	4	5	7	10
P (picowatts)	31.3	12.8	3.3	0.8

d. 10^{-13} watts is $\dfrac{10^{-13}}{10^{-12}} = 10^{-1} = 0.1$ picowatts. Let $P = 0.1$ in the equation from part (b) and solve for d:

$0.1 = 8000d^{-4}$

$0.1d^4 = 8000$

$d^4 = 80,000$

$d = \sqrt[4]{80,000} \approx 16.8$

The aircraft is 16.8 nautical miles away.

e. Refer to the graph in the back of the textbook.

Homework 6.2

1. a. $\sqrt{121} = 11$ since $11^2 = 121$

   b. $\sqrt[3]{27} = 3$ since $3^3 = 27$

   c. $\sqrt[4]{625} = 5$ since $5^4 = 625$

3. a. $\sqrt[5]{32} = 2$ since $2^5 = 32$

   b. $\sqrt[4]{16} = 2$ since $2^4 = 16$

   c. $\sqrt[3]{729} = 9$ since $9^3 = 729$

5. a. $9^{1/2} = \sqrt{9} = 3$

   b. $81^{1/4} = \sqrt[4]{81} = 3$

   c. $64^{1/6} = \sqrt[6]{64} = 2$

Calculator check:

```
9^(1/2)
                    3
81^(1/4)
                    3
64^(1/6)
                    2
```

7. a. $32^{0.2} = 32^{1/5} = \sqrt[5]{32} = 2$

   b. $8^{-1/3} = \dfrac{1}{8^{1/3}} = \dfrac{1}{\sqrt[3]{8}} = \dfrac{1}{2}$

   c. $64^{-0.5} = 64^{-1/2} = \dfrac{1}{64^{1/2}}$

     $= \dfrac{1}{\sqrt{64}} = \dfrac{1}{8}$

Calculator check:

```
32^.2
                    2
8^(-1/3)▶Frac
                  1/2
64^-.5▶Frac
                  1/8
```

9. a. $3^{1/2} = \sqrt{3}$

   b. $4x^{1/3} = 4\sqrt[3]{x}$

   c. $(4x)^{0.2} = (4x)^{1/5} = \sqrt[5]{4x}$

11. a. $6^{-1/3} = \dfrac{1}{6^{1/3}} = \dfrac{1}{\sqrt[3]{6}}$

   b. $3(xy)^{-0.125} = 3(xy)^{-1/8}$

     $= \dfrac{3}{(xy)^{1/8}} = \dfrac{3}{\sqrt[8]{xy}}$

   c. $(x-2)^{1/4} = \sqrt[4]{x-2}$

13. a. $\sqrt{7} = 7^{1/2}$

   b. $\sqrt[3]{2x} = (2x)^{1/3}$

   c. $2\sqrt[5]{z} = 2z^{1/5}$

15. a. $\dfrac{-3}{\sqrt[4]{6}} = \dfrac{-3}{6^{1/4}} = -3\left(6^{-1/4}\right)$

   b. $\sqrt[4]{x-3y} = (x-3y)^{1/4}$

   c. $\dfrac{-1}{\sqrt[5]{1+3b}} = \dfrac{-1}{(1+3b)^{1/5}}$

     $= -(1+3b)^{-1/5}$

17. a. $2^{1/2} \approx 1.414$

b. $\sqrt[3]{75} = 75^{1/3} \approx 4.217$

c. $\sqrt[4]{1.6} = 1.6^{1/4} \approx 1.125$

d. $365^{-1/3} \approx 0.140$

e. $0.006^{-0.2} \approx 2.782$

On the calculator, this looks like:

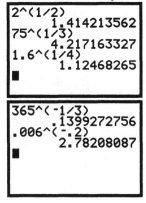

19. Evaluate $w = 4\left(\dfrac{Th}{m}\right)^{1/2}$ at T=293, h=15, and m=4.

$$w = 4\left(\frac{293(15)}{4}\right)^{1/2}$$
$$= 4(1098.75)^{1/2} = 132.6$$

The width of the sonic boom is 132.6 kilometers.

21. a. Use the equation $v = \sqrt{10r}$.

r (meters)	0.2	0.4	0.6	0.8	1.0
v (meters per second)	1.4	2	2.4	2.8	3.2

b. A typical adult man can walk $v = \sqrt{10(0.9)} = \sqrt{9} = 3$ m/s.

c. A typical four-year-old can walk $v = \sqrt{10(0.5)} = \sqrt{5} \approx 2.2$ m/s.

d. Refer to the graph in the back of the textbook.

e. let $v = 4.4$:

$$4.4 = \sqrt{10r}$$
$$(4.4)^2 = 10r$$
$$19.36 = 10r$$
$$1.936 \text{ meters} = r$$

f. On the moon, a typical adult man can walk $v = \sqrt{(1.6)(0.9)} = \sqrt{1.44} = 1.2$ m/s.

23. a. Using the formula
$r = 1.3 \times 10^{-13} A^{1/3}$ with A = 127, the nucleus of an atom of iodine-127 is $1.3 \times 10^{-13}(127)^{1/3}$
$\approx 6.5 \times 10^{-13}$ centimeters. Using the formula for the volume of a sphere, the volume of the nucleus is $V = \frac{4}{3}\pi r^3 = \frac{4}{3}\pi\left(6.5 \times 10^{-13}\right)^3$
$\approx 1.15 \times 10^{-36}$ cm^3.

b. The density is
$\dfrac{2.1 \times 10^{-22} \text{ g}}{1.15 \times 10^{-36} \text{ cm}^3} \approx 1.8 \times 10^{14}$
grams per cubic centimeter.

c. Use $r = 1.3 \times 10^{-13} A^{1/3}$.

Element	Mass number, A	Radius r $\left(\times 10^{-13} \text{ cm}\right)$
Carbon	14	3.1
Potassium	40	4.4
Cobalt	60	5.1
Technetium	99	6.0
Radium	226	7.9

d. Refer to the graph in the back of the textbook.

25. a. In 1950, $t = 1960 - 1950 = 10$ so the number of members was
$M(10) = 72 + 100(10)^{1/3} \approx 287.$

In 1970, $t = 1970 - 1950 = 20$ so the number of members was
$M(20) = 72 + 100(20)^{1/3} \approx 343.$

b. Let $M(t) = 400$ and solve for t:
$$72 + 100t^{1/3} = 400$$
$$100t^{1/3} = 328$$
$$t^{1/3} = 3.28$$
$$t = (3.28)^3 \approx 35.288$$
$1950 + 35 = 1985$, so there were 400 members in the year 1985.

Let $M(t) = 500$ and solve for t:
$$72 + 100t^{1/3} = 500$$
$$100t^{1/3} = 428$$
$$t^{1/3} = 4.28$$
$$t = (4.28)^3 \approx 78.403$$
$1950 + 78 = 2028$, so there will be 500 members in the year 2028.

c. Refer to the graph in the back of the textbook. From the graph we notice that since 1950 the membership has grown each year but by less and less.

27. a. I **b.** III **c.** II

29.

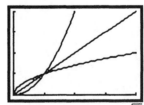

The graph of $y_1 = \sqrt{x}$ is the same as the graph of $y_2 = x^2$ reflected about the line $y_3 = x$.

31.

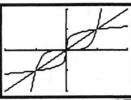

The graph of $y_1 = \sqrt[5]{x}$ is the same as the graph of $y_2 = x^5$ reflected about the line $y_3 = x$.

33. a. $\sqrt{x} = x^{1/2}$

b. $\sqrt{\sqrt{x}} = \sqrt{x^{1/2}} = \left(x^{1/2}\right)^{1/2}$

c. $\sqrt{\sqrt{x}} = \left(x^{1/2}\right)^{1/2}$ from part (a)

$\qquad = x^{1/4}$ multiply exponents

$\qquad = \sqrt[4]{x}$

35. a. $\left(\sqrt[3]{125}\right)^3 = 125$ since

$\qquad \left(\sqrt[3]{125}\right)^3 = \left(125^{1/3}\right)^3 = 125^1 = 125$

b. $\left(\sqrt[4]{2}\right)^4 = 2$ since

$\qquad \left(\sqrt[4]{2}\right)^4 = \left(2^{1/4}\right)^4 = 2^1 = 2$

c. $\left(3\sqrt{7}\right)^2 = 3^2\left(\sqrt{7}\right)^2 = 9 \cdot 7 = 63$

d. $\left(-x^2\sqrt[3]{2x}\right)^3 = (-1)^3\left(x^2\right)^3\left(\sqrt[3]{2x}\right)^3$

$\qquad\qquad = -1 \cdot x^6 \cdot 2x$

$\qquad\qquad = -2x^7$

37. $2\sqrt[3]{x} - 5 = -17$

$\qquad 2\sqrt[3]{x} = -12$

$\qquad \sqrt[3]{x} = -6$

$\qquad \left(\sqrt[3]{x}\right)^3 = (-6)^3$

$\qquad x = -216$

Check:

$2\sqrt[3]{-216} - 5 = 2(-6) - 5$

$\qquad\qquad = -12 - 5$

$\qquad\qquad = -17$

39. $4(x+2)^{1/5} = 12$

$\qquad (x+2)^{1/5} = 3$

$\qquad \left((x+2)^{1/5}\right)^5 = 3^5$

$\qquad x + 2 = 243$

$\qquad x = 241$

Check:

$4(241+2)^{1/5} = 4(243)^{1/5}$

$\qquad\qquad = 4\sqrt[5]{243}$

$\qquad\qquad = 4(3)$

$\qquad\qquad = 12$

41.
$$(2x-3)^{-1/4} = \frac{1}{2}$$
$$\left((2x-3)^{-1/4}\right)^4 = \left(\frac{1}{2}\right)^4$$
$$(2x-3)^{-1} = \frac{1}{16}$$
$$\frac{1}{2x-3} = \frac{1}{16}$$
$$2x-3 = 16$$
$$2x = 19$$
$$x = \frac{19}{2}$$

Check:
$$\left(2\left(\frac{19}{2}\right)-3\right)^{-1/4} = (19-3)^{-1/4}$$
$$= (16)^{-1/4}$$
$$= \frac{1}{16^{1/4}}$$
$$= \frac{1}{\sqrt[4]{16}}$$
$$= \frac{1}{2}$$

43.
$$\sqrt[3]{x^2-3} = 3$$
$$\left(\sqrt[3]{x^2-3}\right)^3 = 3^3$$
$$x^2-3 = 27$$
$$x^2 = 30$$
$$x = \pm\sqrt{30} \approx \pm 5.477$$

Checks:
$$\sqrt[3]{\left(\sqrt{30}\right)^2 - 3} = \sqrt[3]{30-3}$$
$$= \sqrt[3]{27}$$
$$= 3$$
$$\sqrt[3]{\left(-\sqrt{30}\right)^2 - 3} = \sqrt[3]{30-3}$$
$$= \sqrt[3]{27}$$
$$= 3$$

45.
$$\sqrt[3]{2x^2-15x} = 5$$
$$\left(\sqrt[3]{2x^2-15x}\right)^3 = 5^3$$
$$2x^2-15x = 125$$
$$2x^2-15x-125 = 0$$
$$(2x-25)(x+5) = 0$$
$$x = \frac{25}{2} = 12.5 \text{ or } x = -5$$

Check $x = 12.5$:
$$\sqrt[3]{2(12.5)^2 - 15(12.5)}$$
$$= \sqrt[3]{2(156.25) - 187.5}$$
$$= \sqrt[3]{312.5 - 187.5}$$
$$= \sqrt[3]{125}$$
$$= 5$$

Check $x = -5$:
$$\sqrt[3]{2(-5)^2 - 15(-5)} = \sqrt[3]{2(25) + 75}$$
$$= \sqrt[3]{50 + 75}$$
$$= \sqrt[3]{125}$$
$$= 5$$

47. a. Refer to the graph in the back of the textbook.

a	1	2	4	8
V	3.14	25.13	201.06	1608.5

b. Let $V = 816.8$ cubic meters:
$$816.8 = \pi a^3$$
$$\frac{816.8}{\pi} = a^3$$
$$\sqrt[3]{\frac{816.8}{\pi}} = a$$
$$6.38 \approx a$$
The height is approximately 6.38 meters.

49. Substitute the values given for L, R, and s, and solve for T:

$$\left(5.7\times10^{-5}\right)T^4 = \frac{3.9\times10^{33}}{4\pi\left(9.96\times10^{10}\right)^2}$$

$$T^4 = \frac{3.9\times10^{33}}{4\pi\left(9.96\times10^{10}\right)^2\left(5.7\times10^{-5}\right)}$$

$$T^4 \approx 5.4886\times10^{14}$$

$$T = \sqrt[4]{5.4886\times10^{14}} \approx 4840$$

The temperature is about 4840 degrees Kelvin.

51. $r = \sqrt[3]{\dfrac{3V}{4\pi}}$

$r^3 = \dfrac{3V}{4\pi}$ Cube both sides

$\dfrac{4\pi r^3}{3} = V$ Multiply both sides by $\dfrac{4\pi}{3}$

53. $R = \sqrt[4]{\dfrac{8Lvf}{\pi p}}$

$R^4 = \dfrac{8Lvf}{\pi p}$ Raise both sides to the fourth power.

$pR^4 = \dfrac{8Lvf}{\pi}$ Multiply both sides by p.

$p = \dfrac{8Lvf}{\pi R^4}$ Divide both sides by R^4.

55. $\dfrac{\sqrt{x}}{4} - \dfrac{2}{\sqrt{x}} + \dfrac{x}{\sqrt{2}} = \dfrac{1}{4}x^{1/2} - 2x^{-1/2} + \dfrac{1}{\sqrt{2}}x$

57. $\dfrac{6-\sqrt[3]{x}}{2\sqrt[3]{x}} = \dfrac{6}{2\sqrt[3]{x}} - \dfrac{\sqrt[3]{x}}{2\sqrt[3]{x}} = 3x^{-1/3} - \dfrac{1}{2}$

59. $x^{-0.5}\left(x + x^{0.25} - x^{0.5}\right) = x^{-0.5+1} + x^{-0.5+0.25} - x^{-0.5+0.5}$

$$= x^{0.5} + x^{-0.25} - x^0$$

$$= x^{0.5} + x^{-0.25} - 1$$

Homework 6.3

1. a. $81^{3/4} = \left(\sqrt[4]{81}\right)^3 = 3^3 = 27$

b. $125^{2/3} = \left(\sqrt[3]{125}\right)^2 = 5^2 = 25$

c. $625^{0.75} = 625^{3/4} = \left(\sqrt[4]{625}\right)^3 = 5^3 = 125$

3. a. $16^{-3/2} = \dfrac{1}{16^{3/2}} = \dfrac{1}{\left(\sqrt{16}\right)^3} = \dfrac{1}{4^3} = \dfrac{1}{64}$

b. $8^{-4/3} = \dfrac{1}{8^{4/3}} = \dfrac{1}{\left(\sqrt[3]{8}\right)^4} = \dfrac{1}{2^4} = \dfrac{1}{16}$

c. $32^{-1.6} = 32^{-16/10} = 32^{-8/5} = \dfrac{1}{32^{8/5}} = \dfrac{1}{\left(\sqrt[5]{32}\right)^8} = \dfrac{1}{2^8} = \dfrac{1}{256}$

5. a. $x^{4/5} = \sqrt[5]{x^4}$

b. $b^{-5/6} = \dfrac{1}{b^{5/6}} = \dfrac{1}{\sqrt[6]{b^5}}$

c. $(pq)^{-2/3} = \dfrac{1}{(pq)^{2/3}} = \dfrac{1}{\sqrt[3]{(pq)^2}} = \dfrac{1}{\sqrt[3]{p^2q^2}}$

7. a. $3x^{0.4} = 3x^{4/10} = 3x^{2/5} = 3\sqrt[5]{x^2}$

b. $4z^{-4/3} = \dfrac{4}{z^{4/3}} = \dfrac{4}{\sqrt[3]{z^4}}$

c. $-2x^{0.25}y^{0.75} = -2x^{1/4}y^{3/4} = -2\left(xy^3\right)^{1/4} = -2\sqrt[4]{xy^3}$

9. a. $\sqrt[3]{x^2} = x^{2/3}$

b. $2\sqrt[5]{ab^3} = 2\left(ab^3\right)^{1/5} = 2a^{1/5}\left(b^3\right)^{1/5} = 2a^{1/5}b^{3/5}$

c. $\dfrac{-4m}{\sqrt[6]{p^7}} = \dfrac{-4m}{p^{7/6}} = -4mp^{-7/6}$

11. a. $\sqrt[3]{(ab)^2} = \left((ab)^2\right)^{1/3} = \left(a^2b^2\right)^{1/3} = a^{2/3}b^{2/3}$ or $(ab)^{2/3}$

b. $\dfrac{8}{\sqrt[4]{x^3}} = \dfrac{8}{x^{3/4}} = 8x^{-3/4}$

c. $\dfrac{R}{3\sqrt{TK^5}} = \dfrac{R}{3\left(TK^5\right)^{1/2}} = \dfrac{R}{3T^{1/2}K^{5/2}} = \dfrac{1}{3}RT^{-1/2}K^{-5/2}$

13. a. $\sqrt[5]{32^3} = \left(\sqrt[5]{32}\right)^3 = 2^3 = 8$

b. $-\sqrt[3]{27^4} = -\left(\sqrt[3]{27}\right)^4 = -3^4 = -81$

c. $\sqrt[4]{16y^{12}} = \left(16y^{12}\right)^{1/4} = 16^{1/4}\left(y^{12}\right)^{1/4} = \sqrt[4]{16}\,y^3 = 2y^3$

15. a. $-\sqrt{a^8b^{16}} = -\left(a^8b^{16}\right)^{1/2} = -\left(a^8\right)^{1/2}\left(b^{16}\right)^{1/2} = -a^4b^8$

b. $\sqrt[3]{8x^9y^{27}} = \left(8x^9y^{27}\right)^{1/3} = 8^{1/3}\left(x^9\right)^{1/3}\left(y^{27}\right)^{1/3} = \sqrt[3]{8}\,x^3y^9 = 2x^3y^9$

c. $-\sqrt[4]{81a^8b^{12}} = -\left(81a^8b^{12}\right)^{1/4} = -81^{1/4}\left(a^8\right)^{1/4}\left(b^{12}\right)^{1/4} = -\sqrt[4]{81}\,a^2b^3 = -3a^2b^3$

17. a. $12^{5/6} \approx 7.931$

b. $\sqrt[3]{6^4} = 6^{4/3} \approx 10.903$

c. $37^{-2/3} \approx 0.090$

d. $4.7^{2.3} \approx 35.142$

19. a.

t	5	10	15	20
$I(t)$	131	199	254	302

b. Let $I = 300$:

$$300 = 50t^{3/5}$$
$$6 = t^{3/5}$$
$$6^{5/3} = \left(t^{3/5}\right)^{5/3}$$

$19.81 \approx t$
After about 20 days, 300 people will be ill.

c. Refer to the graph in the back of the textbook. The graph passes through at approximately (20, 300), which confirms the result in part (b).

21. All of the graphs are increasing. All of the graphs except the line $y_1 = x$ are concave up. For $x > 1$, each graph increases more quickly than the previous one.

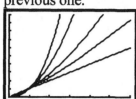

23.a,b. Refer to the graph in the back of the textbook.

c. $L = 1.05(1,600,000)^{0.58} \approx 4165$. The Congo River is about 4165 miles long.

d. Let $L = 1700$:
$$1700 = 1.05A^{0.58}$$
$$\frac{1700}{1.05} = A^{0.58}$$
$$\left(\frac{1700}{1.05}\right)^{1/0.58} = \left(A^{0.58}\right)^{1/0.58}$$
$$A = \left(\frac{1700}{1.05}\right)^{1/0.58} \approx 341,355$$
The Rio Grande's drainage basin is about 341,355 square miles.

25. a. It may be helpful to let $k = 1$ and use your graphing calculator. For example, if $k = 1$ and $p = 1.26$, graph $Y_1 = x^{1.26}$. The resulting graph looks most like II. Match the rest in the same way to find:
Home range size: II
Lung volume: III
Brain mass: I
Respiration rate: IV

b. If $p > 1$, the graph is increasing and concave up. If $0 < p < 1$, the graph is increasing and concave down. If $p < 0$, the graph is decreasing and concave up.

27. a. For any width $d > 1$, *Tricosanthes's* length, $ad^{2.2}$, is larger than *Lagenaria's* length, $ad^{0.81}$. So *Tricosanthes* is relatively thinner and *Lagenaria* is relatively fatter. *Tricosanthes* is the snake guard and *Lagenaria* is the bottle guard.

b. Let $L = 200$ cm and $d = 4$ cm in the formula $L = ad^{2.2}$:
$$a(4)^{2.2} = 200$$
$$a = \frac{200}{4^{2.2}} \approx 9.5$$

c. Let $L = 10$ cm and $d = 7$ cm in the formula $L = ad^{0.81}$:
$$a(7)^{0.81} = 10$$
$$a = \frac{10}{7^{0.81}} \approx 2.07$$

d. The giant bottle guard has a diameter of 20 cm, so according to the formula for bottle guards, it should have a length of $L = 2.07(20)^{0.81} \approx 23.4$ cm. Since it is given in the problem that a giant bottle guard of diameter 20 cm has length 23 cm, we conclude that the formula for the standard bottle guard fits the giant bottle guard quite well.

29. a. Refer to the graph in the back of the textbook.

    b. Where the temperature range is 10°, you would expect the number of species of mammals to be $M = 433.8(10)^{-0.742} \approx 79$. This corresponds to the point $(10, 79)$ on the graph.

    c. Let $M = 50$ species.

$$433.8R^{-0.742} = 50$$

$$R^{-0.742} = 0.11526$$

$$R = (0.11526)^{-1/0.742} \approx 18.4$$

If 50 species are found, you would expect the temperature range to be 18.4° Celsius.

    d. $f(9) = 433.8(9)^{-0.742} \approx 85$

$$f(10) = 433.8(10)^{-0.742} \approx 79$$

$$f(19) = 433.8(19)^{-0.742} \approx 49$$

$$f(20) = 433.8(20)^{-0.742} \approx 47$$

If the temperature range is 9°, there are 85 species. If the temperature range is 10°, there are 79 species. If the temperature range is 19°, there are 49 species. If the temperature range is 20°, there are 47 species. An increase in temperature range from 9° to 10° results in a greater drop in the number of species (6) than an increase from 19° to 20° (2 species). The curve is steeper near 10° than near 20°, so a 1° horizontal change on the graph near 10° results in a greater vertical change than a 1° horizontal change near 20°.

31. a. $4a^{6/5}a^{4/5} = 4a^{6/5+4/5} = 4a^{10/5} = 4a^2$

    b. $9b^{4/3}b^{1/3} = 9b^{4/3+1/3} = 9b^{5/3}$

33. a. $\dfrac{8w^{9/4}}{2w^{3/4}} = 4w^{9/4-3/4} = 4w^{6/4} = 4w^{3/2}$

    b. $\dfrac{12z^{11/3}}{4z^{5/3}} = 3z^{11/3-5/3} = 3z^{6/3} = 3z^2$

35. a. $\dfrac{k^{3/4}}{2k} = \dfrac{k^{3/4-1}}{2} = \dfrac{k^{-1/4}}{2} = \dfrac{1}{2k^{1/4}}$

    b. $\dfrac{4h^{2/3}}{3h} = \dfrac{4h^{2/3-1}}{3} = \dfrac{4h^{-1/3}}{3} = \dfrac{4}{3h^{1/3}}$

37. a. The incubation time for the wren is $I(2.5) = 12.0(2.5)^{0.217}$ ≈ 15 days.
The incubation time for the greylag goose is $I(46) = 12.0(46)^{0.217} \approx 28$ days.

b. The time of incubation multiplied by the rate of water loss equals the total amount of water lost during incubation. Dividing this amount by the mass of the egg, m, gives the fraction of the initial mass that is lost. Therefore, the fraction lost is

$$\frac{I(m) \times W(m)}{m}$$

$$= \frac{12.0m^{0.217} \times 0.015m^{0.742}}{m}$$

$$= \frac{0.18m^{0.959}}{m}$$

$$= 0.18m^{-0.041}$$

c. Since $m^{-0.041} \approx m^0 = 1$, then the fraction of the mass lost during incubation is 0.18, or 18%.

39. $x^{2/3} - 1 = 15$

$$x^{2/3} = 16$$

$$\left(x^{2/3}\right)^{3/2} = 16^{3/2}$$

$$x = \left(\sqrt{16}\right)^3 = 4^3 = 64$$

Check:
$$64^{2/3} - 1 = \left(\sqrt[3]{64}\right)^2 - 1$$

$$= 4^2 - 1$$

$$= 16 - 1$$

$$= 15$$

41. $x^{-2/5} = 9$

$$\left(x^{-2/5}\right)^{-5/2} = 9^{-5/2}$$

$$x = \frac{1}{9^{5/2}} = \frac{1}{\left(\sqrt{9}\right)^5} = \frac{1}{3^5} = \frac{1}{243}$$

Check:
$$\left(\frac{1}{243}\right)^{-2/5} = 243^{2/5}$$

$$= \left(\sqrt[5]{243}\right)^2$$

$$= 3^2$$

$$= 9$$

43. $2\left(5.2 - x^{5/3}\right) = 1.4$

$$5.2 - x^{5/3} = 0.7$$

$$-x^{5/3} = -4.5$$

$$x^{5/3} = 4.5$$

$$x = 4.5^{3/5}$$

$$\approx 2.466$$

Check:
$$2\left(5.2 - 2.466^{5/3}\right) \approx 2(5.2 - 4.501)$$

$$= 2(0.699)$$

$$= 1.398$$

$$\approx 1.4$$

45. a. $p^2 = Ka^3$, so taking the square root of both sides gives $p = \sqrt{Ka^3}$. (We keep only the positive square root the period of a planet's revolution cannot be negative.)

b. Substitute $K = 1.243 \times 10^{-24}$ and $a = 1.417 \times 10^8$:

$$p = \sqrt{\left(1.243 \times 10^{-24}\right)\left(1.417 \times 10^8\right)^3} \approx 1.88.$$ Therefore, Mars takes 1.88 years to revolve around the sun.

Note: Entering the above expression all at once in your calculator looks like the following, where the "E" on the screen comes from the "EE" button on your calculator.

```
√((1.243E-24)*(1
.417E8)^3)
          1.88057362
■
```

47. $f(x) = (3x - 4)^{3/2}$ and $f(x) = 27$ so:

$$27 = (3x - 4)^{3/2}$$

$27^{2/3} = 3x - 4$ Raise both sides to the 2 / 3 power.

$9 = 3x - 4$ Evaluate: $27^{2/3} = \left(\sqrt[3]{27}\right)^2 = 3^2 = 9$

$13 = 3x$

$\dfrac{13}{3} = x$

49. $S(x) = 12x^{-5/4}$ and $S(x) = 20$ so

$20 = 12x^{-5/4}$

$\dfrac{5}{3} = x^{-5/4}$

$\left(\dfrac{5}{3}\right)^{-4/5} = \left(x^{-5/4}\right)^{-4/5}$ Raise both sides to the $-4 / 5$ power.

$x = \left(\dfrac{5}{3}\right)^{-4/5} \approx 0.665$ Evaluate using calculator.

Homework 6.3

51. $2x^{1/2}\left(x - x^{1/2}\right) = 2x^{1/2+1} - 2x^{1/2+1/2} = 2x^{3/2} - 2x$

53. $\frac{1}{2}y^{-1/3}\left(y^{2/3} + 3y^{-5/6}\right) = \frac{1}{2}y^{-1/3+2/3} + \frac{3}{2}y^{-1/3+(-5/6)} = \frac{1}{2}y^{1/3} + \frac{3}{2}y^{-2/6+(-5/6)}$

$= \frac{1}{2}y^{1/3} + \frac{3}{2}y^{-7/6}$

55. Multiply using FOIL:

$\left(2x^{1/4} + 1\right)\left(x^{1/4} - 1\right) = 2x^{2/4} - 2x^{1/4} + x^{1/4} - 1 = 2x^{1/2} - x^{1/4} - 1$

57. Multiply using FOIL:

$\left(a^{3/4} - 2\right)^2 = \left(a^{3/4} - 2\right)\left(a^{3/4} - 2\right) = a^{6/4} - 2a^{3/4} - 2a^{3/4} + 4 = a^{3/2} - 4a^{3/4} + 4$

59. $x^{3/2} + x = x\left(x^{1/2} + 1\right)$

61. $y^{3/4} - y^{-1/4} = y^{-1/4}(y - 1) = \dfrac{y-1}{y^{1/4}}$

63. $a^{1/3} + 3 - a^{-1/3} = a^{-1/3}\left(a^{2/3} + 3a^{1/3} - 1\right) = \dfrac{a^{2/3} + 3a^{1/3} - 1}{a^{1/3}}$

Midchapter 6 Review

1. $4 \cdot 10^{-3} = \dfrac{4}{10^3} = \dfrac{4}{1000} = \dfrac{1}{250}$

3. $6t^{-3} = \dfrac{3}{500}$

$t^{-3} = \dfrac{3}{500} \cdot \dfrac{1}{6}$

$t^{-3} = \dfrac{1}{1000}$

$\dfrac{1}{t^3} = \dfrac{1}{1000}$

$t^3 = 1000$

$t = \sqrt[3]{1000}$

$t = 10$

5. $f(x) = \dfrac{2}{3x^4} = \dfrac{2}{3}x^{-4}$

7. $\dfrac{3w^{-1}}{t^{-7}w^2} = \dfrac{3t^7}{w^1 w^2} = \dfrac{3t^7}{w^{1+2}} = \dfrac{3t^7}{w^3}$

9. $1{,}234{,}000 = 1.234 \times 10^6$

11. $\sqrt[4]{0.0081} = (0.0081)^{1/4} = 0.3$
Note: When entering this expression into your calculator, do not forget to write parentheses around the exponent 1/4.

13. $4096^{1/6} = \sqrt[6]{4096} = 4$

15. $0.00243^{0.4} = 0.00243^{4/10}$

$= 0.00243^{2/5}$

$= \left(\sqrt[5]{0.00243}\right)^2$

$= 0.3^2$

$= 0.09$

17. $-256^{1/4} = -\sqrt[4]{256} = -4$

19. $(-1728)^{1/3} = \sqrt[3]{-1728} = -12$

21. $-2\sqrt[3]{5m} = -2(5m)^{1/3}$

23. $\dfrac{9}{\sqrt[4]{(2p)^3}} = \dfrac{9}{\left((2p)^3\right)^{1/4}}$

$= \dfrac{9}{(2p)^{3/4}}$

$= 9(2p)^{-3/4}$

25. $11h^{1/5} = 11\sqrt[5]{h}$

27. $\dfrac{2}{3}c^{0.8} = \dfrac{2}{3}c^{8/10} = \dfrac{2}{3}c^{4/5} = \dfrac{2}{3}\left(\sqrt[5]{c}\right)^4$

29. $(7t)^3(7t)^{-1} = (7t)^{3+(-1)}$

$= (7t)^2$

$= 7^2 t^2$

$= 49t^2$

31. $\dfrac{\left(2k^{-1}\right)^{-4}}{4k^{-3}} = \dfrac{2^{-4}\left(k^{-1}\right)^{-4}}{4k^{-3}}$

$= \dfrac{2^{-4}k^4}{4k^{-3}}$

$= \dfrac{k^{4-(-3)}}{2^4 \cdot 4}$

$= \dfrac{k^7}{64}$

33. $\dfrac{8a^{-3/4}}{a^{-11/4}} = 8a^{-3/4-(-11/4)}$

$= 8a^{8/4}$

$= 8a^2$

35. $8\sqrt[4]{x+6} = 24$

$\sqrt[4]{x+6} = 3$

$\left(\sqrt[4]{x+6}\right)^4 = 3^4$

$x + 6 = 81$

$x = 75$

37. $\frac{2}{3}(2y+1)^{0.2} = 6$

$\frac{2}{3}(2y+1)^{1/5} = 6$

$(2y+1)^{1/5} = 9$

$\left((2y+1)^{1/5}\right)^5 = 9^5$

$2y+1 = 59,049$

$2y = 59,048$

$y = 29,524$

39.

x	0	1	5	10	20	50	70	100
f(x)	0	1	1.62	2.00	2.46	3.23	3.58	3.98

Refer to the graph in the back of the textbook.

Homework 6.4

1. Distance $= \sqrt{(4-1)^2 + (5-1)^2} = \sqrt{9+16} = \sqrt{25} = 5$

   Midpoint $= \left(\dfrac{1+4}{2}, \dfrac{1+5}{2}\right) = \left(\dfrac{5}{2}, 3\right)$

3. Distance $= \sqrt{(-2-2)^2 + (-1-(-3))^2} = \sqrt{16+4} = \sqrt{20}$

   Midpoint $= \left(\dfrac{2+(-2)}{2}, \dfrac{(-3)+(-1)}{2}\right) = (0, -2)$

5. Distance $= \sqrt{(-2-3)^2 + (5-5)^2} = \sqrt{25+0} = \sqrt{25} = 5$

   Midpoint $= \left(\dfrac{3+(-2)}{2}, \dfrac{5+5}{2}\right) = \left(\dfrac{1}{2}, 5\right)$

7. **a.** Let the harbor be at the point $(0, 0)$. Then Leanne is originally at the point $(-3, -5)$ and the island is at $(-8, 7)$. So the distance between Leanne and the island is

   $$\sqrt{(-8-(-3))^2 + (7-(-5))^2} = \sqrt{25+144} = \sqrt{169} = 13 \text{ miles.}$$

   b. When Leanne is halfway to the island, she is at the midpoint between $(-3, -5)$ and $(-8, 7)$, which is $\left(\dfrac{-8+(-3)}{2}, \dfrac{7+(-5)}{2}\right) = (-5.5, 1)$. The distance between this halfway point and the harbor is

   $$\sqrt{(-5.5-0)^2 + (1-0)^2} = \sqrt{30.25+1} = \sqrt{31.25} \approx 5.6 \text{ miles.}$$

9.

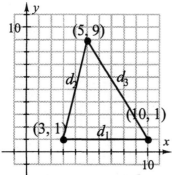

To find the perimeter of this triangle we add together the three distances between vertices.

$$d_1 = \sqrt{(3-10)^2 + (1-1)^2} = \sqrt{49} = 7$$

$$d_2 = \sqrt{(5-3)^2 + (9-1)^2} = \sqrt{68}$$

$$d_3 = \sqrt{(10-5)^2 + (1-9)^2} = \sqrt{89}$$

perimeter $= 7 + \sqrt{68} + \sqrt{89} \approx 24.7$

11.

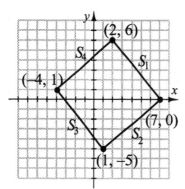

Let S_1, S_2, S_3, and S_4 be the lengths of the four sides of the rectangle as shown. To ensure that this rectangle is a square we must check that $S_1 = S_2 = S_3 = S_4$.

$$S_1 = \sqrt{(7-2)^2 + (0-6)^2} = \sqrt{61}$$

$$S_2 = \sqrt{(1-7)^2 + (-5-0)^2} = \sqrt{61}$$

$$S_3 = \sqrt{(-4-1)^2 + (1-(-5))^2} = \sqrt{61}$$

$$S_4 = \sqrt{(2-(-4))^2 + (6-1)^2} = \sqrt{61}$$

13.

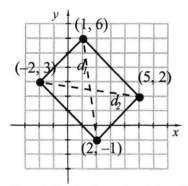

Let d_1 be the length of the diagonal joining points $(1, 6)$ and $(2, -1)$ and let d_2 be the length of the diagonal joining points $(-2, 3)$ and $(5, 2)$ as shown. By the distance formula, $d_1 = \sqrt{(2-1)^2 + (-1-6)^2} = \sqrt{50}$ and $d_2 = \sqrt{(5-(-2))^2 + (2-3)^2} = \sqrt{50}$. Therefore, the diagonals are of equal length.

15.

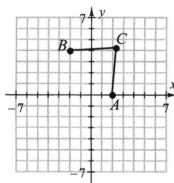

The distance between $A(2, 0)$ and $C\left(\sqrt{5}, 2+\sqrt{5}\right)$ is

$$\overline{AC} = \sqrt{\left(\sqrt{5}-2\right)^2 + \left(2+\sqrt{5}-0\right)^2} = \sqrt{\left(\sqrt{5}-2\right)^2 + \left(2+\sqrt{5}\right)^2}$$

$$= \sqrt{\left(5-4\sqrt{5}+4\right)+\left(4+4\sqrt{5}+5\right)} = \sqrt{18}$$

The distance between $B(-2, 4)$ and $C\left(\sqrt{5}, 2+\sqrt{5}\right)$ is

$$\overline{BC} = \sqrt{\left(\sqrt{5}-(-2)\right)^2 + \left(2+\sqrt{5}-4\right)^2} = \sqrt{\left(\sqrt{5}+2\right)^2 + \left(\sqrt{5}-2\right)^2}$$

$$= \sqrt{\left(5+4\sqrt{5}+4\right)+\left(5-4\sqrt{5}+4\right)} = \sqrt{18}$$

Thus $\overline{AC} = \overline{BC}$.

17.

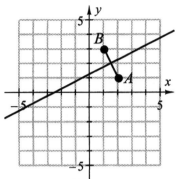

The perpendicular bisector passes through the midpoint of $A(2, 1)$ and $B(1, 3)$, which is $\left(\dfrac{2+1}{2}, \dfrac{1+3}{2}\right) = \left(\dfrac{3}{2}, 2\right)$. The slope of the line passing through A and B is

$m_{AB} = \dfrac{3-1}{1-2} = \dfrac{2}{-1} = -2$. The slope of the perpendicular bisector is the negative

reciprocal of m_{AB}, or $\dfrac{1}{2}$. So the perpendicular bisector has slope $\dfrac{1}{2}$ and passes

through $\left(\dfrac{3}{2}, 2\right)$. Using the point-slope form, we have the equation:

$$y - 2 = \dfrac{1}{2}\left(x - \dfrac{3}{2}\right)$$
$$y - 2 = \dfrac{1}{2}x - \dfrac{3}{4}$$
$$y = \dfrac{1}{2}x + \dfrac{5}{4}$$

19. The equation is a circle with center $(0, 0)$ and radius 5. Refer to the graph in the back of the textbook.

21. Divide both sides of $4x^2 + 4y^2 = 16$ by 4 so that $x^2 + y^2 = 4$. The equation is a circle with center $(0, 0)$ and radius 2. Refer to the graph in the back of the textbook.

23. The equation is a circle with center $(4, -2)$ and radius 3. Refer to the graph in the back of the textbook.

25. The equation is a circle with center $(-3, 0)$ and radius $\sqrt{10} \approx 3.2$. Refer to the graph in the back of the textbook.

27. Prepare to complete the square by writing the equation in the form:

$$\left(x^2 + 2x + \underline{}\right) + \left(y^2 - 4y + \underline{}\right) = 6$$

Complete the square in x by adding $\left[\frac{1}{2}(2)\right]^2 = 1$ to both sides of the equation and

complete the square in y by adding $\left[\frac{1}{2}(-4)\right]^2 = 4$ to both sides of the equation:

$$\left(x^2 + 2x + \underline{1}\right) + \left(y^2 - 4y + \underline{4}\right) = 6 + \underline{1} + \underline{4}$$

Factor the trinomials on the left and combine the numbers on the right:

$$(x+1)^2 + (y-2)^2 = 11$$

Center: $(-1, 2)$; Radius: $\sqrt{11} \approx 3.3$.

29. Prepare to complete the square by writing the equation in the form:

$$\left(x^2 + 8x + \underline{}\right) + y^2 = 4$$

Complete the square in x by adding $\left[\frac{1}{2}(8)\right]^2 = 16$ to both sides of the equation:

$$\left(x^2 + 8x + \underline{16}\right) + y^2 = 4 + \underline{16}$$

Factor the trinomial on the left and combine the numbers on the right:

$$(x+4)^2 + y^2 = 20$$

Center: $(-4, 0)$; Radius: $\sqrt{20} \approx 4.5$.

31. $\left(x - (-2)\right)^2 + (y - 5)^2 = \left(2\sqrt{3}\right)^2$

$\quad (x+2)^2 + (y-5)^2 = 12$ Simplify.

$\quad x^2 + 4x + 4 + y^2 - 10y + 25 = 12$ Expand.

$\quad x^2 + y^2 + 4x - 10y + 17 = 0$ Simplify.

33. To find the radius we compute the distance from the center to any point on the circle.

$$\text{radius} = \sqrt{\left(4 - \frac{3}{2}\right)^2 + \left(-3 - (-4)\right)^2} = \sqrt{\frac{25}{4} + 1} = \sqrt{\frac{29}{4}}$$

Thus, the equation of the circle is:

$$\left(x - \frac{3}{2}\right)^2 + \left(y - (-4)\right)^2 = \left(\sqrt{\frac{29}{4}}\right)^2$$

$\quad \left(x - \frac{3}{2}\right)^2 + (y + 4)^2 = \frac{29}{4}$ Simplify.

$\quad x^2 - 3x + \frac{9}{4} + y^2 + 8y + 16 = \frac{29}{4}$ Expand.

$\quad x^2 + y^2 - 3x + 8y + 11 = 0$ Simplify.

35. The center of circle is the midpoint of the diameter. Therefore,

$$\text{center} = \left(\frac{1+3}{2}, \frac{5-1}{2}\right) = (2,\ 2)$$

The radius is half the length of the diameter. Thus,

$$\text{radius} = \frac{1}{2}\sqrt{(3-1)^2 + (-1-5)^2} = \frac{\sqrt{40}}{2}$$

The equation of the circle is:

$$(x-2)^2 + (y-2)^2 = \left(\frac{\sqrt{40}}{2}\right)^2$$

$(x-2)^2 + (y-2)^2 = 10$ Simplify.

$x^2 - 4x + 4 + y^2 - 4y + 4 = 10$ Expand.

$x^2 + y^2 - 4x - 4y - 2 = 0$ Simplify.

37. Since the *x*-axis is tangent to the circle, the length of the vertical line segment originating at the center and terminating at the *x*-axis equals the radius. The length of this segment is the distance from (–3, –1) to (–3, 0), which is

$\sqrt{(-3-(-3))^2 + (0-(-1))^2} = 1$. So the radius is 1 and the equation of the circle is:

$$\left(x-(-3)\right)^2 + \left(y-(-1)\right)^2 = 1^2$$

$(x+3)^2 + (y+1)^2 = 1$ Simplify.

$x^2 + 6x + 9 + y^2 + 2y + 1 = 1$ Expand.

$x^2 + y^2 + 6x + 2y + 9 = 0$ Simplify.

39. Plug each point into the equation $x^2 + y^2 + ax + by + c = 0$ and then simplify:

$(2, 3)$: $\qquad 2^2 + 3^2 + a(2) + b(3) + c = 0 \qquad \rightarrow \qquad 2a + 3b + c = -13$ (1)

$(3, 2)$: $\qquad 3^2 + 2^2 + a(3) + b(2) + c = 0 \qquad \rightarrow \qquad 3a + 2b + c = -13$ (2)

$(-4, -5)$: $\quad (-4)^2 + (-5)^2 + a(-4) + b(-5) + c = 0 \quad \rightarrow \quad -4a - 5b + c = -41$ (3)

This gives us a system of three equations, which we can solve using the methods of Section 2.3 or by using matrices as in Section 2.4. Below, we use the methods taught in Section 2.3:

Multiply equation (1) by -1 and add the result to equation (2):

$$-2a - 3b - c = 13 \quad \text{(1a)}$$
$$\underline{3a + 2b + c = -13 \quad \text{(2)}}$$
$$a - b = 0$$
$$a = b \quad \text{(4)}$$

Substitute b for a in equation (1) so that $2a + 3a + c = -13$ or $5a + c = -13$. (5)

Substitute b for a in equation (3) so that $-4a - 5a + c = -41$ or $-9a + c = -41$. (6)

Multiply (5) by -1 and add the result to (6):

$$-5a - c = 13 \quad \text{(5a)}$$
$$\underline{-9a + c = -41 \quad \text{(6)}}$$
$$-14a = -28$$
$$a = 2$$

Substitute $a = 2$ into equation (5) to get $5(2) + c = -13$. Solving this equation for c gives us $c = -23$. Since $a = b$ from equation (4), we have $b = 2$. Thus, the equation of the circle is $x^2 + y^2 + 2x + 2y - 23 = 0$.

Homework 6.5

1. **a.** $\sqrt{18} = \sqrt{9}\sqrt{2}$ Factor out perfect square.
   $\phantom{\sqrt{18}} = 3\sqrt{2}$ Simplify.

   b. $\sqrt[3]{24} = \sqrt[3]{8}\sqrt[3]{3}$ Factor out perfect cube.
   $\phantom{\sqrt[3]{24}} = 2\sqrt[3]{3}$ Simplify.

   c. $-\sqrt[4]{64} = -\sqrt[4]{16}\sqrt[4]{4}$ Factor out perfect fourth root.
   $\phantom{-\sqrt[4]{64}} = -2\sqrt[4]{4}$ Simplify.

3. **a.** $\sqrt{60,000} = \sqrt{10,000}\sqrt{6}$ Factor out perfect square.
   $\phantom{\sqrt{60,000}} = 100\sqrt{6}$ Simplify.

   b. $\sqrt[3]{900,000} = \sqrt[3]{1,000}\sqrt[3]{900}$ Factor out perfect cube.
   $\phantom{\sqrt[3]{900,000}} = 10\sqrt[3]{900}$ Simplify.

   c. $\sqrt[3]{\dfrac{-40}{27}} = \sqrt[3]{\dfrac{-8}{27}}\sqrt[3]{5}$ Factor out perfect cube.
   $\phantom{\sqrt[3]{\dfrac{-40}{27}}} = -\dfrac{2}{3}\sqrt[3]{5}$ Simplify.

5. **a.** $\sqrt[3]{x^{10}} = \sqrt[3]{x^9}\sqrt[3]{x}$ Factor out perfect cube.
   $\phantom{\sqrt[3]{x^{10}}} = x^3\sqrt[3]{x}$ Simplify.

   b. $\sqrt{27z^3} = \sqrt{9z^2}\sqrt{3z}$ Factor out perfect squares.
   $\phantom{\sqrt{27z^3}} = 3z\sqrt{3z}$ Simplify.

   c. $\sqrt[4]{48a^9} = \sqrt[4]{16a^8}\sqrt[4]{3a}$ Factor out perfect fourth powers.
   $\phantom{\sqrt[4]{48a^9}} = 2a^2\sqrt[4]{3a}$ Simplify.

7. **a.** $-\sqrt{18s}\sqrt{2s^3} = -\sqrt{36s^4}$ Apply property (1).
   $\phantom{-\sqrt{18s}\sqrt{2s^3}} = -6s^2$ Simplify.

   b. $\sqrt[3]{7h^2}\sqrt[3]{-49h} = \sqrt[3]{-343h^3}$ Apply property (1).
   $\phantom{\sqrt[3]{7h^2}\sqrt[3]{-49h}} = -7h$ Simplify.

   c. $\sqrt{16-4x^2} = \sqrt{4}\sqrt{4-x^2}$ Factor out perfect squares.
   $\phantom{\sqrt{16-4x^2}} = 2\sqrt{4-x^2}$ Simplify.

9. a. $\sqrt[3]{8A^3 + A^6} = \sqrt[3]{A^3}\sqrt[3]{8 + A^3}$ Factor out perfect cube.

$$= A\sqrt[3]{8 + A^3} \qquad \text{Simplify.}$$

b. $\dfrac{\sqrt{45x^3y^3}}{\sqrt{5y}} = \sqrt{\dfrac{45x^3y^3}{5y}}$ Apply property (2).

$$= \sqrt{9x^3y^2} \qquad \text{Simplify.}$$
$$= \sqrt{9x^2y^2}\sqrt{x} \qquad \text{Factor out perfect squares.}$$
$$= 3xy\sqrt{x} \qquad \text{Simplify.}$$

c. $\dfrac{\sqrt[3]{8b^7}}{\sqrt[3]{a^6b^2}} = \sqrt[3]{\dfrac{8b^7}{a^6b^2}}$ Apply property (2).

$$= \sqrt[3]{\dfrac{8b^5}{a^6}} \qquad \text{Simplify.}$$
$$= \sqrt[3]{\dfrac{8b^3}{a^6}}\sqrt[3]{b^2} \qquad \text{Factor out perfect cubes.}$$
$$= \dfrac{2b}{a^2}\sqrt[3]{b^2} \qquad \text{Simplify.}$$

11. $3\sqrt{7} + 2\sqrt{7} = 5\sqrt{7}$ Combine like terms.

13. $4\sqrt{3} - \sqrt{27} = 4\sqrt{3} - \sqrt{9}\sqrt{3}$ Factor out perfect squares.

$$= 4\sqrt{3} - 3\sqrt{3} \qquad \text{Simplify.}$$
$$= \sqrt{3} \qquad \text{Combine like terms.}$$

15. $\sqrt{50x} + \sqrt{32x} = \sqrt{25}\sqrt{2x} + \sqrt{16}\sqrt{2x}$ Factor out perfect squares.

$$= 5\sqrt{2x} + 4\sqrt{2x} \qquad \text{Simplify.}$$
$$= 9\sqrt{2x} \qquad \text{Combine like terms.}$$

17. $3\sqrt[3]{16} - \sqrt[3]{2} - 2\sqrt[3]{54} = 3\sqrt[3]{8}\sqrt[3]{2} - \sqrt[3]{2} - 2\sqrt[3]{27}\sqrt[3]{2}$ Factor out perfect cubes.

$$= 6\sqrt[3]{2} - \sqrt[3]{2} - 6\sqrt[3]{2} \qquad \text{Simplify.}$$
$$= -\sqrt[3]{2} \qquad \text{Combine like terms.}$$

19. $2(3 - \sqrt{5}) = (2)3 - (2)\sqrt{5} = 6 - 2\sqrt{5}$ Distribute.

21. $\sqrt{2}\left(\sqrt{6}+\sqrt{10}\right)=\left(\sqrt{2}\right)\sqrt{6}+\left(\sqrt{2}\right)\sqrt{10}$ Distribute.

$\qquad\qquad = \sqrt{12}+\sqrt{20}$ Apply property (1).

$\qquad\qquad = \sqrt{4}\sqrt{3}+\sqrt{4}\sqrt{5}$ Factor out perfect squares.

$\qquad\qquad = 2\sqrt{3}+2\sqrt{5}$ Simplify.

23. $\sqrt[3]{2}\left(\sqrt[3]{20}-2\sqrt[3]{12}\right)=\left(\sqrt[3]{2}\right)\sqrt[3]{20}-\left(\sqrt[3]{2}\right)2\sqrt[3]{12}$ Distribute.

$\qquad\qquad = \sqrt[3]{40}-2\sqrt[3]{24}$ Apply property (1).

$\qquad\qquad = \sqrt[3]{8}\sqrt[3]{5}-2\sqrt[3]{8}\sqrt[3]{3}$ Factor out perfect cubes.

$\qquad\qquad = 2\sqrt[3]{5}-4\sqrt[3]{3}$ Simplify.

25. $\left(\sqrt{x}-3\right)\left(\sqrt{x}+3\right)=\sqrt{x}\sqrt{x}+3\sqrt{x}-3\sqrt{x}-9$ FOIL.

$\qquad\qquad = \sqrt{x^2}+3\sqrt{x}-3\sqrt{x}-9$ Apply property (1).

$\qquad\qquad = x+3\sqrt{x}-3\sqrt{x}-9$ Simplify.

$\qquad\qquad = x-9$ Combine like terms.

27. $\left(\sqrt{2}-\sqrt{3}\right)\left(\sqrt{2}+2\sqrt{3}\right)=\sqrt{2}\sqrt{2}+2\sqrt{2}\sqrt{3}-\sqrt{2}\sqrt{3}-2\sqrt{3}\sqrt{3}$ FOIL.

$\qquad\qquad = \sqrt{4}+2\sqrt{6}-\sqrt{6}-2\sqrt{9}$ Apply property (1).

$\qquad\qquad = 2+2\sqrt{6}-\sqrt{6}-6$ Simplify.

$\qquad\qquad = \sqrt{6}-4$ Combine like terms.

29. $\left(\sqrt{5}-\sqrt{2}\right)^2=\left(\sqrt{5}-\sqrt{2}\right)\left(\sqrt{5}-\sqrt{2}\right)$

$\qquad\qquad = \sqrt{5}\sqrt{5}-\sqrt{2}\sqrt{5}-\sqrt{2}\sqrt{5}+\sqrt{2}\sqrt{2}$ FOIL.

$\qquad\qquad = \sqrt{25}-\sqrt{10}-\sqrt{10}+\sqrt{4}$ Apply property (1).

$\qquad\qquad = 5-\sqrt{10}-\sqrt{10}+2$ Simplify.

$\qquad\qquad = 7-2\sqrt{10}$ Combine like terms.

31. $\left(\sqrt{a}-2\sqrt{b}\right)^2=\left(\sqrt{a}-2\sqrt{b}\right)\left(\sqrt{a}-2\sqrt{b}\right)$

$\qquad\qquad = \sqrt{a}\sqrt{a}-2\sqrt{a}\sqrt{b}-2\sqrt{a}\sqrt{b}+4\sqrt{b}\sqrt{b}$ FOIL.

$\qquad\qquad = \sqrt{a^2}-2\sqrt{ab}-2\sqrt{ab}+4\sqrt{b^2}$ Apply property (1).

$\qquad\qquad = a-2\sqrt{ab}-2\sqrt{ab}+4b$ Simplify.

$\qquad\qquad = a-4\sqrt{ab}+4b$ Combine like terms.

33. Let $x = 1 + \sqrt{3}$ in $x^2 - 2x - 2$ and simplify the result:

$$x^2 - 2x - 3 = \left(1 + \sqrt{3}\right)^2 - 2\left(1 + \sqrt{3}\right) - 2$$
$$= \left(1 + \sqrt{3}\right)\left(1 + \sqrt{3}\right) - 2\left(1 + \sqrt{3}\right) - 2$$
$$= 1 + \sqrt{3} + \sqrt{3} + \sqrt{9} - 2 - 2\sqrt{3} - 2$$
$$= 1 + 2\sqrt{3} + 3 - 2 - 2\sqrt{3} - 2$$
$$= 0$$

So $x = 1 + \sqrt{3}$ is a solution to the equation $x^2 - 2x - 2 = 0$.

35. Let $x = -3 + 3\sqrt{2}$ in $x^2 + 6x - 9$ and simplify the result:

$$x^2 + 6x - 9 = \left(-3 + 3\sqrt{2}\right)^2 + 6\left(-3 + 3\sqrt{2}\right) - 9$$
$$= \left(-3 + 3\sqrt{2}\right)\left(-3 + 3\sqrt{2}\right) + 6\left(-3 + 3\sqrt{2}\right) - 9$$
$$= 9 - 9\sqrt{2} - 9\sqrt{2} + 9\sqrt{4} - 18 + 18\sqrt{2} - 9$$
$$= 9 - 18\sqrt{2} + 18 - 18 + 18\sqrt{2} - 9$$
$$= 0$$

So $x = -3 + 3\sqrt{2}$ is a solution to the equation $x^2 + 6x - 9 = 0$.

37. $\dfrac{6}{\sqrt{3}} = \dfrac{6\sqrt{3}}{\sqrt{3}\sqrt{3}}$ Multiply numerator and denominator by $\sqrt{3}$.

$\qquad = \dfrac{6\sqrt{3}}{3}$ Simplify.

$\qquad = 2\sqrt{3}$ Simplify.

39. $\sqrt{\dfrac{7x}{18}} = \dfrac{\sqrt{7x}}{\sqrt{18}}$ Apply property (2).

$\qquad = \dfrac{\sqrt{7x}}{\sqrt{9}\sqrt{2}}$ Factor out perfect squares.

$\qquad = \dfrac{\sqrt{7x}\sqrt{2}}{3\sqrt{2}\sqrt{2}}$ Multiply numerator and denominator by $\sqrt{2}$.

$\qquad = \dfrac{\sqrt{14x}}{6}$ Simplify.

41. $\sqrt{\dfrac{2a}{b}} = \dfrac{\sqrt{2a}}{\sqrt{b}}$ Apply property (2).

$\qquad = \dfrac{\sqrt{2a}\sqrt{b}}{\sqrt{b}\sqrt{b}}$ Multiply numerator and denominator by $\sqrt{b}$.

$\qquad = \dfrac{\sqrt{2ab}}{b}$ Apply property (1) and simplify.

43. $\dfrac{2\sqrt{3}}{\sqrt{2k}} = \dfrac{2\sqrt{3}\sqrt{2k}}{\sqrt{2k}\sqrt{2k}}$ Multiply numerator and denominator by $\sqrt{2k}$.

$\qquad\quad = \dfrac{2\sqrt{6k}}{2k}$ Apply property (1) and simplify.

$\qquad\quad = \dfrac{\sqrt{6k}}{k}$ Simplify.

45. $\dfrac{4}{(1+\sqrt{3})} = \dfrac{4(1-\sqrt{3})}{(1+\sqrt{3})(1-\sqrt{3})}$ Multiply numerator and denominator by conjugate.

$\qquad\quad = \dfrac{4(1-\sqrt{3})}{1^2 - (\sqrt{3})^2}$ Simplify denominator.

$\qquad\quad = \dfrac{4(1-\sqrt{3})}{1-3}$ Simplify.

$\qquad\quad = \dfrac{4(1-\sqrt{3})}{-2}$

$\qquad\quad = -2(1-\sqrt{3})$

47. $\dfrac{x}{x-\sqrt{3}} = \dfrac{x(x+\sqrt{3})}{(x-\sqrt{3})(x+\sqrt{3})}$ Multiply numerator and denominator by conjugate.

$\qquad\quad = \dfrac{x(x+\sqrt{3})}{x^2 - (\sqrt{3})^2}$ Simplify.

$\qquad\quad = \dfrac{x(x+\sqrt{3})}{x^2 - 3}$

49. $\dfrac{\sqrt{6}-3}{2-\sqrt{6}} = \dfrac{(\sqrt{6}-3)(2+\sqrt{6})}{(2-\sqrt{6})(2+\sqrt{6})}$ Multiply numerator and denominator by conjugate.

$\qquad\quad = \dfrac{2\sqrt{6} + (\sqrt{6})^2 - 6 - 3\sqrt{6}}{2^2 - (\sqrt{6})^2}$ Simplify.

$\qquad\quad = \dfrac{2\sqrt{6} + 6 - 6 - 3\sqrt{6}}{4-6}$

$\qquad\quad = \dfrac{\sqrt{6}}{2}$

51. a. Graph $y = \sqrt{x^2}$ using the standard window:

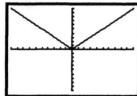

This is identical to the graph of $y = |x|$ since $\sqrt{x^2} = |x|$.

b. Graph $y = \sqrt[3]{x^3}$ using the standard window:

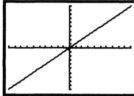

This is identical to the graph of $y = x$ since $\sqrt[3]{x^3} = x$.

53. a. $\sqrt{4x^2} = 2|x|$

b. $\sqrt{(x-5)^2} = |x-5|$

c. $\sqrt{x^2 - 6x + 9} = \sqrt{(x-3)^2}$
$$= |x-3|$$

Homework 6.6

1. $\sqrt{x} - 5 = 3$

   $\sqrt{x} = 8$ Isolate radical.

   $\left(\sqrt{x}\right)^2 = 8^2$ Square both sides.

   $x = 64$

   Check:

   $\sqrt{64} - 5 = 3$?

   $3 = 3$ Yes.

   The solution to the original equation is 64.

3. $\sqrt{y+6} = 2$

   $\left(\sqrt{y+6}\right)^2 = 2^2$ Square both sides.

   $y + 6 = 4$

   $y = -2$ Subtract 6 from both sides.

   Check:

   $\sqrt{-2+6} = 2$?

   $2 = 2$ Yes.

   The solution to the original equation is –2.

5. $4\sqrt{z} - 8 = -2$

   $4\sqrt{z} = 6$ Add 8 to both sides.

   $\sqrt{z} = \dfrac{3}{2}$ Divide both sides by 4.

   $\left(\sqrt{z}\right)^2 = \left(\dfrac{3}{2}\right)^2$ Square both sides.

   $z = \dfrac{9}{4}$

   Check:

   $4\sqrt{\left(\dfrac{9}{4}\right)} - 8 = -2$?

   $4\left(\dfrac{3}{2}\right) - 8 = -2$?

   $-2 = -2$ Yes.

   The solution to the original equation is $\dfrac{9}{4}$.

7. $5 + 2\sqrt{6 - 2w} = 13$

  $2\sqrt{6 - 2w} = 8$ Subtract 5 from both sides.

   $\sqrt{6 - 2w} = 4$ Divide both sides by 2.

  $\left(\sqrt{6 - 2w}\right)^2 = 4^2$ Square both sides.

    $6 - 2w = 16$

     $-2w = 10$ Subtract 6 from both sides.

      $w = -5$ Divide both sides by ± 2.

Check:

 $5 + 2\sqrt{6 - 2(-5)} = 13$?

   $5 + 2\sqrt{16} = 13$

     $13 = 13$ Yes.

The solution to the original equation is –5.

9. Let $T = 10.54$ seconds and solve for L:

$$10.54 = 2\pi\sqrt{\frac{L}{32}}$$

$$\frac{10.54}{2\pi} = \sqrt{\frac{L}{32}} \qquad \text{Divide both sides by } 2\pi.$$

$$\left(\frac{10.54}{2\pi}\right)^2 = \left(\sqrt{\frac{L}{32}}\right)^2 \qquad \text{Square both sides.}$$

$$2.814 \approx \frac{L}{32}$$

$$90 \approx L \qquad \text{Multiply both sides by 32.}$$

This answer checks in the original equation, so the pendulum is about 90 feet.

11. Solve $T = 2\pi\sqrt{\dfrac{L}{g}}$ for L.

$$\frac{T}{2\pi} = \sqrt{\frac{L}{g}} \qquad \text{Divide both sides by } 2\pi.$$

$$\left(\frac{T}{2\pi}\right)^2 = \left(\sqrt{\frac{L}{g}}\right)^2 \qquad \text{Square both sides.}$$

$$\frac{T^2}{4\pi^2} = \frac{L}{g}$$

$$\frac{gT^2}{4\pi^2} = L \qquad \text{Multiply both sides by } g.$$

13. Solve $r = \sqrt{t^2 - s^2}$ for s.

$$r^2 = \left(\sqrt{t^2 - s^2}\right)^2 \qquad \text{Square both sides.}$$
$$r^2 = t^2 - s^2$$
$$r^2 - t^2 = -s^2 \qquad \text{Solve for } s^2.$$
$$t^2 - r^2 = s^2$$
$$\pm\sqrt{t^2 - r^2} = s \qquad \text{Take square root of both sides.}$$

15.
$$3z + 4 = \sqrt{3z + 10}$$
$$(3z + 4)^2 = \left(\sqrt{3z + 10}\right)^2 \qquad \text{Square both sides.}$$
$$9z^2 + 24z + 16 = 3z + 10$$
$$9z^2 + 21z + 6 = 0 \qquad \text{Get zero on one side.}$$
$$3\left(3z^2 + 7z + 2\right) = 0 \qquad \text{Factor.}$$
$$3(3z + 1)(z + 2) = 0$$

Setting each factor equal to 0 gives $z = -\dfrac{1}{3}$ or $z = -2$.

Checks:

$$3\left(-\frac{1}{3}\right) + 4 = \sqrt{3\left(-\frac{1}{3}\right) + 10} \ ? \qquad 3(-2) + 4 = \sqrt{3(-2) + 10} \ ?$$
$$-1 + 4 = \sqrt{-1 + 10} \ ? \qquad\qquad -6 + 4 = \sqrt{-6 + 10} \ ?$$
$$3 = 3 \ \text{ Yes.} \qquad\qquad\qquad -2 = 2 \ \text{ No.}$$

Thus, the solution to the original equation is $z = -\dfrac{1}{3}$.

17.
$$2x + 1 = \sqrt{10x + 5}$$
$$(2x + 1)^2 = \left(\sqrt{10x + 5}\right)^2 \qquad \text{Square both sides.}$$
$$4x^2 + 4x + 1 = 10x + 5$$
$$4x^2 - 6x - 4 = 0 \qquad \text{Get zero on one side.}$$
$$2\left(2x^2 - 3x - 2\right) = 0 \qquad \text{Factor.}$$
$$2(2x + 1)(x - 2) = 0$$

Setting each factor equal to 0 gives $x = -\dfrac{1}{2}$ or $x = 2$.

Checks:

$$2\left(-\frac{1}{2}\right) + 1 = \sqrt{10\left(-\frac{1}{2}\right) + 5} \ ? \qquad 2(2) + 1 = \sqrt{10(2) + 5} \ ?$$
$$-1 + 1 = \sqrt{-5 + 5} \ ? \qquad\qquad 4 + 1 = \sqrt{20 + 5} \ ?$$
$$0 = 0 \ \text{Yes.} \qquad\qquad\qquad 5 = 5 \ \text{Yes.}$$

Thus, the solutions to the original equation are $x = -\dfrac{1}{2}$ and $x = 2$.

19.
$$\sqrt{y + 4} = y - 8$$
$$\left(\sqrt{y + 4}\right)^2 = (y - 8)^2 \qquad \text{Square both sides.}$$
$$y + 4 = y^2 - 16y + 64$$
$$0 = y^2 - 17y + 60 \qquad \text{Get zero on one side.}$$
$$0 = (y - 12)(y - 5) \qquad \text{Factor.}$$

Setting each factor equal to 0 gives $y = 12$ or $y = 5$.

Checks:

$$\sqrt{12 + 4} = 12 - 8 \ ? \qquad \sqrt{5 + 4} = 5 - 8 \ ?$$
$$4 = 4 \ \text{Yes.} \qquad\qquad 3 = -3 \ \text{No.}$$

Thus, the solution to the original equation is $y = 12$.

21.
$$\sqrt{2y - 1} = \sqrt{3y - 6}$$
$$\left(\sqrt{2y - 1}\right)^2 = \left(\sqrt{3y - 6}\right)^2 \qquad \text{Square both sides.}$$
$$2y - 1 = 3y - 6 \qquad\qquad \text{Solve the linear equation.}$$
$$5 = y$$

Check:

$$\sqrt{2(5) - 1} = \sqrt{3(5) - 6} \ ?$$
$$3 = 3 \ \text{Yes.}$$

Thus, the solution to the original equation is $y = 5$.

23.

$$\sqrt{x-3}\sqrt{x} = 2$$

$$\left(\sqrt{x-3}\sqrt{x}\right)^2 = 2^2 \qquad \text{Square both sides.}$$

$$(x-3)\cdot x = 4$$

$$x^2 - 3x = 4$$

$$x^2 - 3x - 4 = 0 \qquad \text{Get zero on one side.}$$

$$(x-4)(x+1) = 0 \qquad \text{Factor.}$$

Setting each factor equal to 0 gives $x = 4$ or $x = -1$.

Checks:

$$\sqrt{4-3}\sqrt{4} = 2 \text{ ?} \qquad\qquad \sqrt{-1-3}\sqrt{-1} = 2 \text{ ?}$$

$$2 = 2 \text{ Yes.} \qquad \text{(nonreal number)} = 2 \text{ No.}$$

Thus, the solution to the original equation is $x = 4$.

25.

$$\sqrt{y+4} = \sqrt{y+20} - 2$$

$$\left(\sqrt{y+4}\right)^2 = \left(\sqrt{y+20} - 2\right)^2 \qquad \text{Square both sides.}$$

$$y+4 = y+20 - 4\sqrt{y+20} + 4 \qquad \text{Isolate radical.}$$

$$4\sqrt{y+20} = 20$$

$$\sqrt{y+20} = 5$$

$$\left(\sqrt{y+20}\right)^2 = 5^2 \qquad \text{Square both sides.}$$

$$y+20 = 25 \qquad \text{Solve the linear equation.}$$

$$y = 5$$

Check:

$$\sqrt{5+4} = \sqrt{5+20} - 2 \text{ ?}$$

$$3 = 3 \text{ Yes.}$$

Thus, the solution to the original equation is $y = 5$.

27.
$$\sqrt{x} + \sqrt{2} = \sqrt{x+2}$$
$$\left(\sqrt{x} + \sqrt{2}\right)^2 = \left(\sqrt{x+2}\right)^2 \qquad \text{Square both sides.}$$
$$x + 2\sqrt{2x} + 2 = x + 2 \qquad \text{Isolate radical.}$$
$$2\sqrt{2x} = 0$$
$$\sqrt{2x} = 0$$
$$\left(\sqrt{2x}\right)^2 = 0^2 \qquad \text{Square both sides.}$$
$$2x = 0$$
$$x = 0$$

Check:
$$\sqrt{0} + \sqrt{2} = \sqrt{0+2} \ ?$$
$$\sqrt{2} = \sqrt{2} \ \text{Yes.}$$

Thus, the solution to the original equation is $x = 0$.

29.
$$\sqrt{5+x} + \sqrt{x} = 5$$
$$\sqrt{5+x} = 5 - \sqrt{x} \qquad \text{Isolate one of the radicals.}$$
$$\left(\sqrt{5+x}\right)^2 = \left(5 - \sqrt{x}\right)^2 \qquad \text{Square both sides.}$$
$$5 + x = 25 - 10\sqrt{x} + x$$
$$10\sqrt{x} = 20 \qquad \text{Isolate radical.}$$
$$\sqrt{x} = 2$$
$$\left(\sqrt{x}\right)^2 = 2^2 \qquad \text{Square both sides.}$$
$$x = 4$$

Check:
$$\sqrt{5+4} + \sqrt{4} = 5 \ ?$$
$$3 + 2 = 5 \ ?$$
$$5 = 5 \ \text{Yes.}$$

Thus, the solution to the original equation is $x = 4$.

31. The distance between $A(2, -1)$ and (x, y) is $d_1 = \sqrt{(x-2)^2 + (y+1)^2}$.

The distance between $B(-2, 3)$ and (x, y) is $d_2 = \sqrt{(x+2)^2 + (y-3)^2}$.

The distance between $C(2, 5)$ and (x, y) is $d_3 = \sqrt{(x-2)^2 + (y-5)^2}$.

Since (x, y) is the orthocenter of the triangle, these three distances are equal. To find the values of x and y, write a few equations by setting the distances equation to one another. Using the first two distances, we have:

$$\sqrt{(x-2)^2 + (y+1)^2} = \sqrt{(x+2)^2 + (y-3)^2} \qquad \text{Set } d_1 \text{ equal to } d_2.$$

$$\left(\sqrt{(x-2)^2 + (y+1)^2}\right)^2 = \left(\sqrt{(x+2)^2 + (y-3)^2}\right)^2 \qquad \text{Square both sides.}$$

$$(x-2)^2 + (y+1)^2 = (x+2)^2 + (y-3)^2 \qquad \text{Multiply binomials.}$$

$$x^2 - 4x + 4 + y^2 + 2y + 1 = x^2 + 4x + 4 + y^2 - 6y + 9$$

$$-8x + 8y = 8 \qquad \text{Combine like terms.}$$

$$-x + y = 1 \qquad (1) \qquad \text{Divide both sides by 8.}$$

This gives us one equation involving x and y. Find another equation in the same way:

$$\sqrt{(x-2)^2 + (y+1)^2} = \sqrt{(x-2)^2 + (y-5)^2} \qquad \text{Set } d_1 \text{ equal to } d_3.$$

$$\left(\sqrt{(x-2)^2 + (y+1)^2}\right)^2 = \left(\sqrt{(x-2)^2 + (y-5)^2}\right)^2 \qquad \text{Square both sides.}$$

$$(x-2)^2 + (y+1)^2 = (x-2)^2 + (y-5)^2 \qquad \text{Multiply binomials.}$$

$$x^2 - 4x + 4 + y^2 + 2y + 1 = x^2 - 4x + 4 + y^2 - 10y + 25$$

$$12y = 24 \qquad \text{Combine like terms.}$$

$$y = 2 \qquad (2) \qquad \text{Solve.}$$

Substitute $y = 2$ in equation (1) so that $-x + 2 = 1$ and, therefore, $x = 1$. The orthocenter of the triangle is the point $(1, 2)$.

33. a. Francine is traveling at 1 mile per minute on the highway. Therefore, *t* minutes after turning onto the highway she has traveled *t* miles.

b. Let *d(t)* be Francine's distance from the college *t* minutes after turning onto the highway. By the Pythagorean theorem, $d(t) = \sqrt{8^2 + t^2} = \sqrt{64 + t^2}$.

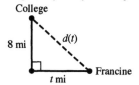

c.

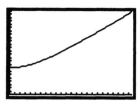

d. The range of KGVC is 15 miles so we solve the equation $\sqrt{64 + t^2} = 15$ for *t*.

$$\left(\sqrt{64 + t^2}\right)^2 = 15^2 \quad \text{Square both sides}$$

$64 + t^2 = 225$

$t^2 = 161$ Subtract 64 from both sides

$t = \sqrt{161} = 12.7$ Take the square root of both sides

This value checks in the original equation, so Francine will be out of range in 12.7 minutes.

35. a. The UFO descends at a rate of 10 feet per second from an altitude of 700 feet. Therefore, after t seconds, the altitude of the UFO is $700 - 10t$ feet.

b. You are running at 15 feet per second. Therefore, after t seconds, you have run $15t$ feet.

c.

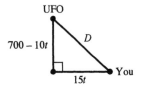

d. By the Pythagorean theorem, $D(t) = \sqrt{(700 - 10t)^2 + (15t)^2}$.

e. $D(10) = \sqrt{(700 - 10(10))^2 + (15(10))^2} = \sqrt{360,000 + 22,500} = 618.5$ feet.

f.

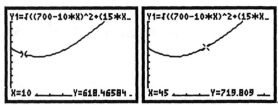

Use *trace* or *value* ("EVAL" on some calculators) to evaluate D when $t = 10$ and when $t = 45$. This gives $D(10) = 618.5$, which confirms our answer to part (e), and $D(45) = 719.8$, which means that the distance between you and the UFO when $t = 45$ seconds is 719.8 feet.

g. Using *trace* to locate the lowest point on the graph of $D(t)$, we find that the distance is smallest when t is approximately 22 seconds.

h. To find when the UFO reaches the ground we solve for t in the equation
$$700 - 10t = 0$$
$$700 = 10t$$
$$70 = t$$
The UFO hits the ground after t seconds, so $0 \le t \le 70$, and a more realistic Xmax would be 70.

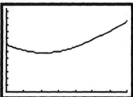

Chapter 6 Review

1. a. $(-3)^{-4} = \dfrac{1}{(-3)^4} = \dfrac{1}{81}$

   b. $4^{-3} = \dfrac{1}{4^3} = \dfrac{1}{64}$

3. a. $(3m)^{-5} = \dfrac{1}{(3m)^5} = \dfrac{1}{3^5 m^5} = \dfrac{1}{243 m^5}$

   b. $-7y^{-8} = -\dfrac{7}{y^8}$

5. a. $6c^{-7} \cdot 3^{-1} c^4 = \dfrac{6c^4}{3c^7} = \dfrac{2}{c^{7-4}} = \dfrac{2}{c^3}$

   b. $\dfrac{11z^{-7}}{3^{-2} z^{-5}} = \dfrac{11\left(3^2\right) z^5}{z^7} = \dfrac{99}{z^{7-5}} = \dfrac{99}{z^2}$

7. a. The speed of light is 186,000 mi/s and 1 mi = 5280 ft, therefore, the speed of light in ft/s is $186000 \cdot 5280 = 9.8208 \times 10^8$ ft/s. The time it takes light to travel

   1 ft is $\dfrac{1}{9.8208 \times 10^8} = 1.018 \times 10^{-9} = 0.000000001018$ seconds.

   b. 92,956,000 miles $= 92,956,000 \cdot 5280 = 4.90808 \times 10^{11}$ feet. The time it takes light to travel to the earth is $\dfrac{4.90808 \times 10^{11}}{9.8208 \times 10^8} \approx 500$ seconds, which is equivalent to $8\dfrac{1}{3}$ minutes, or 8 minutes and 20 seconds.

9. a. From problem 59 in Section 6.1, the average distance from the earth to the sun is 1.5×10^{11} meters. Using the relationship $time = \dfrac{distance}{rate}$, the time it would take

to reach the sun is $\dfrac{1.5 \times 10^{11}}{3 \times 10^7} = 5000$ seconds. Dividing 5000 by 60, this time is

equivalent to $83\dfrac{1}{3}$ minutes.

b. Also from problem 59 in Section 6.1, the distance from the sun to Proxima Centauri is 3.99×10^{16} meters. The time it would take to travel from the sun to

this star is $\dfrac{3.99 \times 10^{16}}{3 \times 10^7} = 1.33 \times 10^9$ seconds. There are $(365)(24)(3600) =$

31,536,000 seconds in a year, so 1.33×10^9 seconds $= \dfrac{1.33 \times 10^9}{31,536,000} \approx 42$ years.

(Note that if a person needed to first travel by the sun en route to Proxima Centauri, the additional time would not change this last figure by a significant amount.)

c. $\dfrac{3 \times 10^7}{3 \times 10^8} = \dfrac{1}{10}$. Thus, spacecraft may be able to travel at one-tenth the speed of light in the 21rst century.

11. a. $25m^{1/2} = 25\sqrt{m}$

b. $8n^{-1/3} = \dfrac{8}{n^{1/3}} = \dfrac{8}{\sqrt[3]{n}}$

13. a. $(3q)^{-3/4} = \dfrac{1}{(3q)^{3/4}} = \dfrac{1}{\left((3q)^3\right)^{1/4}} = \dfrac{1}{\sqrt[4]{(3q)^3}}$

b. $7(uv)^{3/2} = 7\left((uv)^3\right)^{1/2} = 7\sqrt{(uv)^3} = 7\sqrt{u^3 v^3}$

15. a. $2\sqrt[3]{x^2} = 2x^{2/3}$

b. $\dfrac{1}{4}\sqrt[4]{x} = \dfrac{1}{4}x^{1/4}$

17. a. $\dfrac{6}{\sqrt[4]{b^3}} = \dfrac{6}{b^{3/4}} = 6b^{-3/4}$

b. $\dfrac{-1}{3\sqrt[3]{b}} = \dfrac{-1}{3b^{1/3}} = -\dfrac{1}{3}b^{-1/3}$

19. Solve for m in the formula $m = \dfrac{M}{\sqrt{1 - \dfrac{v^2}{c^2}}}$, where $M = 80$ kg and $v = 0.7c$.

$$m = \frac{80}{\sqrt{1 - \dfrac{(0.7c)^2}{c^2}}} = \frac{80}{\sqrt{1 - \dfrac{0.49c^2}{c^2}}} = \frac{80}{\sqrt{1 - 0.49}} = 112$$

Thus, the original mass of the man is 112 kg.

21. a. Compute q with $m = 100$ and $w = 1600$.

$$q = 0.6(100)^{1/4}(1600)^{3/4} = 0.6(100)^{1/4}(100)^{3/4}(16)^{3/4}$$

$$= 0.6(100)^{1/4+3/4}8 = 480$$

Thus, they can produce 480 saddlebags.

b. Solve for w with $m = 100$ and $q = 200$.

$$200 = 0.6(100)^{1/4}w^{3/4}$$

$$\frac{200}{0.6(100)^{1/4}} = w^{3/4}$$

$$\left(\frac{200}{0.6(100)^{1/4}}\right)^{4/3} = \left(w^{3/4}\right)^{4/3} = w$$

$$498 = w$$

Thus, they need 498 hours of labor.

23. a.

t	0	2	4	6	8	10	12	14	16	18	20
$M(t)$	0	50.5	84.9	115.0	142.7	168.7	193.4	217.1	240.0	262.2	283.7

Refer to the graph in the back of the textbook.

b. To find the membership in 1990, we compute $M(20)$, since 1990 is 20 years after 1970: $M(20) = 30(20)^{3/4} = 284$. The membership was 284 in 1990.

c. $810 = 30t^{3/4}$

$27 = t^{3/4}$

$(27)^{4/3} = \left(t^{3/4}\right)^{4/3}$

$81 = t$

The year will be $1970 + 81 = 2051$.

25. a. If an old vat cost $5000, the function is $N = 5000r^{0.6}$. Refer to the graph in the back of the textbook.

b. The accountant should budget $5000(1.8)^{0.6} = \$7{,}114.32$.

27. a. When $x = 1$, $C = a$, so a is the cost of producing the first ship.

b. $C = 12x^{-1/8} = \dfrac{12}{x^{1/8}} = \dfrac{12}{\sqrt[8]{x}}$

c. $C = \dfrac{12}{\sqrt[8]{2}} \approx 11$. When two ships are built, the cost per ship is about 11 million. Since the cost of building the first ship is 12 million, the cost per ship has gone down by $\dfrac{12-11}{12} = 0.083$, or about 8.3%. When four ships are built, the cost per ship is $C = \dfrac{12}{\sqrt[8]{4}} \approx 10.09$ million. From building two ships to building four ships, the cost per ship has gone down by $\dfrac{11-10.09}{11} = 0.083$, or about 8.3%.

d. The average cost decreases by about 8.3% from building n ships to building $2n$ ships.

29. a. $Q(16) = 4(16)^{5/2} = 4\left(\sqrt{16}\right)^5 = 4 \cdot 4^5 = 4 \cdot 1024 = 4096$

b. $Q\left(\dfrac{1}{4}\right) = 4\left(\dfrac{1}{4}\right)^{5/2} = 4\left(\sqrt{\dfrac{1}{4}}\right)^5 = 4 \cdot \left(\dfrac{1}{2}\right)^5 = 4 \cdot \dfrac{1}{32} = \dfrac{1}{8}$

c. $Q(3) = 4(3)^{5/2} = 4\sqrt{3^5} = 4\sqrt{3^4 \cdot 3} = 4 \cdot 3^2\sqrt{3} = 36\sqrt{3} \approx 62.35$

d. $Q(100) = 4(100)^{5/2} = 4\left(\sqrt{100}\right)^5 = 4 \cdot 10^5 = 4 \cdot 100{,}000 = 400{,}000$

31. $2\sqrt{w} - 5 = 21$

$\quad 2\sqrt{w} = 26 \qquad$ Isolate radical.

$\quad \sqrt{w} = 13$

$\quad \left(\sqrt{w}\right)^2 = 13^2 \qquad$ Square both sides.

$\quad w = 169$

Check:

$2\sqrt{169} - 5 = 21$?

$21 = 21$; Yes.

The solution to the original equation is 169.

33. $12 - \sqrt{5v+1} = 3$

$-\sqrt{5v+1} = -9$ Isolate radical.

$\sqrt{5v+1} = 9$

$\left(\sqrt{5v+1}\right)^2 = 9^2$ Square both sides.

$5v + 1 = 81$ Solve the linear equation.

$5v = 80$

$v = 16$

Check:

$12 - \sqrt{5(16)+1} = 3$?

$12 - \sqrt{81} = 3$?

$3 = 3$; Yes.

The solution to the original equation is 16.

35. $x - 3\sqrt{x} + 2 = 0$

$x + 2 = 3\sqrt{x}$ Add $3\sqrt{x}$ to both sides.

$(x+2)^2 = \left(3\sqrt{x}\right)^2$ Square both sides.

$x^2 + 4x + 4 = 9x$

$x^2 - 5x + 4 = 0$ Subtract $9x$ from both sides.

$(x-1)(x-4) = 0$ Factor.

$x = 1$ or $x = 4$ Set each factor equal to 0.

Check:

$1 - 3\sqrt{1} + 2 = 3 - 3 = 0$ Yes; 1 is a solution.

$4 - 3\sqrt{4} + 2 = 6 - 3 \cdot 2 = 0$ Yes; 4 is a solution.

Thus, the solutions of the original equation are 1 and 4.

37. $(x+7)^{1/2} + x^{1/2} = 7$

$(x+7)^{1/2} = 7 - x^{1/2}$ Subtract $x^{1/2}$ from both sides.

$\left((x+7)^{1/2}\right)^2 = \left(7 - x^{1/2}\right)^2$ Square both sides.

$x + 7 = 49 - 14x^{1/2} + x$

$14x^{1/2} = 42$ Rearrange and combine like terms.

$x^{1/2} = 3$ Divide both sides by 14.

$\left(x^{1/2}\right)^2 = 3^2$ Square both sides.

$x = 9$

Check:

$(9+7)^{1/2} + 9^{1/2} = 4 + 3 = 7$ Yes; 9 is a solution.

Thus, the solution to the original equation is 9.

39. $\sqrt[3]{x+1} = 2$

$\left(\sqrt[3]{x+1}\right)^3 = 2^3$ Cube both sides.

$x + 1 = 8$

$x = 7$ Subtract 1 from both sides.

Check:

$\sqrt[3]{7+1} = \sqrt[3]{8} = 2$ Yes; 7 is a solution.

Thus, the solution to the original equation is 7.

41. $(x-1)^{-3/2} = \dfrac{1}{8}$

$\dfrac{1}{(x-1)^{3/2}} = \dfrac{1}{8}$ Rewrite without negative exponents.

$8 = (x-1)^{3/2}$ Fundamental property of proportions.

$\sqrt[3]{8} = \sqrt[3]{(x-1)^{3/2}}$ Take the cube root of both sides.

$2 = (x-1)^{1/2}$

$2^2 = \left((x-1)^{1/2}\right)^2$ Square both sides.

$4 = x - 1$

$5 = x$ Add 1 to both sides.

Check:

$(5-1)^{-3/2} = 4^{-3/2} = \dfrac{1}{4^{3/2}} = \dfrac{1}{8}$ Yes; 5 is a solution.

Thus, the solution to the original equation is 5.

43. $t = \sqrt{\dfrac{2v}{g}}$

$t^2 = \dfrac{2v}{g}$ Square both sides.

$gt^2 = 2v$ Multiply both sides by g.

$g = \dfrac{2v}{t^2}$ Divide both sides by t^2.

45. $R = \dfrac{1 + \sqrt{p^2 + 1}}{2}$

$2R - 1 = \sqrt{p^2 + 1}$ Isolate $\sqrt{p^2 + 1}$.

$4R^2 - 4R + 1 = p^2 + 1$ Square both sides.

$4R^2 - 4R = p^2$ Subtract 1 from both sides.

$\pm 2\sqrt{R^2 - R} = p$ Take the square root of both sides.

47. The car traveling east will be $50t$ miles from the intersection t hours after passing the intersection. The car traveling north will be $5 + 40t$ miles from the intersection t hours after the eastbound car passes the intersection. By the Pythagorean theorem, the distance between the two cars t hours after the eastbound car passes the intersection is $\sqrt{(50t)^2 + (5 + 40t)^2}$ miles. The cars will be 200 miles apart when

$200 = \sqrt{(50t)^2 + (5 + 40t)^2}$

$40000 = 2500t^2 + 25 + 400t + 1600t^2$ Square both sides.

$0 = 4100t^2 + 400t - 39975$ Combine like terms.

$t = \dfrac{-400 \pm \sqrt{400^2 - 4 \cdot 4100 \cdot (-39975)}}{2(4100)}$ Apply quadratic formula.

$t = -3.17$ or 3.07

A negative answer doesn't make sense, therefore, the cars are 200 miles apart after 3.07 hours.

49. Compute the length of each side.

length of $\overline{AB} = \sqrt{(5-(-1))^2 + (4-2)^2} = \sqrt{36+4} = \sqrt{40} = 2\sqrt{10}$

length of $\overline{BC} = \sqrt{(1-5)^2 + (-4-4)^2} = \sqrt{16+64} = \sqrt{80} = 4\sqrt{5}$

length of $\overline{AC} = \sqrt{(1-(-1))^2 + (-4-2)^2} = \sqrt{4+36} = \sqrt{40} = 2\sqrt{10}$

Thus, the perimeter equals $2\sqrt{10} + 2\sqrt{10} + 4\sqrt{5} = 4\sqrt{10} + 4\sqrt{5} \approx 21.59$. Since
$\overline{AB}^2 + \overline{AC}^2 = \left(2\sqrt{10}\right)^2 + \left(2\sqrt{10}\right)^2 = 40+40 = 80 = \overline{BC}^2$, $\triangle ABC$ is a right triangle.

51. The circle has center $(0, 0)$ and radius 9. Refer to the graph in the back of the textbook.

53. The circle has center $(2, -4)$ and radius 3. Refer to the graph in the back of the textbook.

55. The equation is $(x-5)^2 + (y-(-2))^2 = \left(4\sqrt{2}\right)^2$ which simplifies to
$(x-5)^2 + (y+2)^2 = 32$.

57. The center is at the midpoint of a diameter.

center $= \left(\dfrac{-2+4}{2}, \dfrac{3+5}{2}\right) = (1,4)$

The radius is half the length of the diameter.

radius $= \dfrac{1}{2}\sqrt{(4-(-2))^2 + (5-3)^2} = \dfrac{1}{2}\sqrt{36+4} = \dfrac{\sqrt{40}}{2} = \dfrac{2\sqrt{10}}{2} = \sqrt{10}$

The equation is $(x-1)^2 + (y-4)^2 = \left(\sqrt{10}\right)^2$ which simplifies to
$(x-1)^2 + (y-4)^2 = 10$.

59. a. $\sqrt{\dfrac{125p^9}{a^4}} = \dfrac{\sqrt{25p^8}\sqrt{5p}}{\sqrt{a^4}} = \dfrac{5p^4\sqrt{5p}}{a^2}$

  b. $\sqrt[3]{\dfrac{24v^2}{w^6}} = \dfrac{\sqrt[3]{8}\sqrt[3]{3v^2}}{\sqrt[3]{w^6}} = \dfrac{2\sqrt[3]{3v^2}}{w^2}$

61. a. $\sqrt[3]{8a^3 - 16b^6} = \sqrt[3]{8}\sqrt[3]{a^3 - 2b^6} = 2\sqrt[3]{a^3 - 2b^6}$

  b. $\sqrt[3]{8a^3}\sqrt[3]{-16b^6} = 2a\sqrt[3]{-16b^6} = 2a\sqrt[3]{-8b^6}\sqrt[3]{2} = 2a\cdot\left(-2b^2\right)\cdot\sqrt[3]{2} = -4ab^2\sqrt[3]{2}$

63. a. $\left(x-2\sqrt{x}\right)^2=\left(x-2\sqrt{x}\right)\left(x-2\sqrt{x}\right)=x^2-2x\sqrt{x}-2x\sqrt{x}+4\sqrt{x}\sqrt{x}$
$$=x^2-4x\sqrt{x}+4x$$

b. $\left(x-2\sqrt{x}\right)\left(x+2\sqrt{x}\right)=x^2-2x\sqrt{x}+2x\sqrt{x}-4\sqrt{x}\sqrt{x}=x^2-4x$

65. a $\dfrac{7}{\sqrt{5y}}=\dfrac{7\sqrt{5y}}{\sqrt{5y}\sqrt{5y}}=\dfrac{7\sqrt{5y}}{5y}$

b. $\dfrac{6d}{\sqrt{2d}}=\dfrac{6d\sqrt{2d}}{\sqrt{2d}\sqrt{2d}}=\dfrac{6d\sqrt{2d}}{2d}=3\sqrt{2d}$

67. a. $\dfrac{-3}{\sqrt{a}+2}=\dfrac{-3\left(\sqrt{a}-2\right)}{\left(\sqrt{a}+2\right)\left(\sqrt{a}-2\right)}=\dfrac{-3\sqrt{a}+6}{a-4}$

b. $\dfrac{-3}{\sqrt{z}-4}=\dfrac{-3\left(\sqrt{z}+4\right)}{\left(\sqrt{z}-4\right)\left(\sqrt{z}+4\right)}=\dfrac{-3\sqrt{z}-12}{z-16}$

Chapter Seven: Exponential and Logarithmic Functions

Homework 7.1

1. a.

Weeks	0	1	2	3	4
Bacteria	300	600	1200	2400	4800

    b. $P(t) = 300(2)^t$

    c. Refer to the graph in the back of the textbook.

    d. After 8 weeks, there will be $P(8) = 300(2)^8 = 76{,}800$ bacteria. After 5 days $= \frac{5}{7}$ weeks, there will be

$$P\left(\frac{5}{7}\right) = 300(2)^{5/7} = 492 \text{ bacteria.}$$

3. a.

Weeks	0	6	12	18	24
Bees	20,000	50,000	125,000	312,500	781,250

    b. $P(t) = 20{,}000(2.5)^{t/6}$

    c. Refer to the graph in the back of the textbook.

    d. After 4 weeks, there are $P(4) = 20{,}000(2.5)^{4/6} = 36{,}840$ insects. After 20 weeks, there are $P(20) = 20{,}000(2.5)^{20/6} = 424{,}128$ insects.

5. a.

Years	0	1	2	3	4
Account balance	4000	4320	4665.60	5038.85	5441.96

    b. $P(t) = 4000(1.08)^t$

    c. Refer to the graph in the back of the textbook.

    d. After 2 years, there is $P(2) = 4000(1.08)^2 = \$4665.60$. After 10 years, there is $P(10) = 4000(1.08)^{10} = \$8635.70$.

7. a.

Years since 1963	Value of house
0	20,000
5	25,525.63
10	32,577.89
15	41,578.56
20	53,065.95

    b. $P(t) = 20{,}000(1.05)^t$

    c. Refer to the graph in the back of the textbook.

    d. In 1975, $t = 12$, and the value is $P(12) = \$35{,}917.13$. In 1990, $t = 27$, and the value is $P(27) = \$74{,}669.13$.

9. a.

Weeks	0	2	4	6	8
Mosquitoes	250,000	187,500	140,625	105,469	79,102

b. $P(t) = 250,000(0.75)^{t/2}$

c. Refer to the graph in the back of the textbook.

d. After 3 weeks of spraying, there were $P(3) = 162,380$ mosquitoes. After 8 weeks of spraying, there were $P(8) = 79,102$ mosquitoes.

11. a.

Feet	0	4	8	12	16
% of light	100	85	72.25	61.41	52.20

b. $L(d) = (0.85)^{d/4}$

c. Refer to the graph in the back of the textbook.

d. The amount of sunlight that can penetrate to a depth of 20 feet is $L(20) = 44\%$. The amount of sunlight that can penetrate to a depth of 45 feet is $L(45) = 16\%$.

13. a.

Years	0	10	20	30	40
Pounds of plutonium-238	50	46.1	42.6	39.3	36.3

b. $P(t) = 50(0.992)^t$

c. Refer to the graph in the back of the textbook.

d. After 10 years there will be $P(10) = 46.1$ pounds of plutonium-238 left.
After 100 years there will be $P(100) = 22.4$ pounds left.

15. a. To evaluate $P(t)$, we raise 3 to a power and then multiply the result by 2. To evaluate $Q(t)$, we raise 6 to the power.

b.

t	0	1	2	3
$P(t)$	2	6	18	54
$Q(t)$	1	6	36	216

17. a. Let n represent the number that would have originally hatched to produce 130 one-year-olds. Then $0.356n = 130$ and

$$n = \frac{130}{0.356} = 365 \text{ robins.}$$

b. Let N represent the number of the original 365 robins that are alive after t years. Then

$$N(t) = 365(0.356)^t.$$

c. Refer to the graph in the back of the textbook.

d. The model predicts that
$$N(9) = 365(0.356)^9 \approx 0 \text{ robins}$$
will survive for 9 years.

19. growth factor = 1.5

t	0	1	2	3	4
P	8	12	18	27	40.5

21. growth factor = 1.2

x	0	1	2	3	4
Q	20	24	28.8	34.56	41.472

23. decay factor = 0.8

w	0	1	2	3	4
N	120	96	76.8	61.44	49.152

25.
$$768 = 12a^3$$
$$64 = a^3$$
$$\sqrt[3]{64} = a$$
$$4 = a$$

27.
$$14{,}929.92 = 5000a^6$$
$$2.985984 = a^6$$
$$\pm\sqrt[6]{2.985984} = a$$
$$\pm 1.2 = a$$

29.
$$1253 = 260(1+r)^{12}$$
$$\frac{1253}{260} = (1+r)^{12}$$
$$\pm\left(\frac{1253}{260}\right)^{1/12} = 1+r$$
$$\pm\left(\frac{1253}{260}\right)^{1/12} - 1 = r$$
$$r \approx 0.14 \text{ or } r \approx -2.14$$

31.
$$56.27 = 78(1-r)^8$$
$$\frac{56.27}{78} = (1-r)^8$$
$$\pm\left(\frac{56.27}{78}\right)^{1/8} = 1-r$$
$$r = 1 \pm \left(\frac{56.27}{78}\right)^{1/8}$$
$$r \approx 1.96 \text{ or } r \approx 0.04$$

33. a. After 6 days, the population of species A is $30(1.3) = 39$. To find the daily growth factor a, use $1.3 = a^6$. So $a = 1.3^{1/6} = 1.0447$.

b. After 4 days, the population of species B is $30(1.2) = 36$. To find the daily growth factor a, use $1.2 = a^4$. Then $a = 1.2^{1/4} = 1.0466$.

c. Species B multiplies faster since it has the greater daily growth factor.

35. a. $P(t) = 9{,}579{,}700(1+r)^t$

b. $11{,}196{,}700 = 9{,}579{,}700(1+r)^{10}$, so
$$r = \left(\frac{11{,}196{,}700}{9{,}579{,}700}\right)^{1/10} - 1$$
$$= 0.0157 = 1.57\%$$

37. a. $20{,}000 = 10{,}000(1+r)^{20}$, so
$$r = \left(\frac{20{,}000}{10{,}000}\right)^{1/20} - 1 = 2^{1/20} - 1$$
$$= 0.0353 = 3.53\%$$

b. $700{,}000 = 350{,}000(1+r)^{20}$, so
$$r = \left(\frac{700{,}000}{350{,}000}\right)^{1/20} - 1 = 2^{1/20} - 1$$
$$= 0.0353 = 3.53\%$$

c. No

d. 3.53%.

39. a. $L(t) = mt + b$, $m = \dfrac{6-3}{2} = 1.5$,
and $b = 3$. So $L(t) = 1.5t + 3$.

t	0	2	4	6	8
$L(t)$	3	6	9	12	15

Refer to the graph in the back of the textbook.

b. $E(t) = E_0 a^t$ and $E_0 = 3$ so
$E(t) = 3a^t$. When $t = 2$, $E(t) = 6$,
so $6 = 3a^2$. This gives $a^2 = 2$
and $a = \sqrt{2} = 2^{1/2}$. Thus,
$E(t) = 3\left(2^{1/2}\right)^t = 3(2)^{t/2}$.

t	0	2	4	6	8
$E(t)$	3	6	12	24	48

Refer to the graph in the back of the textbook.

41. a. $m = \dfrac{720 - 400}{5} = \dfrac{320}{5} = 64$ geese
per year.

b. Use $P(t) = P_0 a^t$ with $P_0 = 400$
and $P(5) = 720$. This gives
$720 = 400a^5$ so the annual
growth factor is
$a = \left(\dfrac{720}{400}\right)^{1/5} \approx 1.125$. The annual
percent increase is 12.5%.

Homework 7.2

1. a. $f(0) = 26(1.4)^0 = 26$ so the y-intercept is $(0, 26)$. The graph is increasing.

b. $g(0) = 1.2(0.84)^0 = 1.2$ so the y-intercept is $(0, 1.2)$. The graph is decreasing.

c. $h(0) = 75\left(\dfrac{4}{5}\right)^0 = 75$ so the y-intercept is $(0, 75)$. The graph is decreasing.

d. $k(0) = \dfrac{2}{3}\left(\dfrac{9}{8}\right)^0 = \dfrac{2}{3}$ so the y-intercept is $\left(0, \dfrac{2}{3}\right)$. The graph is increasing.

3. a. $g(t) = 3t^{0.4}$: power

b. $h(t) = 4(0.3)^t$: exponential

c. $D(x) = 6x^{1/2}$: power

d. $E(x) = 4x + x^4$: neither

5. Make two tables similar to the following. Refer to the graph in the back of the textbook.

a. $f(x) = 3^x$

x	$f(x)$
-2	1/9
-1	1/3
0	1
1	3
2	9

b. $g(x) = \left(\dfrac{1}{3}\right)^x$

x	$g(x)$
-2	9
-1	3
0	1
1	1/3
2	1/9

7. Make two tables similar to the following. Refer to the graph in the back of the textbook.

a. $h(t) = 4^{-t}$

t	$h(t)$
-2	16
-1	4
0	1
1	1/4
2	1/16

b. $q(t) = -4^t$

t	$q(t)$
-2	$-1/16$
-1	$-1/4$
0	-1
1	-4
2	-16

9. (b), (c), and (d) have identical graphs because $\left(\frac{1}{6}\right)^x = 6^{-x} = \frac{1}{6^x}$.

11. **a.** Refer to the graph in the back of the textbook.

 b. To find the smallest and largest function values using your graph, use *value* (EVAL on some calculators) to locate the y-values that correspond to $x = -5$ and $x = 5$. Answer: $0.27 \le y \le 3.71$

13. **a.** Refer to the graph in the back of the textbook.

 b. To find the smallest and largest function values using your graph, use *value* (EVAL on some calculators) to locate the y-values that correspond to $x = -5$ and $x = 5$. Answer: $0.33 \le y \le 3.05$

15. $25^{4/3} = \left(5^2\right)^{4/3} = 5^{8/3}$ so the equation is $5^{x+2} = 5^{8/3}$. Set the exponents equal and solve:
$$x + 2 = \frac{8}{3} \text{ so } x = \frac{2}{3}.$$

17. $\frac{\sqrt{3}}{9} = 3^{1/2} \cdot 3^{-2} = 3^{-3/2}$ so the equation is $3^{2x-1} = 3^{-3/2}$. Set the exponents equal and solve:
$$2x - 1 = -\frac{3}{2} \text{ so } 2x = -\frac{1}{2} \text{ and }$$
$$x = -\frac{1}{4}.$$

19. $4 \cdot 2^{x-3} = 2^2 \cdot 2^{x-3} = 2^{x-1}$ and $8^{-2x} = \left(2^3\right)^{-2x} = 2^{-6x}$ so the equation is $2^{x-1} = 2^{-6x}$. Set the exponents equal and solve:
$$x - 1 = -6x \text{ so } 7x = 1 \text{ and } x = \frac{1}{7}.$$

21. $27^{4x+2} = \left(3^3\right)^{4x+2} = 3^{12x+6}$ and $81^{x-1} = \left(3^4\right)^{x-1} = 3^{4x-4}$ so the equation is $3^{12x+6} = 3^{4x-4}$. Set the exponents equal and solve:
$$12x + 6 = 4x - 4 \text{ so } 8x = -10 \text{ and}$$
$$x = -\frac{5}{4}.$$

23. $1000 = 10^3$ so the equation is $10^{x^2-1} = 10^3$. Set the exponents equal and solve: $x^2 - 1 = 3$ so $x^2 = 4$ and $x = \pm 2$.

25. **a.** $N(t) = 26(2)^{t/6}$

 b. Refer to the graph in the back of the textbook.

 c. $106{,}496 = 26(2)^{t/6}$
$4096 = 2^{t/6}$ Divide by 26.
$2^{12} = 2^{t/6}$ Since $2^{12} = 4096$.
$12 = \frac{t}{6}$ Set exponents equal.
$72 = t$
Hospitals should expect to be treating 106,496 cases after 72 days. To verify graphically, graph the line $y = 106{,}496$ in the same window as your graph from (b) and use the *intersect* feature on your calculator. The intersection occurs at the point (72, 106,496).

27. a. Note that if the television loses 30% of its value every 2 years, then it retains $70\% = 0.7$ of its value every two years. Thus,

$$V(t) = 700(0.7)^{t/2}.$$

b. Refer to the graph in the back of the textbook.

c. $343 = 700(0.7)^{t/2}$

$0.49 = 0.7^{t/2}$ Divide by 700.

$0.7^2 = 0.7^{t/2}$ Since $0.7^2 = 0.49$.

$2 = \dfrac{t}{2}$ Set exponents equal.

$4 = t$

A $700 television set will depreciate to $343 after 4 years. To verify graphically, graph the line $y = 343$ in the same window as your graph from (b) and use the *intersect* feature on your calculator. The intersection occurs at the point (4, 343).

29. Graph $Y_1 = 3^{x-1}$ and $Y_2 = 4$. (Don't forget to put parentheses around $x - 1$.) Use the *intersection* feature to find the value of x where $Y_1 = Y_2$:

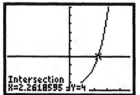

The solution is $x = 2.26$.

31. Graph $Y_1 = 4^{-x}$ and $Y_2 = 7$. Use the *intersection* feature to find the value of x where $Y_1 = Y_2$:

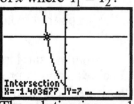

The solution is $x = -1.40$.

33. a. The y-values increase by a factor of 2 for each unit increase in x. Therefore, this is an exponential function with growth factor 2.

b. This is not an exponential function because P does not increase by the same factor for each unit increase of t.

c. This is not an exponential function because N does not increase by the same factor for each unit increase of x.

d. The R-values decrease by a factor of 3 for each unit increase in h. Therefore, this is an exponential function with decay factor $\dfrac{1}{3}$.

35. a. $P_0 = 300$

b. The table below confirms that this is exponential. For every unit increase in x, $f(x)$ increases by a factor of 2.

x	0	1	2
$f(x)$	300	600	1200

c. By part (b), the growth factor is $a = 2$.

d. $f(x) = 300(2)^x$

37. a. $S_0 = 150$

b. Using the points (1, 82) and (2, 45), the decay factor is $a = \dfrac{45}{82} \approx 0.55$. (Using any of the other consecutive points will give approximately the same value.)

c. $S(d) = 150(0.55)^d$

39. a. $N(1) = ab^{1-1} = a$, so from the table, $a = 1{,}600{,}000$. The decay factor is $b = \dfrac{339{,}200}{1{,}600{,}000} = 0.212$. So, $N(x) = 1{,}600{,}000(0.212)^{x-1}$.

b., c., d.

Order	Number	Average length	Total length
1	1,600,000	1	1,600,000
2	339,200	2.3	780,160
3	71,910	5.29	380,404
4	15,245	12.167	185,486
5	3232	27.9841	90,445
6	685	64.3634	44,089
7	145	148.036	21,465
8	31	340.483	10,555
9	7	783.110	5482
10	1	1801.15	1801

c. $L(1) = ab^{1-1} = a$, so from the table, $a = 1$. The decay factor is $b = \dfrac{2.3}{1} = 2.3$. So, $L(x) = 2.3^{x-1}$.

d. The total length of all stream channels is about 3,120,000 miles. (The sum of the last column.)

41.

x	$f(x) = x^2$	$g(x) = 2^x$
-2	4	$\dfrac{1}{4}$
-1	1	$\dfrac{1}{2}$
0	0	1
1	1	2
2	4	4
3	9	8
4	16	16
5	25	32
6	36	64

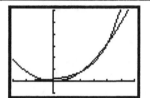

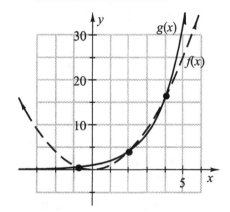

a. 3 (the intersection points are marked on the above graph)

b. $x \approx -0.8$, $x = 2$, $x = 4$

c. $(-0.8, 2) \cup (4, \infty)$

d. $g(x)$

Homework 7.3

1. a. $\log_7(49) = 2$ since $7^2 = 49$

   b. $\log_2(32) = 5$ since $2^5 = 32$

3. a. $\log_3 \sqrt{3} = \frac{1}{2}$ since $3^{1/2} = \sqrt{3}$

   b. $\log_3 \frac{1}{3} = -1$ since $3^{-1} = \frac{1}{3}$

5. a. $\log_4 4 = 1$ since $4^1 = 4$

   b. $\log_6 1 = 0$ since $6^0 = 1$

7. a. $\log_8 8^5 = 5$ since $8^5 = 8^5$

   b. $\log_7 7^6 = 6$ since $7^6 = 7^6$

9. a. $\log_{10} 0.1 = -1$ since
   $$10^{-1} = \frac{1}{10} = 0.1$$

   b. $\log_{10} 0.001 = -3$ since
   $$10^{-3} = \frac{1}{10^3} = \frac{1}{1000} = 0.001$$

11. The base is t and the exponent is $\frac{3}{2}$. A logarithm is equal to an exponent, so $\log_t 16 = \frac{3}{2}$.

13. The base is 0.8 and the exponent is 1.2. A logarithm is equal to an exponent, so $\log_{0.8} M = 1.2$.

15. The base is x and the exponent is $5t$. A logarithm is equal to an exponent, so $\log_x(W - 3) = 5t$.

17. The base is 3 and the exponent is $-0.2t$. A logarithm is equal to an exponent, so $\log_3(2N_0) = -0.2t$.

19. a. Convert to logarithmic form:
   $$x = \log_4 2.5$$

   b. $x \approx 0.7$ (Try $4^{0.6}, 4^{0.7}$, and $4^{0.8}$ in your calculator.)

21. a. Convert to logarithmic form:
   $$x = \log_{10} 0.003$$

   b. $x \approx -2.5$ (Try $10^{-2.4}, 10^{-2.5}$, and $10^{-2.6}$ in your calculator.)

23. Let $x = \log_{10} 7$. Converting to exponential form, $10^x = 7$. Graph $y_1 = 10^x$ and $y_2 = 7$ with Xmin $= -3$, Xmax $= 3$, Ymin $= -4$, and Ymax $= 10$. Use the *intersection* feature to find out where $y_1 = y_2$.

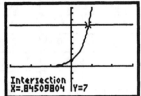

Intersection
X=.84509804 Y=7

Thus, $\log_{10} 7 \approx 0.85$.

25. Let $x = \log_3 67.9$. Converting to exponential form, $3^x = 67.9$. Graph $y_1 = 3^x$ and $y_2 = 67.9$ with Xmin $= -1$, Xmax $= 5$, Ymin $= -10$, and Ymax $= 80$. Use the *intersection* feature to find out where $y_1 = y_2$.

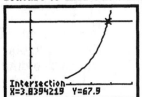

Intersection
X=3.8394219 Y=67.9

Thus, $\log_3 67.9 \approx 3.84$.

27. For each of the following, use the "log" button on your calculator. This key is the logarithm base 10.

    a. 1.7348

    b. 3.3700

    c. −1.1367

    d. −2.2118

29. Convert to logarithmic form:
$$-3x = \log_{10} 5$$
Divide by 3 and evaluate:
$$x = -\frac{\log_{10} 5}{3} = -0.23$$

31. Divide by 25 and reduce:
$$10^{0.2x} = \frac{80}{25} = \frac{16}{5}$$
Convert to logarithmic form:
$$0.2x = \log_{10} \frac{16}{5}$$
Divide by 0.2 and evaluate:
$$x = \frac{\log_{10} \frac{16}{5}}{0.2} = 2.53$$

33. Add 11.6 to both sides:
$$2(10^{1.4x}) = 23.8$$
Divide both sides by 2:
$$10^{1.4x} = 11.9$$
Convert to logarithmic form:
$$1.4x = \log_{10} 11.9$$
Divide by 1.4 and evaluate:
$$x = \frac{\log_{10} 11.9}{1.4} = 0.77$$

35. Add 14.7 to both sides:
$$3(10^{-1.5x}) = 31.8$$
Divide both sides by 3:
$$10^{-1.5x} = 10.6$$
Convert to logarithmic form:
$$-1.5x = \log_{10} 10.6$$
Divide by −1.5 and evaluate:
$$x = \frac{\log_{10} 10.6}{-1.5} = -0.68$$

37. Divide both sides by 80 and reduce:
$$1 - 10^{-0.2x} = \frac{13}{16}$$
Subtract 1 from both sides and then multiply both sides by −1:
$$10^{-0.2x} = \frac{3}{16}$$
Convert to logarithmic form:
$$-0.2x = \log_{10} \frac{3}{16}$$
Divide by −0.2 and evaluate:
$$x = \frac{\log_{10} \frac{3}{16}}{-0.2} = 3.63$$

39. $a = 29,028 \text{ ft} \cdot \dfrac{1 \text{ mi}}{5280 \text{ ft}}$
$= 5.4977$ miles. For this altitude, the atmospheric pressure is
$$P = 30(10)^{-(0.09)(5.4977)} = 9.60 \text{ inches}$$
of mercury. Graphical verification:

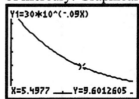

41. Let $P = 20.2$ and solve for a:
$$20.2 = 30(10)^{-0.09a}$$
Divide both sides by 30:
$$\frac{20.2}{30} = 10^{-0.09a}$$
Convert to logarithmic form:
$$-0.09a = \log_{10}\left(\frac{20.2}{30}\right)$$
Divide by –0.09 and evaluate:
$$a = 1.91 \text{ miles}$$
Graphical verification:

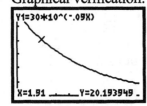

43. Since $P(0) = 30(10)^0 = 30$, the atmospheric pressure at sea level is 30 inches of mercury. Half of this is 15, so $15 = 30(10)^{-0.09a}$.
Divide both sides by 30:
$$\frac{1}{2} = 10^{-0.09a}$$
Convert to logarithmic form:
$$-0.09a = \log_{10}\frac{1}{2}$$
Divide by –0.09 and evaluate:
$$a = 3.34 \text{ miles}$$
Graphical verification:

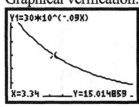

45. a. In 1970, $t = 10$, so the population was
$$P(10) = 15,717,000(10)^{(0.0104)(10)}$$
$$= 19,969,613$$

b. 1980: $P(20) = 25,372,873$
1990: $P(30) = 32,238,116$
2000: $P(40) = 40,960,915$

c. From part (a), $P(10)$ is just about 20,000,000, so the population will reach 20 million around 1970.

d. Solve the equation $30,000,000$
$$= 15,717,000(10)^{0.0104t} \text{ for } t:$$
Divide by 15,717,000:
$$\frac{30,000,000}{15,717,000} = 10^{0.0104t}.$$
Convert to logarithmic form:
$$0.0104t = \log_{10}\frac{30,000,000}{15,717,000}$$
Divide by 0.0104 and evaluate:
$t \approx 27$, so in 1987 the population reached 30,000,000.

e. On a graphing calculator, use Xmin = 0, Xmax = 40, Ymin = 0, and Ymax = 50,000,000.

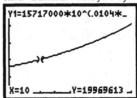

47. $\log_2\left(\log_4 16\right) = \log_2 2 = 1$

49. $\log_{10}\left[\log_3\left(\log_5 125\right)\right] = \log_{10}\left[\log_3 3\right]$
$$= \log_{10} 1$$
$$= 0$$

51. $\log_2\left[\log_2\left(\log_3 81\right)\right] = \log_2\left[\log_2 4\right]$
$$= \log_2 2$$
$$= 1$$

53. $\log_b\left(\log_b b\right) = \log_b 1 = 0$

Midchapter 7 Review

1. a. $2^{1/4} \approx 1.1892$

b. $P(t) = 5000\left(2^{t/4}\right)$

c. Refer to the graph in the back of the textbook.

3. a.

Years since 1992	0	1	2	3	4
Value of house	135,000	148,500	163,350	179,685	197,654

b. $V(t) = 135,000(1.10)^t$

c. Refer to the graph in the back of the textbook.

d. In 1999, $t = 7$, so the value of the house was $V(7) = 135,000(1.10)^7$ $= \$263,076.81$.

5. a.

Years since 1995	0	1	2	3	4
Agriculture majors	200	160	128	102	82

b. $N(t) = 200\left(\dfrac{4}{5}\right)^t$

c. Refer to the graph in the back of the textbook.

d. In 2000, $t = 5$, so the number of agriculture majors was
$$N(5) = 200\left(\frac{4}{5}\right)^5 \approx 66$$

7. Refer to the graph in the back of the textbook.

9. Since $625 = 5^4$, the equation is $5^{x-3} = 5^4$. Equate the exponents: $x - 3 = 4$, so $x = 7$.

11. Graph the equations $Y_1 = 2^x$ and $Y_2 = 3$ using Xmin $= -3$, Xmax $= 3$, Ymin $= -2$, Ymax $= 7$. Use the *intersect* feature on your calculator to find where $Y_1 = Y_2$.

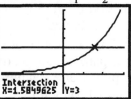

Intersection
X=1.5849625 Y=3

The solution is $x \approx 1.58$.

13. $\log_6 \dfrac{1}{36} = -2$ since $6^{-2} = \dfrac{1}{36}$

15. The base is 64 and a logarithm is equal to an exponent, so
$$\log_{64}\left(\frac{1}{2}\right) = -\frac{1}{6}.$$

17. Divide both sides by 5: $10^x = 6$. Convert to logarithmic form and evaluate: $x = \log_{10} 6 \approx 0.7782$

19. a. $N(t) = 20(9)^{t/5}$

b. Refer to the graph in the back of the textbook.

c. Let $N = 14,580$ and solve for t:
$14,580 = 20(9)^{t/5}$
Divide both sides by 20:
$729 = 9^{t/5}$
Since $729 = 9^3$, the equation can be written as $9^3 = 9^{t/5}$. Equate the exponents: $3 = \dfrac{t}{5}$ so $t = 15$. Thus, 14,580 cosmetologists will have tried the new product after 15 weeks. This answer is verified graphically by the point $(15, 14,580)$.

Homework 7.4

1. a. $f(x) = 2^x$:

x	-2	-1	0	1	2
$f(x)$	$\frac{1}{4}$	$\frac{1}{2}$	1	2	4

The inverse of $f(x)$ is the logarithmic function $g(x) = \log_2 x$. Note that to find ordered pairs for g, you only need to swap the x and y values from the ordered pairs of f.

x	$\frac{1}{4}$	$\frac{1}{2}$	1	2	4
$g(x)$	-2	-1	0	1	2

b. Refer to the graph in the back of the textbook.

3. a. $f(x) = \left(\frac{1}{3}\right)^x$

x	-2	-1	0	1	2
$f(x)$	9	3	1	$\frac{1}{3}$	$\frac{1}{9}$

The inverse of $f(x)$ is the logarithmic function $g(x) = \log_{1/3} x$. Note that to find ordered pairs for g, you only need to swap the x and y values from the ordered pairs of f.

x	9	3	1	$\frac{1}{3}$	$\frac{1}{9}$
$g(x)$	-2	-1	0	1	2

b. Refer to the graph in the back of the textbook.

5. a. $f(487) = \log_{10} 487 \approx 2.6875$
Note: the "log" button on your calculator gives you the logarithm base 10.

b. $f(2.16) = \log_{10} 2.16 \approx 0.3345$

7. a. $f(-7) = \log_{10}(-7)$ is undefined, since there is no real number a such that 10^a is negative.

b. $6f(28) = 6\log_{10} 28 \approx 8.6829$

9. a. $18 - 5f(3) = 18 - 5\log_{10} 3$
≈ 15.6144

b. $\dfrac{2}{5 + f(0.6)} = \dfrac{2}{5 + \log_{10} 0.6}$
≈ 0.4186
Note: To evaluate the above expression in your calculator, be sure to write parentheses around the terms in the denominator.

11. $\dfrac{1}{8500}\log_{10}\left(\dfrac{3600}{8500 - 3600}\right)$
$\approx -0.0000158 = -1.58 \times 10^{-5}$
To evaluate this expression all at once in your calculator, your screen should look like this:

```
(1/8500)*log(360
0/(8500-3600))
     -1.57521858E-5
■
```

234

13. $\sqrt{\dfrac{\log_{10} 0.93}{0.006\log_{10} 0.02}} \approx 1.76$

Calculator screen:

```
√(log(.93)/(.006
*log(.02)))
           1.758345885
```

15. The base is 16 and a logarithm is equal to an exponent, so $16^w = 256$.

17. The base is b and a logarithm is equal to an exponent, so $b^{-2} = 9$.

19. The base is 10 and a logarithm is equal to an exponent, so $10^{-2.3} = A$.

21. The base is 4 and a logarithm is equal to an exponent, so $4^{2q-1} = 36$.

23. The base is u and a logarithm is equal to an exponent, so $u^w = v$.

25. Convert to exponential form:
$b^3 = 8$. So $b = 8^{1/3} = 2$.

27. Convert to exponential form:
$2^{-1} = y$. So $y = \dfrac{1}{2}$.

29. Convert to exponential form:
$b^{1/2} = 10$. So $b = 10^2 = 100$.

31. Convert to exponential form and solve the linear equation:
$2^5 = 3x - 1$
$32 = 3x - 1$
$33 = 3x$
$11 = x$

33. Subtract 5 from both sides:
$3\log_7 x = 2$.
Divide by 3:
$\log_7 x = \dfrac{2}{3}$
Convert to exponential form:
$x = 7^{2/3} \approx 3.66$.

35. $\log_{10} x = 1.41$
Convert to exponential form and evaluate: $x = 10^{1.41} \approx 25.70$

37. $\log_{10} x = 0.52$
Convert to exponential form and evaluate: $x = 10^{0.52} \approx 3.31$

39. $\log_{10} x = -1.3$
Convert to exponential form and evaluate: $x = 10^{-1.3} \approx 0.0501$

41. According to Example 7, when $H^+ = 6.3 \times 10^{-4}$, the
$pH = -\log_{10}(6.3 \times 10^{-4}) \approx 3.2$.

43. From Example 7, we have the equation $1.9 = -\log_{10}\left[H^+\right]$. Multiply both sides by -1: $-1.9 = \log_{10}\left[H^+\right]$. Convert to exponential form and evaluate: $H^+ = 10^{-1.9} \approx 1.26 \times 10^{-2}$

45. From Example 8,
$D = 10\log_{10}\left(\dfrac{10^{-2}}{10^{-12}}\right)$
$= 10\log_{10}\left(10^{10}\right) = 100$ decibels

47. From Example 8, we have the equation: $188 = 10\log_{10}\left(\dfrac{I}{10^{-12}}\right)$.

Divide both sides by 10:

$18.8 = \log_{10}\left(\dfrac{I}{10^{-12}}\right)$

Convert to exponential form:

$\dfrac{I}{10^{-12}} = 10^{18.8}$.

Multiply both sides by 10^{-12}:

$I = 10^{18.8} \cdot 10^{-12} = 6.31 \times 10^{6}$ watts per square meter

49. Refer to Example 8. Let I_{who} be the intensity at the concert. Then

$120 = 10\log_{10}\left(\dfrac{I_{who}}{10^{-12}}\right)$.

Divide by 10: $12 = \log_{10}\left(\dfrac{I_{who}}{10^{-12}}\right)$

Convert: $\dfrac{I_{who}}{10^{-12}} = 10^{12}$.

Then $I_{who} = 1$ watt / m^2.

Let I_{pain} be the intensity of the threshold of pain for the human ear.

Then $90 = 10\log_{10}\left(\dfrac{I_{pain}}{10^{-12}}\right)$.

Divide by 10: $9 = \log_{10}\left(\dfrac{I_{pain}}{10^{-12}}\right)$

Convert: $\dfrac{I_{pain}}{10^{-12}} = 10^{9}$.

Then $I_{pain} = 10^{-3}$ watts / m^2.

Since $\dfrac{I_{who}}{I_{pain}} = \dfrac{1}{10^{-3}} = 1000$, the concert by The Who was 1000 times more intense than a sound at the threshold of pain for the human ear.

51. Refer to Example 9. Convert $M = \log_{10}\left(\dfrac{A}{A_0}\right)$ to exponential form:

$\dfrac{A}{A_0} = 10^{M}$ or $A = A_0 10^{M}$. For the 1964 Alaskan earthquake, $M = 8.4$ and for the 4.0 earthquake, $M = 4.0$. Comparing the amplitudes of these earthquakes, we have the ratio:

$\dfrac{A_0 10^{8.4}}{A_0 10^{4.0}} = 10^{4.4} \approx 25{,}119$. So the Alaskan earthquake was 25,119 times stronger than one measuring 4.0 on the Richter scale.

53. If $M = \log_{10}\left(\dfrac{A}{A_0}\right) = 4.2$, then

$\dfrac{A}{A_0} = 10^{4.2}$ and the amplitude is $A = A_0 10^{4.2}$. An earthquake 3 times as strong would have magnitude

$M = \log_{10}\left(\dfrac{3A}{A_0}\right) = \log_{10}\left(\dfrac{3A_0 10^{4.2}}{A_0}\right)$

$= \log_{10}\left(3 \cdot 10^{4.2}\right) = 4.7$

55. a.

Distance from bed (feet)	0.2	0.4	0.6	0.8	1.0	1.2
Velocity, Hoback River (ft/sec)	1.91	2.22	2.40	2.53	2.63	2.71
Velocity, Pole Creek (ft/sec)	1.51	1.70	1.82	1.90	1.96	2.01

b. For Hoback River, if you double the distance from the bed from 0.2 feet to 0.4 feet, the velocity increases by $2.22 - 1.91 = 0.31$ feet per second. We find this same result no matter what distance we start with.
For Pole Creek, if you double the distance from the bed from 0.2 feet to 0.4 feet, the velocity increases by $1.70 - 1.51 = 0.19$ feet per second. We find this same (approximate) result no matter what distance we start with.

c. Refer to the graph in the back of the textbook.

d. 20% of the total depth is $0.20(1.2) = 0.24$ feet and 80% of the total depth is $0.80(1.2) = 0.96$ feet. For Hoback River,
$v(0.24) = 2.63 + 1.03 \log_{10}(0.24)$
$= 1.9916$ ft/sec, and
$v(0.96) = 2.63 + 1.03 \log_{10}(0.96)$
$= 2.6117$ ft/sec. Thus the average velocity of Hoback River is
$\dfrac{1.9916 + 2.6117}{2} \approx 2.3$ ft/sec.
For Pole Creek,
$v(0.24) = 1.96 + 0.65 \log_{10}(0.24)$
$= 1.5571$ ft/sec, and
$v(0.96) = 1.96 + 0.65 \log_{10}(0.96)$
$= 1.9485$ ft/sec. Thus the average velocity of Pole Creek is
$\dfrac{1.5571 + 1.9485}{2} \approx 1.75$ ft/sec.

57. a. $f(4) = 3^4 = 81$

b. $g[f(4)] = g(81) = \log_3 81 = 4$

c. $g(x)$ is the inverse, or "undoing" function of $f(x)$, so
$g(3^x) = \log_3 3^x = x$

d. $\log_3 3^{1.8} = 1.8$

e. $\log_3 3^a = a$

59. a. IV
b. V
c. I
d. II
e. III
f. VI

61. a. "take the base-6 logarithm of x"
b. "raise 5 to the power of x"

63. a. The graph reaches a height of 4 when $y = 4$, so solve $4 = \log_{10} x$. Convert to exponential form:
$x = 10^4 = 10,000$.

b. Convert $8 = \log_{10} x$ to exponential form:
$x = 10^8 = 100,000,000$.

65. a.

x	x^2	$\log_{10} x$	$\log_{10} x^2$
1	1	0	0
2	4	0.3010	0.6021
3	9	0.4771	0.9542
4	16	0.6021	1.2041
5	25	0.6990	1.3979
6	36	0.7782	1.5563

b. $\log_{10} x^2$ is always twice $\log_{10} x$;
$\log_{10} x^2 = 2 \log_{10} x$

Homework 7.5

1. a. $\log_b 2x = \log_b 2 + \log_b x$ by Property (1)

b. $\log_b \dfrac{x}{y} = \log_b x - \log_b y$ by Property (2)

c. $\log_b \dfrac{xy}{z} = \log_b xy - \log_b z$ by Property (2)
$\qquad\qquad = \log_b x + \log_b y - \log_b z$ by Property (1)

d. $\log_b x^3 = 3\log_b x$ by Property (3)

3. a. $\log_b \sqrt{x} = \log_b x^{1/2} = \dfrac{1}{2}\log_b x$ by Property (3)

b. $\log_b \sqrt[3]{x^2} = \log_b x^{2/3} = \dfrac{2}{3}\log_b x$ by Property (3)

c. $\log_b x^2 y^3 = \log_b x^2 + \log_b y^3$ by Property (1)
$\qquad\qquad = 2\log_b x + 3\log_b y$ by Property (3)

d. $\log_b \dfrac{x^{1/2} y}{z^2} = \log_b x^{1/2} y - \log_b z^2$ by Property (2)

$\qquad\qquad = \log_b x^{1/2} + \log_b y - \log_b z^2$ by Property (1)
$\qquad\qquad = \dfrac{1}{2}\log_b x + \log_b y - 2\log_b z$ by Property (3)

5. a. $\log_{10} \sqrt[3]{\dfrac{xy^2}{z}} = \log_{10}\left(\dfrac{xy^2}{z}\right)^{1/3} = \dfrac{1}{3}\log_{10}\left(\dfrac{xy^2}{z}\right)$ by Property (3)

$\qquad\qquad = \dfrac{1}{3}\left(\log_{10} xy^2 - \log_{10} z\right)$ by Property (2)

$\qquad\qquad = \dfrac{1}{3}\left(\log_{10} x + \log_{10} y^2 - \log_{10} z\right)$ by Property (1)

$\qquad\qquad = \dfrac{1}{3}\left(\log_{10} x + 2\log_{10} y - \log_{10} z\right)$ by Property (3)

$\qquad\qquad = \dfrac{1}{3}\log_{10} x + \dfrac{2}{3}\log_{10} y - \dfrac{1}{3}\log_{10} z$ Distribute

b. $\log_{10}\sqrt{\dfrac{2L}{R^2}} = \log_{10}\left(\dfrac{2L}{R^2}\right)^{1/2} = \dfrac{1}{2}\log_{10}\left(\dfrac{2L}{R^2}\right)$ by Property (3)

$\qquad\qquad = \dfrac{1}{2}\left(\log_{10} 2L - \log_{10} R^2\right)$ by Property (2)

$\qquad\qquad = \dfrac{1}{2}\left(\log_{10} 2 + \log_{10} L - \log_{10} R^2\right)$ by Property (1)

$\qquad\qquad = \dfrac{1}{2}\left(\log_{10} 2 + \log_{10} L - 2\log_{10} R\right)$ by Property (3)

$\qquad\qquad = \dfrac{1}{2}\log_{10} 2 + \dfrac{1}{2}\log_{10} L - \log_{10} R$ Distribute

c. $\log_{10} 2\pi\sqrt{\dfrac{l}{g}} = \log_{10} 2\pi + \log_{10}\sqrt{\dfrac{l}{g}}$ by Property (1)

$\qquad\qquad = \log_{10} 2 + \log_{10} \pi + \log_{10}\left(\dfrac{l}{g}\right)^{1/2}$ by Property (1)

$\qquad\qquad = \log_{10} 2 + \log_{10} \pi + \dfrac{1}{2}\log_{10}\dfrac{l}{g}$ by Property (3)

$\qquad\qquad = \log_{10} 2 + \log_{10} \pi + \dfrac{1}{2}\log_{10} l - \dfrac{1}{2}\log_{10} g$ by Property (2)

d. $\log_{10} 2y\sqrt[3]{\dfrac{x}{y}} = \log_{10} 2y + \log_{10}\sqrt[3]{\dfrac{x}{y}}$ by Property (1)

$\qquad\qquad = \log_{10} 2 + \log_{10} y + \log_{10}\left(\dfrac{x}{y}\right)^{1/3}$ by Property (1)

$\qquad\qquad = \log_{10} 2 + \log_{10} y + \dfrac{1}{3}\log_{10}\dfrac{x}{y}$ by Property (3)

$\qquad\qquad = \log_{10} 2 + \log_{10} y + \dfrac{1}{3}\log_{10} x - \dfrac{1}{3}\log_{10} y$ by Property (2)

$\qquad\qquad = \log_{10} 2 + \dfrac{2}{3}\log_{10} y + \dfrac{1}{3}\log_{10} x$ Combine like terms.

7. a. $\log_b 6 = \log_b(2 \cdot 3)$
$= \log_b 2 + \log_b 3$
$= 0.6931 + 1.0986$
$= 1.7917$

b. $\log_b \dfrac{2}{5} = \log_b 2 - \log_b 5$
$= 0.6931 - 1.6094$
$= -0.9163$

9. a. $\log_b 9 = \log_b 3^2$
$= 2\log_b 3$
$= 2(1.0986)$
$= 2.1972$

b. $\log_b \sqrt{50} = \log_b 5\sqrt{2}$
$= \log_b 5 + \log_b \sqrt{2}$
$= \log_b 5 + \dfrac{1}{2}\log_b 2$
$= 1.6094 + \dfrac{1}{2}(0.6931)$
$= 1.95595$

11. $2^x = 7$
$\log_{10} 2^x = \log_{10} 7$
$x\log_{10} 2 = \log_{10} 7$
$x = \dfrac{\log_{10} 7}{\log_{10} 2}$
$x \approx 2.8074$

13. $3^{x+1} = 8$
$\log_{10} 3^{x+1} = \log_{10} 8$
$(x+1)\log_{10} 3 = \log_{10} 8$
$x+1 = \dfrac{\log_{10} 8}{\log_{10} 3}$
$x = \dfrac{\log_{10} 8}{\log_{10} 3} - 1$
$x \approx 0.8928$

15. $4^{x^2} = 15$
$\log_{10} 4^{x^2} = \log_{10} 15$
$x^2 \log_{10} 4 = \log_{10} 15$
$x^2 = \dfrac{\log_{10} 15}{\log_{10} 4}$
$x = \pm\sqrt{\dfrac{\log_{10} 15}{\log_{10} 4}}$
$x \approx \pm 1.3977$

17. $4.26^{-x} = 10.3$
$\log_{10} 4.26^{-x} = \log_{10} 10.3$
$-x\log_{10} 4.26 = \log_{10} 10.3$
$x = -\dfrac{\log_{10} 10.3}{\log_{10} 4.26}$
$x \approx -1.6092$

19. $25 \cdot 3^{2.1x} = 47$
$3^{2.1x} = \dfrac{47}{25}$
$\log_{10} 3^{2.1x} = \log_{10} \dfrac{47}{25}$
$2.1x\log_{10} 3 = \log_{10} \dfrac{47}{25}$
$x = \dfrac{\log_{10} \dfrac{47}{25}}{2.1\log_{10} 3}$
$x = \dfrac{\log_{10} 47 - \log_{10} 25}{2.1\log_{10} 3}$
$x \approx 0.2736$

21.
$$3600 = 20 \cdot 8^{-0.2x}$$
$$180 = 8^{-0.2x}$$
$$\log_{10} 180 = \log_{10} 8^{-0.2x}$$
$$\log_{10} 180 = -0.2x \log_{10} 8$$
$$\frac{\log_{10} 180}{-0.2 \log_{10} 8} = x$$
$$-12.4864 \approx x$$

23. a. $w(t) = 0.01(3)^{t/16}$

b. $0.5 = 0.01(3)^{t/16}$
$$50 = 3^{t/16}$$
$$\log_{10} 50 = \log_{10} 3^{t/16}$$
$$\log_{10} 50 = \frac{t}{16} \log_{10} 3$$
$$\frac{16 \log_{10} 50}{\log_{10} 3} = t$$
$$56.97 \approx t$$
The culture will weigh 0.5 gram after 56.97 hours. To check graphically, find the intersection of the line $y = 0.5$ and $w(t)$:
Xmin = 0, Xmax = 70,
Ymin = 0, Ymax = 0.6

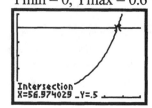

Intersection
X=56.974029 _Y=.5

25. a. Note that if the concentration decreases by 20%, then there is 80% left in the bloodstream. Therefore, the concentration is
$$C(t) = 0.7(0.80)^t.$$

b. $0.4 = 0.7(0.80)^t$
$$\frac{4}{7} = 0.80^t$$
$$\log_{10} \frac{4}{7} = \log_{10} 0.80^t$$
$$\log_{10} \frac{4}{7} = t \log_{10} 0.80$$
$$\frac{\log_{10} \frac{4}{7}}{\log_{10} 0.80} = t$$
$$2.5 \approx t$$
The second dose should be administered after 2.5 hours.

c. To check graphically, find the intersection of the line $y = 0.4$ and $C(t)$:
Xmin = 0, Xmax = 5,
Ymin = 0, Ymax = 0.7

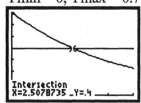

Intersection
X=2.5078735 _Y=.4

27. a. $\log_2(4\cdot 8) = \log_2 32 = 5$;
$(\log_2 4)(\log_2 8) = (2)(3) = 6$;
The expressions are not equal.

b. $\log_2(16+16) = \log_2 32 = 5$;
$\log_2 16 + \log_2 16 = 4+4 = 8$;
The expressions are not equal.

29. a. $\log_{10}\left(\frac{240}{10}\right) = \log_{10} 24$;

$\dfrac{\log_{10} 240}{\log_{10} 10} = \dfrac{\log_{10} 240}{1} = \log_{10} 240$
The expressions are not equal.

b. $\log_{10}\left(\frac{1}{2}\cdot 80\right) = \log_{10} 40 \approx 1.6021$;

$\frac{1}{2}\log_{10}(80) \approx 0.9515$;
The expressions are not equal.

31. Refer to the graph in the back of the textbook. The graphs are the same since by Property (1),
$\log(2x) = \log 2 + \log x$.

33. Refer to the graph in the back of the textbook. The graphs are the same since by Property (2),

$\log\left(\frac{1}{x}\right) = \log 1 - \log x = 0 - \log x$

$\qquad = -\log x$

35. a. $P(t) = P_0(1.037)^t$

b. The population has doubled when $P(t) = 2P_0$. Solve the equation:

$2P_0 = P_0(1.037)^t$

$2 = 1.037^t$

$\log_{10} 2 = \log_{10} 1.037^t$

$\log_{10} 2 = t\log_{10} 1.037$

$t = \dfrac{\log_{10} 2}{\log_{10} 1.037} \approx 19.08$

The population doubles after 19.08 years.

c. Refer to the graph in the back of the textbook or in part (d) below.

d. To verify that the population doubles from 500 to 1000 after 19.08 years, find the intersection of the equation $P(t) = 500(1.037)^t$ and the line $y = 1000$:

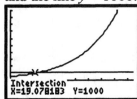

Intersection
X=19.078183 Y=1000

In the same way, one can verify that the population doubles from 1000 to 2000 and from 2000 to 4000 after 19.08 years.

37. a. The amount of potassium-42 left after t hours is $P(t) = 400(0.946)^t$. To find out when half of the 400 milligrams has decayed, let $P(t) = 200$:

$$200 = 400(0.946)^t$$

$$\frac{1}{2} = 0.946^t$$

$$\log_{10} \frac{1}{2} = t \log_{10} 0.946$$

$$t = \frac{\log_{10} \frac{1}{2}}{\log_{10} 0.946} \approx 12.49 \text{ hours}$$

b. Three-fourths will be gone in two half-lives, or 24.97 hours. Seven-eighths will be gone in three half-lives, or 37.46 hours.

39. a. $\log_b 8 - \log_b 2 = \log_b \frac{8}{2} = \log_b 4$

b. $2 \log_b x + 3 \log_b y = \log_b x^2 + \log_b y^3 = \log_b x^2 y^3$

c. $-2 \log_b x = \log_b x^{-2} = \log_b \frac{1}{x^2}$

41. a. $\frac{1}{2}(\log_{10} y + \log_{10} x - 3\log_{10} z) = \frac{1}{2}(\log_{10} xy - \log_{10} z^3) = \frac{1}{2} \log_{10} \frac{xy}{z^3}$

$$= \log_{10}\left(\frac{xy}{z^3}\right)^{1/2} = \log_{10} \sqrt{\frac{xy}{z^3}}$$

b. $\frac{1}{2} \log_b 16 + 2(\log_b 2 - \log_b 8) = \log_b 16^{1/2} + 2\log_b \frac{2}{8} = \log_b 4 + \log_b \left(\frac{1}{4}\right)^2$

$$= \log_b 4 + \log_b \frac{1}{16} = \log_b \frac{4}{16} = \log_b \frac{1}{4}$$

43. $\log_{10} x + \log_{10}(x + 21) = 2$

$\log_{10}[x(x + 21)] = 2$ Apply Property (1).

$\log_{10}(x^2 + 21x) = 2$

$x^2 + 21x = 10^2$ Change to exponential form.

$x^2 + 21x - 100 = 0$

$(x + 25)(x - 4) = 0$ Factor.

This gives the values $x = -25$ or $x = 4$. However, for $\log_{10} x$ to be defined, x must be positive, so the only solution to the original equation is $x = 4$.

45. $\log_8(x+5) - \log_8 2 = 1$

$\log_8 \dfrac{x+5}{2} = 1$ Apply Property (2).

$\dfrac{x+5}{2} = 8^1$ Change to exponential form.

$x+5 = 16$

$x = 11$

For this value of x, the logarithms in the original equation are defined, so the solution to the original equation is $x = 11$.

47. $\log_{10}(x+2) + \log_{10}(x-1) = 1$

$\log_{10}[(x+2)(x-1)] = 1$ Apply Property (1).

$\log_{10}(x^2 + x - 2) = 1$

$x^2 + x - 2 = 10^1$ Change to exponential form.

$x^2 + x - 12 = 0$

$(x+4)(x-3) = 0$ Factor.

This gives the values $x = -4$ and $x = 3$. However, if $x = -4$, then $x + 2 < 0$ so $\log_{10}(x+2)$ is undefined. Thus, the only solution to the original equation is $x = 3$.

49. $\log_3(x-2) - \log_3(x+1) = 3$

$\log_3 \dfrac{x-2}{x+1} = 3$ Apply Property (2).

$\dfrac{x-2}{x+1} = 3^3$ Change to exponential form.

$x - 2 = 27(x+1)$

$x - 2 = 27x + 27$

$-26x = 29$

$x = -\dfrac{29}{26}$

For this value of x, $x - 2 < 0$ and thus $\log_{10}(x-2)$ is undefined. Therefore, the original equation has no solutions.

51. a. with annual compounding:
$$Y_1 = 5000(1+0.12)^t;$$

with quarterly compounding:
$$Y_2 = 5000\left(1+\frac{0.12}{4}\right)^{4t}.$$
Refer to the graph in the back of the textbook.

b. with annual compounding:
$$Y_1 = 5000(1+0.12)^{10}$$
$$= \$15,529.24$$

with quarterly compounding:
$$Y_2 = 5000\left(1+\frac{0.12}{4}\right)^{4\cdot10}$$
$$= \$16,310.19$$
These answers are verified on the graph by the points
(10, 15,529.24) and
(10, 16,310.19).

c. The graphs are not the same. Using Table Start = 0 and ΔTbl = 0.5, we find that Y_1 and Y_2 first differ by $250 for about $t = 5.5$.

X	Y₁	Y₂
3	7024.6	7128.8
3.5	7434.2	7562.9
4	7867.6	8023.5
4.5	8326.3	8512.2
5	8811.7	9030.6
5.5	9325.4	9580.5
6	9869.1	10164

X=5.5

53. $1900 = 1000\left(1+\frac{r}{12}\right)^{12\cdot5}$

$$\left(\frac{1900}{1000}\right)^{1/60} = 1+\frac{r}{12}$$

$$r = 12\left[\left(\frac{1900}{1000}\right)^{1/60} - 1\right] = 0.129$$
or 12.9%

55. $3 = \left(1+\frac{0.10}{365}\right)^{365t}$

$$\log_{10} 3 = \log_{10}\left(1+\frac{0.10}{365}\right)^{365t}$$

$$\log_{10} 3 = 365t\log_{10}\left(1+\frac{0.10}{365}\right)$$

$$t = \frac{\log_{10} 3}{365\log_{10}\left(1+\frac{0.10}{365}\right)} = 10.988$$
$$= 10 \text{ years, } 361 \text{ days}$$

57. $A = A_0(10^{kt} - 1)$

$$10^{kt} = \frac{A}{A_0} + 1$$

$$\log_{10} 10^{kt} = \log_{10}\left(\frac{A}{A_0} + 1\right)$$

$$kt = \log_{10}\left(\frac{A}{A_0} + 1\right)$$

$$t = \frac{1}{k}\log_{10}\left(\frac{A}{A_0} + 1\right)$$

59. $w = pv^q$

$$\frac{w}{p} = v^q$$
Change to logarithmic form:
$$q = \log_v \frac{w}{p}$$

61. $t = T\log_{10}\left(1+\frac{A}{k}\right)$

$$\frac{t}{T} = \log_{10}\left(1+\frac{A}{k}\right)$$
Change to exponential form:
$$10^{t/T} = 1+\frac{A}{k}$$

$$A = k(10^{t/T} - 1)$$

Homework 7.6

1.

x	-10	-5	0	5	10	15	20
f(x)	0.14	0.37	1	2.72	7.39	20.09	54.60

Refer to the graph in the back of the textbook.

3.

x	-10	-5	0	5	10	15	20
f(x)	20.09	4.48	1	0.22	0.05	0.01	0.00

Refer to the graph in the back of the textbook.

5. a. $x = \log_e 1.9 = \ln 1.9 = 0.6419$

   b. $x = \ln 45 = 3.8067$

   c. $x = \ln 0.3 = -1.2040$

7. a. $x = e^{1.42} = 4.1371$

   b. $x = e^{0.63} = 1.8776$

   c. $x = e^{-2.6} = 0.0743$

9. a. $N(t) = 6000e^{0.04t}$

   b.

t	0	5	10	15
N(t)	6000	7328	8951	10,933

t	20	25	30
N(t)	13,353	16,310	19,921

   c. Refer to the graph in the back of the textbook.

   d. After 24 hours, there were
$N(24) = 6000e^{0.04 \cdot 24} \approx 15,670$
bacteria present.

   e. Let $N(t) = 100,000$:
$100,000 = 6000e^{0.04t}$
Divide by 6000 and reduce:
$\dfrac{50}{3} = e^{0.04t}$.
Change to logarithmic form:
$$0.04t = \log_e\left(\frac{50}{3}\right) = \ln\left(\frac{50}{3}\right)$$
$$t = \frac{1}{0.04}\ln\left(\frac{50}{3}\right) \approx 70.3$$
There will be 100,000 bacteria present after about 70.3 hours.

11. a. Refer to the graph in the back of the textbook.

   b. When $t = 0.6$ cm, the intensity is
$1000e^{-(0.1)(0.6)} = 941.8$ lumens.

   c. $800 = 1000e^{-0.1t}$
Divide by 1000: $0.8 = e^{-0.1t}$
Change to logarithmic form:
$-0.1t = \ln 0.8$
$t = \dfrac{\ln 0.8}{-0.1} \approx 2.2$ cm

13. $P(t) = 20e^{0.4t} = 20\left(e^{0.4}\right)^t$

$\approx 20(1.49)^t$

Since $a > 1$, the function is increasing.

15. $P(t) = 6500e^{-2.5t} = 6500\left(e^{-2.5}\right)^t$

$\approx 6500(0.082)^t$

Since $0 < a < 1$, the function is decreasing.

17. **a.**

x	0	0.5	1	1.5	2	2.5
e^x	1	1.6487	2.7183	4.4817	7.3891	12.1825

b. $\dfrac{1.6487}{1} = 1.6487,$

$\dfrac{2.7183}{1.6487} = 1.6488,$

$\dfrac{4.4817}{2.7183} = 1.6487,$

$\dfrac{7.3891}{4.4817} = 1.6487,$

$\dfrac{12.1825}{7.3891} = 1.6487$

Each ratio equals $e^{0.5} \approx 1.6487$

19. a.

x	0	0.6931	1.3863	2.0794
e^x	1	2	4	8

x	2.7726	3.4657	4.1589
e^x	16	32	64

b. $0.6931 - 0 = 0.6931,$
$1.3863 - 0.6931 = 0.6932,$
$2.0794 - 1.3863 = 0.6931,$
$2.7726 - 2.0794 = 0.6932,$
$3.4657 - 2.7726 = 0.6931,$
$4.1589 - 3.4657 = 0.6932;$
Each difference in x-values is approximately 0.6931. Since $e^{0.6931} \approx 2$, by the first law of exponents, each function value is approximately double the previous. For example, consider the last value in the table:
$e^{4.1589} = e^{0.6932} \cdot e^{3.4657} = 2e^{3.4657}$

21. $6.21 = 2.3e^{1.2x}$
Divide by 2.3:
$2.7 = e^{1.2x}$
Change to logarithmic form:
$1.2x = \ln 2.7$
$x = \dfrac{\ln 2.7}{1.2} \approx 0.8277$

23. $6.4 = 20e^{0.3x} - 1.8$
$8.2 = 20e^{0.3x}$

$0.41 = e^{0.3x}$
Change to logarithmic form:
$0.3x = \ln 0.41$

$x = \dfrac{\ln 0.41}{0.3} \approx -2.9720$

25. $46.52 = 3.1e^{1.2x} + 24.2$

$22.32 = 3.1e^{1.2x}$

$7.2 = e^{1.2x}$

Change to logarithmic form:

$1.2x = \ln 7.2$

$x = \dfrac{\ln 7.2}{1.2} \approx 1.6451$

27. $16.24 = 0.7e^{-1.3x} - 21.7$

$37.94 = 0.7e^{-1.3x}$

$54.2 = e^{-1.3x}$

Change to logarithmic form:

$-1.3x = \ln 54.2$

$x = \dfrac{\ln 54.2}{-1.3} \approx -3.0713$

29. $y = e^{kt};\ kt = \ln y;\ t = \dfrac{\ln y}{k}$

31. $y = k\left(1 - e^{-t}\right)$

$\dfrac{y}{k} = 1 - e^{-t}$

$e^{-t} = 1 - \dfrac{y}{k} = \dfrac{k-y}{k}$

$-t = \ln\left(\dfrac{k-y}{k}\right)$

$t = -\ln\left(\dfrac{k-y}{k}\right) = \ln\left(\dfrac{k-y}{k}\right)^{-1}$

$\quad = \ln\left(\dfrac{k}{k-y}\right)$

33. $T = T_0 \ln(k+10)$

$\dfrac{T}{T_0} = \ln(k+10)$

$e^{T/T_0} = k + 10$

$k = e^{T/T_0} - 10$

35. a.

n	0.39	3.9	39	390
$\ln n$	-0.942	1.361	3.664	5.966

b. $\ln 3.9 - \ln 0.39 = 2.3026$

$\ln 39 - \ln 3.9 = 2.3026$

$\ln 390 - \ln 39 = 2.3026$

Each x-value is 10 times the previous, so by the first log rule, each function value is $\ln 10 \approx 2.3026$ greater than the previous. For example, consider the first two values in the table:

$\ln 3.9 - \ln 0.39 = \ln \dfrac{3.9}{0.39}$

$\quad = \ln 10$

$\quad \approx 2.3026$

37. a.

n	2	4	8	16
$\ln n$	0.693	1.386	2.079	2.773

b. $\dfrac{\ln 2}{\ln 2} = 1,\ \dfrac{\ln 4}{\ln 2} = 2,$

$\dfrac{\ln 8}{\ln 2} = 3,\ \dfrac{\ln 16}{\ln 2} = 4;$

Each quotient equals k, where

$\dfrac{\ln 2^k}{\ln 2} = \dfrac{k \ln 2}{\ln 2} = k.$

39. a. $N(t) = 100 \cdot 2^t = 100(e^{\ln 2})^t$

$\quad = 100e^{(\ln 2)t} = 100e^{0.6931t}$

b. Xmin = 0, Xmax = 4,
Ymin = 0, Ymax = 800

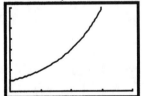

The two graphs are the same.

41. a. $N(t) = 1200(0.6)^t = 1200(e^{\ln 0.6})^t$
$= 1200e^{(\ln 0.6)t} = 1200e^{-0.5108t}$

b. Xmin = 0, Xmax = 6,
Ymin = 0, Ymax = 1400

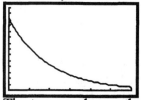

The two graphs are the same.

43. a. $N(t) = 10(1.15)^t = 10(e^{\ln 1.15})^t$

$= 10e^{(\ln 1.15)t} = 10e^{0.1398t}$

b. Xmin = 0, Xmax = 20,
Ymin = 0, Ymax = 100

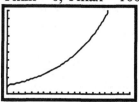

The two graphs are the same.

45. a. $P_0 = 20,000$

b. $P(t) = 20,000e^{kt}$. In 1990:
$35,000 = 20,000e^{k \cdot 10}$
$e^{10k} = \dfrac{35,000}{20,000} = \dfrac{7}{4}$
$10k = \ln\left(\dfrac{7}{4}\right)$
$k = \dfrac{\ln\left(\frac{7}{4}\right)}{10} = 0.05596$
The growth factor is
$e^{0.05596} \approx 1.0576$.

c. $P(t) = 20,000e^{0.05596t}$

d. $P(30) = 20,000e^{0.05596 \cdot 30}$
$= 107,182$

47. a. $P(t) = 500e^{kt}$. After 2 years:
$385 = 500e^{k \cdot 2}$
$e^{2k} = \dfrac{385}{500} = 0.77$
$2k = \ln 0.77$
$k = \dfrac{\ln 0.77}{2} \approx -0.1307$
The decay factor is
$e^{-0.1307} \approx 0.8775$.

b. $N(t) = 500e^{-0.1307t}$

c. $N(10) = 500e^{-0.1307 \cdot 10} = 135.3$
About 135.3 milligrams are left.

49. a. $A(t) = 500e^{0.095t}$

b. $1000 = 500e^{0.095t}$

$2 = e^{0.095t}$

$0.095t = \ln 2$

$t = \dfrac{\ln 2}{0.095} \approx 7.3$ years

c. $2000 = 1000e^{0.095t}$

$2 = e^{0.095t}$

$0.095t = \ln 2$

$t = \dfrac{\ln 2}{0.095} \approx 7.3$ years

d. Refer to the graph in the back of the textbook.

e. Choose, for example, the point (5, 804). The point with vertical coordinate 2(804) = 1608 has t-coordinate 12.3. The difference of these two t-coordinates is 12.3 − 5 = 7.3.

51. a. $0.788N_0 = N_0 e^{-0.000124t}$

$0.788 = e^{-0.000124t}$

$-0.000124t = \ln 0.788$

$t = \dfrac{\ln 0.788}{-0.000124} \approx 1921$ years

b. $\dfrac{1}{2}N_0 = N_0 e^{-0.000124t}$

$\dfrac{1}{2} = e^{-0.000124t}$

$-0.000124t = \ln \dfrac{1}{2}$

$t = \dfrac{\ln \dfrac{1}{2}}{-0.000124} = 5589.9$ years

53. a. After 8 days, there are $\dfrac{1}{2}N_0$ grams left. After 16 days, there are $\dfrac{1}{2}\left(\dfrac{1}{2}N_0\right) = \dfrac{1}{4}N_0$ grams left. After 32 days, there are $\dfrac{1}{4}\left(\dfrac{1}{4}N_0\right) = \dfrac{1}{16}N_0$ grams left.

b. Choose, for example, $N_0 = 1$. Refer to the graph in the back of the textbook.

c. $\dfrac{1}{2}N_0 = N_0 e^{k \cdot 8}$

$\dfrac{1}{2} = e^{k \cdot 8}$

$8k = \ln \dfrac{1}{2}$

$k = \dfrac{1}{8}\ln \dfrac{1}{2} \approx -0.0866$

$N(t) = N_0 e^{-0.0866t}$

55. a. Xmin = 0, Xmax = 100, Ymin = 0, Ymax = 130

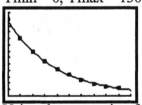

Using the regression feature on the calculator, we have approximately $y = 116(0.975)^t$.

b. $G(t) = 116\left(e^{\ln 0.975}\right)^t$

$= 116e^{(\ln 0.975 t)} = 116e^{-0.0253t}$

c. $\dfrac{1}{2} = e^{-0.0253t}$

$-0.0253t = \ln \dfrac{1}{2}$

$t = \dfrac{\ln \dfrac{1}{2}}{-0.0253} \approx 27.4$ minutes

Chapter 7 Review

1. a.

Years After 1974	Number of Degrees
0	8
5	$1.5(8) = 12$
10	$1.5(12) = 18$
15	$1.5(18) = 27$
20	$1.5(27) = 40.5$

b. $N(t) = 8(1.5)^{t/5}$

c. Refer to the graph in the back of the textbook.

d. $N(10) = 8(1.5)^{10/5} = 18$

$N(21) = 8(1.5)^{21/5} = 43.9 \approx 44$

3. a.

Hours After 8 a.m.	Amount in body (mg)
0	100
1	$0.85 \times 100 = 85$
2	$0.85 \times 85 = 72.25$
3	$0.85 \times 72.25 = 61.4125$
4	$0.85 \times 61.4125 = 52.2006$
5	$0.85 \times 52.2006 = 44.3705$

b. $A(t) = 100(0.85)^t$

c. Refer to the graph in the back of the textbook.

d. $A(4) = 100(0.85)^4 = 52.20$ mg

$A(10) = 100(0.85)^{10} = 19.69$ mg

5. Refer to the graph in the back of the textbook.

7. Refer to the graph in the back of the textbook.

9. $3^{x+2} = 9^{1/3} = (3^2)^{1/3} = 3^{2/3}$;

Equate exponents: $x + 2 = \dfrac{2}{3}$;

$x = \dfrac{2}{3} - 2 = -\dfrac{4}{3}$

11. $4^{2x+1} = 8^{x-3}$; $(2^2)^{2x+1} = (2^3)^{x-3}$;

$2^{4x+2} = 2^{3x-9}$; Equate exponents:

$4x + 2 = 3x - 9$; $x = -11$

13. $\log_2 16 = 4$ since $2^4 = 16$

15. $\log_3 \dfrac{1}{3} = -1$ since $3^{-1} = \dfrac{1}{3}$

17. $\log_{10} 10^{-3} = -3$

19. $2^{x-2} = 3$

21. $\log_{0.3}(x+1) = -2$

23. $y = \log_3 \dfrac{1}{3} = \log_3 3^{-1} = -1$

25. Convert to exponential form:

$b^2 = 16$. Since the base of a logarithm must be positive, the only solution is $b = 4$.

27. $4 \cdot 10^{1.3x} = 20.4;\ 10^{1.3x} = 5.1;$
Take the log of both sides:
$$\log_{10} 10^{1.3x} = \log_{10} 5.1$$
$$1.3x = \log_{10} 5.1$$
$$x = \frac{\log_{10} 5.1}{1.3} \approx 0.5443$$

29. $3\left(10^{-0.7x}\right) + 6.1 = 9$
$$3\left(10^{-0.7x}\right) = 2.9$$
$$10^{-0.7x} = \frac{2.9}{3}$$
$$\log_{10} 10^{-0.7x} = \log_{10} \frac{2.9}{3}$$
$$-0.7x = \log_{10} \frac{2.9}{3}$$
$$x = -\frac{1}{0.7} \log_{10} \frac{2.9}{3} \approx 0.0210$$

31. $\frac{1}{2.3}\left(\log_{10} 12,000 - \log_{10} 9,000\right)$
$= 0.0543$

33. $1.2 \log_{10}\left(\frac{6400}{6400 - 2000}\right) = 0.195$

35. $\log_b\left(\frac{xy^{1/3}}{z^2}\right) = \log_b(xy^{1/3}) - \log_b z^2$
$$= \log_b x + \log_b y^{1/3} - \log_b z^2$$
$$= \log_b x + \frac{1}{3}\log_b y - 2\log_b z$$

37. $\log_{10}\left(x \sqrt[3]{\frac{x}{y}}\right) = \log_{10} \frac{x^{4/3}}{y^{1/3}}$
$$= \log_{10} x^{4/3} - \log_{10} y^{1/3}$$
$$= \frac{4}{3}\log_{10} x - \frac{1}{3}\log_{10} y$$

39. $\frac{1}{3}(\log_{10} x - 2\log_{10} y)$
$$= \frac{1}{3}(\log_{10} x - \log_{10} y^2)$$
$$= \frac{1}{3}\log_{10} \frac{x}{y^2} = \log_{10} \sqrt[3]{\frac{x}{y^2}}$$

41. $\frac{1}{3}\log_{10} 8 - 2(\log_{10} 8 - \log_{10} 2)$
$$= \log_{10} 8^{1/3} - 2\left(\log_{10} \frac{8}{2}\right)$$
$$= \log_{10} 2 - \log_{10} 4^2$$
$$= \log_{10} \frac{2}{16} = \log_{10} \frac{1}{8}$$

43. $\log_3 x + \log_3 4 = 2;\ \log_3 4x = 2;$
$4x = 3^2 = 9;\ x = \frac{9}{4}$

45. $\log_{10}(x - 1) + \log_{10}(x + 2) = 3$
$$\log_{10}[(x - 1)(x + 2)] = 3$$
$$\log_{10}(x^2 + x - 2) = 3$$
$$x^2 + x - 2 = 10^3$$
$$x^2 + x - 1002 = 0$$
$$x = \frac{-1 \pm \sqrt{1 + 4008}}{2}$$
This gives the values $x = 31.16$ and $x = -32.16$. But since x must be greater than one for $\log_{10}(x - 1)$ to be defined, the only solution to the original equation is $x = 31.16$.

47. $3^{x-2} = 7$
$$\log_{10} 3^{x-2} = \log_{10} 7$$
$$(x - 2)\log_{10} 3 = \log_{10} 7$$
$$x = \frac{\log_{10} 7}{\log_{10} 3} + 2 = 3.77$$

49. $1200 = 24 \cdot 6^{-0.3x}$

$50 = 6^{-0.3x}$

$\log_{10} 50 = \log_{10} 6^{-0.3x}$

$\log_{10} 50 = -0.3x \log_{10} 6$

$x = \dfrac{\log_{10} 50}{-0.3 \log_{10} 6} \approx -7.278$

51. $\dfrac{N}{N_0} = 10^{kt}$

$\log_{10} \dfrac{N}{N_0} = \log_{10} 10^{kt}$

$\log_{10} \dfrac{N}{N_0} = kt$

$t = \dfrac{1}{k} \log_{10} \dfrac{N}{N_0}$

53. a. $P(80) = 3800 \cdot 2^{-80/20} = 238$

b. $120 = 3800 \cdot 2^{-t/20}$

$2^{-t/20} = \dfrac{120}{3800} = \dfrac{3}{95}$

$\log_{10} 2^{-t/20} = \log_{10} \dfrac{3}{95}$

$-\dfrac{t}{20} \log_{10} 2 = \log_{10} \dfrac{3}{95}$

$t = -\dfrac{20 \log_{10} \frac{3}{95}}{\log_{10} 2} = 100$ years

The population will dip below 120 people in the year 2010.

55. a. $C(t) = 90(1.06)^t$

b. 10 months = $\dfrac{5}{6}$ year, so

$C\left(\dfrac{5}{6}\right) = 90(1.06)^{5/6} \approx \$94.48$

c. $120 = 90(1.06)^t$

$\dfrac{4}{3} = 1.06^t$

$\log_{10} \dfrac{4}{3} = \log_{10} 1.06^t$

$\log_{10} \dfrac{4}{3} = t \log_{10} 1.06$

$t = \dfrac{\log_{10} \frac{4}{3}}{\log_{10} 1.06} \approx 4.94$ years

57. Convert to logarithmic form:
$x = \ln 4.7 = 1.548$

59. Convert to exponential form:
$x = e^{6.02} \approx 411.58$

61. $e^{0.6x} = \dfrac{4.73}{1.2}$

Convert to logarithmic form:

$0.6x = \ln \dfrac{4.73}{1.2}$

$x = \dfrac{1}{0.6} \ln \dfrac{4.73}{1.2} = 2.286$

63. a. Refer to the graph in the back of the textbook.

t	0	2	4	6	8	10
V(t)	0	63.2	86.5	95.0	98.2	99.3

b. The graph is increasing and concave down. In the long run, $V(t)$ approaches 100.

c. $75 = 100\left(1 - e^{-0.5t}\right)$

$\dfrac{3}{4} = 1 - e^{-0.5t}$

$e^{-0.5t} = \dfrac{1}{4}$

$-0.5t = \ln\dfrac{1}{4}$

$t = -2\ln\dfrac{1}{4} \approx 2.77$ seconds

65. $y = 12e^{-kt} + 6$

$y - 6 = 12e^{-kt}$

$\dfrac{y-6}{12} = e^{-kt}$

$t = -\dfrac{1}{k}\ln\left(\dfrac{y-6}{12}\right)$

This answer can also be written as

$t = \dfrac{1}{k}\ln\left(\dfrac{y-6}{12}\right)^{-1} = \dfrac{1}{k}\ln\left(\dfrac{12}{y-6}\right)$

67. $N(t) = 600(0.4)^t = 600(e^{\ln 0.4})^t$

$= 600e^{-0.9163t}$

69. a.

d	*W(d)*
1	2¢
2	4¢
3	8¢
4	16¢
5	32¢

There is definitely a pattern: See part (b):

b. $W(d) = 2^d$

c. $W(15) = \$327.68$
$W(30) = \$10,737,418.24$

Chapter Eight: Polynomial and Rational Functions

Homework 8.1

1. $(3x-2)(4x^2+x-2) = 3x(4x^2+x-2)-2(4x^2+x-2)$

   $= 12x^3+3x^2-6x-8x^2-2x+4 = 12x^3-5x^2-8x+4$

3. $(x-2)(x-1)(x-3) = (x-2)(x^2-4x+3) = x(x^2-4x+3)-2(x^2-4x+3)$

   $= x^3-4x^2+3x-2x^2+8x-6 = x^3-6x^2+11x-6$

5. $(2a^2-3a+1)(3a^2+2a-1) = 2a^2(3a^2+2a-1)-3a(3a^2+2a-1)+1(3a^2+2a-1)$

   $= 6a^4+4a^3-2a^2-9a^3-6a^2+3a+3a^2+2a-1 = 6a^4-5a^3-5a^2+5a-1$

7. $(y-2)(y+2)(y+4)(y+1) = [(y-2)(y+2)][(y+4)(y+1)] = (y^2-4)(y^2+5y+4)$

   $= y^2(y^2+5y+4)-4(y^2+5y+4) = y^4+5y^3+4y^2-4y^2-20y-16$

   $= y^4+5y^3-20y-16$

9. We need to find the constant, 1rst degree, and 2nd degree terms of the product. Multiply each term in the first polynomial by only the terms in the second polynomial that will result in a 2nd degree or lower product.

   $2(3)+2(2x)+2(-x^2)-x(3)-x(2x)+3x^2(3)$

   $= 6+4x-2x^2-3x-2x^2+9x^2 = 6+x+5x^2$

11. The product of the polynomials will not have any 1rst degree or 3rd degree terms. Therefore, we need to find the constant, 2nd degree, and 4th degree terms of the product. Multiply each term in the first polynomial by only the terms in the second polynomial that will result in a 4th degree or lower product.

    $1(4+x^2-2x^4)-2x^2(4+x^2)-x^4(4)$

    $= 4+x^2-2x^4-8x^2-2x^4-4x^4 = 4-7x^2-8x^4$

13. Multiply each term in the first polynomial by only the terms in the second polynomial that will result in a 2nd degree product.

    $4(2x^2)+2x(-3x)-x^2(2) = 8x^2-6x^2-2x^2 = 0x^2$

15. Multiply each term in the first polynomial by only the terms in the second polynomial that will result in a 3rd degree product.

$$3x\left(-3x^2\right) + x^3(1) = -9x^3 + x^3 = -8x^3$$

17. $(x + y)^3 = (x + y)(x + y)^2 = (x + y)(x^2 + 2xy + y^2)$
$= x(x^2 + 2xy + y^2) + y(x^2 + 2xy + y^2) = x^3 + 2x^2y + xy^2 + x^2y + 2xy^2 + y^3$
$= x^3 + 3x^2y + 3xy^2 + y^3$

19. $(x + y)(x^2 - xy + y^2) = x(x^2 - xy + y^2) + y(x^2 - xy + y^2)$
$= x^3 - x^2y + xy^2 + x^2y - xy^2 + y^3 = x^3 + y^3$

21. a. The formula begins with x^3 and ends with y^3. As you proceed from term to term, the powers on x decrease while the powers on y increase, and on each term the sum of the powers is 3. The coefficients of the two middle terms are both 3.

b. The formula is the same as for $(x + y)^3$, except that the terms alternate in sign.

23. $(1 + 2z)^3 = 1^3 + 3(1)^2(2z) + 3(1)(2z)^2 + (2z)^3 = 1 + 6z + 12z^2 + 8z^3$

25. $\left(1 - 5\sqrt{t}\right)^3 = 1^3 - 3(1)^2\left(5\sqrt{t}\right) + 3(1)\left(5\sqrt{t}\right)^2 - \left(5\sqrt{t}\right)^3 = 1 - 15\sqrt{t} + 75t - 125t\sqrt{t}$

27. $(x - 1)\left(x^2 + x + 1\right) = x\left(x^2 + x + 1\right) - 1\left(x^2 + x + 1\right)$
$= x^3 + x^2 + x - x^2 - x - 1 = x^3 - 1$

29. $(2x + 1)\left(4x^2 - 2x + 1\right) = 2x\left(4x^2 - 2x + 1\right) + 1\left(4x^2 - 2x + 1\right)$
$= 8x^3 - 4x^2 + 2x + 4x^2 - 2x + 1 = 8x^3 + 1$

31. $(3a - 2b)\left(9a^2 + 6ab + 4b^2\right) = 3a\left(9a^2 + 6ab + 4b^2\right) - 2b\left(9a^2 + 6ab + 4b^2\right)$
$= 27a^3 + 18a^2b + 12ab^2 - 18a^2b - 12ab^2 - 8b^3 = 27a^3 - 8b^3$

33. $x^3 + 27 = x^3 + 3^3 = (x + 3)\left(x^2 - 3x + 3^2\right) = (x + 3)\left(x^2 - 3x + 9\right)$

35. $a^3 - 8b^3 = a^3 - (2b)^3 = (a - 2b)\left(a^2 + 2ab + (2b)^2\right) = (a - 2b)\left(a^2 + 2ab + 4b^2\right)$

37. $x^3y^3 - 1 = (xy)^3 - 1^3 = (xy - 1)\left((xy)^2 + xy + 1\right) = (xy - 1)\left(x^2y^2 + xy + 1\right)$

39. $27a^3 + 64b^3 = (3a)^3 + (4b)^3 = (3a + 4b)\left((3a)^2 - (3a)(4b) + (4b)^2\right)$
$= (3a + 4b)\left(9a^2 - 12ab + 16b^2\right)$

41. $125a^3b^3 - 1 = (5ab)^3 - 1^3 = (5ab - 1)\left((5ab)^2 + 5ab + 1^2\right)$

$\quad = (5ab - 1)\left(25a^2b^2 + 5ab + 1\right)$

43. a. The base and the top of the box each have area x^2. Each of the four sides of the box have area $8x$. The total surface area is $S(x) = 2x^2 + 32x$.

b. $S(18) = 2(18)^2 + 32(18) = 1224$ square inches

45. a. The area of the shaded region is equal to the area of the rectangular face minus the areas of the circles:

$$A(x) = (3x)(2x) - \pi\left(\frac{x}{2}\right)^2 - \pi x^2 = 6x^2 - \frac{\pi x^2}{4} - \pi x^2 = 6x^2 - \frac{5}{4}\pi x^2 = \left(6 - \frac{5\pi}{4}\right)x^2$$

b. $A(8) = \left(6 - \frac{5\pi}{4}\right)8^2 = 132.67$ square inches

47. a. The total volume is the sum of the volumes of the cylinder and the hemisphere:

$$V = \pi r^2 h + \frac{1}{2}\left(\frac{4}{3}\right)\pi r^3 = \pi r^2 h + \frac{2}{3}\pi r^3$$

b. The total height of the silo is $h + r$, so $h = 5r - r = 4r$.

$$V(r) = \pi r^2(4r) + \frac{2}{3}\pi r^3 = 4\pi r^3 + \frac{2}{3}\pi r^3 = \frac{14}{3}\pi r^3$$

49. a. $500(1 + r)^2;\ 500(1 + r)^3;\ 500(1 + r)^4$

b. $500(1 + r)^2 = 500\left(1 + 2r + r^2\right) = 500 + 1000r + 500r^2 = 500r^2 + 1000r + 500$

$500(1 + r)^3 = 500\left(1 + 3r + 3r^2 + r^3\right) = 500r^3 + 1500r^2 + 1500r + 500$

$500(1 + r)^4 = 500(1 + r)(1 + r)^3$

$\qquad = 500(1 + r)\left(1 + 3r + 3r^2 + r^3\right)$

$\qquad = 500\left(1 + 3r + 3r^2 + r^3 + r + 3r^2 + 3r^3 + r^4\right)$

$\qquad = 500\left(r^4 + 4r^3 + 6r^2 + 4r + 1\right)$

$\qquad = 500r^4 + 2000r^3 + 3000r^2 + 2000r + 500$

c. $500(1.08)^2 = \$583.20;\ 500(1.08)^3 = \$629.86;\ 500(1.08)^4 = \$680.24$

51. a. $l = 16 - 2x$; $w = 12 - 2x$; $h = x$

  b. $V(x) = lwh = x(16 - 2x)(12 - 2x)$

  c. Since x must be positive and $2x$ cannot exceed 12, $0 < x < 6$.

  d.

x	1	2	3	4	5
V	140	192	180	128	60

  e. Refer to the graph in the back of the textbook.

  f. Use TRACE or *maximum* to find that the high point on the graph, and thus the maximum volume, occurs where $x = 2.26$ inches. The maximum volume is the y-value of the high point on the graph, or $V(2.26) = 194.07$ cubic inches.

53. a. When
$$3x^2 - \frac{1}{3}x^3 = x^2\left(3 - \frac{x}{3}\right) = 0,$$
$x = 0$ or $x = 9$.

  b. When $x = 0$ or $x = 9$, the blood pressure remains constant, so these values should not be included. Since a person cannot take a negative amount of medication, x must be positive. An amount greater than 9 milliliters gives a negative value of R, which would result in an increased blood pressure. Therefore, $0 < x < 9$.

  c. Refer to the graph in the back of the textbook.

  d. The patient's blood pressure should drop by
$$f(2) = 3(2)^2 - \left(\frac{1}{3}\right)(2)^3$$
$$= \frac{28}{3} \text{ points.}$$

  e. At the highest point on the graph, $f(x) = 36$ points, so this is the maximum drop in blood pressure.

  f. The graph indicates, and computation confirms, that $f(3) = 3^2(3 - 1) = 18$, which is half of the 36 points found in part (e). Thus, 3 milliliters should be administered. (There is also a higher value, $x = 8.2$ ml, such that $f(x) = 18$, but there's no point in administering more medication if less will do.)

55. a. Refer to the graph in back of the textbook.

   b. In 1960, $P(0) = 900$.
   In 1975, $P(15) = 11,145$.
   In 1994, $P(34) = 15,078$

   c. From 1960 to 1961, the population grew by $P(1) - P(0) = 2241 - 900 = 1341$. From 1975 to 1976, the population grew by $P(16) - P(15) = 11,316 - 11,145 = 171$. From 1994 to 1995, the population grew by $P(35) - P(34) = 15,705 - 15,078 = 627$

   d. The graph is flattest at $t = 21$, or in 1981.

57. a. For $x = 0$, $y = 20$, so the upper section shifted by 20 cm.

   b. Graphing y versus x using a window with Xmin $= -150$, Xmax $= 0$, Ymin $= -5$, and Ymax $= 20$, and then locating the point where $y = 0$, shows that $x = -100$ at this point. This is confirmed algebraically: $f(-100) = -0.00004(-100)^3 - 0.006(-100)^2 + 20 = 0$. Hence the ramp runs from -100 cm to 0 cm and is 100 cm long.

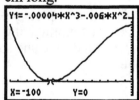

Homework 8.2

1. a. The duck's ground speed is
$50 - v$, so $t = \dfrac{150}{50 - v}$.

b.

v	0	5	10	15	20	25	30	35	40	45	50
t	3	3.33	3.75	4.29	5	6	7.5	10	15	30	–

As the headwind increases, so does the travel time.

c. The vertical asymptote, $v = 50$, signifies that as the wind speed approaches 50 miles per hour, the duck's travel time gets larger and larger without bound. The degree of the denominator exceeds that of the numerator, so $t = 0$ is a horizontal asymptote. Refer to the graph in the back of the textbook.

3. a. Percentages range from 0 to 100, so $0 \le p < 100$. (Note that $p \ne 100$ or else the denominator would be zero.)

b.

P	0	15	25	40	50	75	80	90	100
C	0	12.7	24	48	72	216	288	648	–

c. Refer to the graph in the back of the textbook. Let $C = 108$.
$$108 = \frac{72p}{100 - p}$$
$$10{,}800 - 108p = 72p$$
$$10{,}800 = 180p$$
$$60 = p$$
Thus, 60% of the population can be immunized. This answer can be found graphically by locating the point that has vertical coordinate 108. (Use TRACE or graph the line $y = 108$ and use the intersect feature on your calculator.)

d. The curve rises above 1728 when for $p > 96$. This number can also be obtained algebraically by letting $C = 1728$ and solving for p as in part (c).

e. The vertical asymptote, $p = 100$, indicates that immunizing 100% is infinitely expensive and so cannot be done.

5. a. $C(n) = \dfrac{20,000 + 8n}{n} = 8 + \dfrac{20,000}{n}$

b.

n	100	200	400	500	1000	2000	4000	5000	8000
C	208	108	58	48	28	18	13	12	10.5

c. Refer to the graph in the back of the textbook.

d. From the table, when the cost is $18, the number of calculators produced should be 2000.

e. $8 + \dfrac{20,000}{n} < 12$; $4n > 20,000$;

$n > 5000$. (This result can also be found from the point (5000, 12) in the table and from noting that the C-values of the graph are less than 12 for n greater than 5000.)

f. The numerator's and denominator's degrees in $\dfrac{20,000 + 8n}{n}$ are equal with leading coefficients ratio of 8. So $C = 8$ is the horizontal asymptote. $8 is what the cost per calculator approaches as n gets large.

7. a. Let $R(x)$ represent the cost of reordering x computers. Then

$R(x) = \dfrac{300}{x}(15x + 10)$

$= 4500 + \dfrac{3000}{x}$

The cost of storage is $6x$, so the inventory cost is

$C(x) = R(x) + 6x$

$= 4500 + \dfrac{3000}{x} + 6x$

b.

x	10	20	30	40	50
C	4860	4770	4780	4815	4860

x	60	70	80	90	100
C	4910	4963	5018	5073	5130

c. Refer to the graph in the back of the textbook. The smallest value of C is the second coordinate of the lowest point on the graph: $4768.33. To find this minimum point, you can use TRACE, or, for better accuracy, use the *minimum* feature on your graphing calculator (called FMIN on some models).

d. The minimum comes at $x = 22.36$, which is the x-value of the point found in part (c). Since computers come as whole units, the inventory cost will be minimized when $x = 22$. Orders per year is $\dfrac{300}{22} = 13.6$, so there will be 14 orders per year. (One of the orders will be smaller than the rest.)

e. Refer to the graph in the back of the textbook. $y = 6x + 4500$ is a "slant" asymptote for the cost function.

9. a. The base and the top of the box each have area x^2, and each of the four sides have area xh. Therefore, the surface area is $2x^2 + 4xh$. Let this expression equal 96 and solve for h:

$$2x^2 + 4xh = 96$$

$$4xh = 96 - 2x^2$$

$$h = \frac{96 - 2x^2}{4x} = \frac{24}{x} - \frac{x}{2}$$

b. $V(x) = lwh = x^2h = 24x - \dfrac{x^3}{2}$

c.

x	1	2	3	4	5	6	7
h	23.5	11	6.5	4	2.3	1	-0.07
V	23.5	44	58.5	64	57.5	36	-3.5

If $x = 7$, then the base of the box has area 49 sq. cm and the top of the box has area 49 sq. cm., and the total of these areas is greater than the total surface area, 96 sq. cm. This makes it impossible to find positive values of h and V.

d. Refer to the graph in the back of the textbook. At the highest point on the curve, $V(x) = 64$. (Use TRACE or the *maximum* feature of your graphing calculator.)

e. $V(4) = 64\ cm^3$, so $x = 4$ cm gives the maximum volume.

f. Refer to the graph in the back of the textbook. When $x = 4$, $h = 4$. This can be found using *value* (EVAL on some calculators) and letting $x = 4$. Algebraically, we have
$$h = \frac{24}{4} - \frac{4}{2} = 4 \text{ cm.}$$

11. a.

v	-100	-75	-50	-25	0
P	338.15	358.92	382.41	409.19	440

v	25	50	75	100
P	475.83	518.01	568.4	629.66

b. Refer to the graph in the back of the textbook.

c. For a pitch of 415 hertz:
$$415 = \frac{440(332)}{332 - v}$$
$$137{,}780 - 415v = 146{,}080$$
$$v = \frac{146{,}080 - 137{,}780}{-415} = -20$$
The train is receding at 20 m/s.

For a pitch of $553\frac{1}{3}$ hertz:
$$553\frac{1}{3} = \frac{440(332)}{332 - v}$$
$$183{,}706\frac{2}{3} - \left(553\frac{1}{3}\right)v = 146{,}080$$
$$v = \frac{146{,}080 - 183{,}706\frac{2}{3}}{-553\frac{1}{3}} \approx 68$$
The train is approaching at about 68 m/s. These answers can be found graphically by locating the first coordinates of the points with second coordinates 415 and $553\frac{1}{3}$.

d. $456.5 = \dfrac{440(332)}{332 - v}$
$$151{,}558 - 456.5v = 146{,}080$$
$$v = \frac{146{,}080 - 151{,}558}{-456.5} = 12.$$
From the graph, we see that when $v > 12$, $P > 456.5$. The train must be approaching at over 12 m/s.

e. The vertical asymptote is $v = 332$. As the train's speed approaches 332 m/s (the speed of sound), the pitch becomes infinitely high.

13. $x + 3$ equals zero for $x = -3$, so that is the vertical asymptote. The degree of the denominator exceeds that of the numerator, so $y = 0$ is the horizontal asymptote. To complete the sketch, plot a few points on either side of the vertical asymptote. For example, when $x = -2$, $y = 1$, so $(-2, 1)$ is a point on the graph. The y-intercept is $\left(0, \frac{1}{3}\right)$. Refer to the graph in the back of the textbook.

15. $x^2 - 5x + 4 = (x - 4)(x - 1)$ equals zero for $x = 1$ and $x = 4$, so those are the two vertical asymptotes. The degree of the denominator exceeds that of the numerator, so $y = 0$ is the horizontal asymptote. To complete the graph, plot several points: for example, $(2, -1)$, $\left(5, \frac{1}{2}\right)$, and the y-intercept $\left(0, \frac{1}{2}\right)$. Refer to the graph in the back of the textbook.

17. The vertical asymptote is $x = -3$ (where the denominator is zero). The degree of the numerator and denominator are equal, and the ratio of lead coefficients is 1, so $y = 1$ is the horizontal asymptote. The numerator equals zero for $x = 0$, so $(0, 0)$ is the x-intercept (and also the y-intercept in this case). Plot a few more points: for example, $(-2, -2)$ and $\left(1, \frac{1}{4}\right)$. Refer to the graph in the back of the textbook.

19. The vertical asymptote is $x = -2$ (where the denominator is zero). The numerator and denominator have equal degree and the ratio of lead coefficients is 1, so $y = 1$ is the horizontal asymptote. The numerator equals zero for $x = -1$, so the x-intercept is $(-1, 0)$. The y-intercept is $\left(0, \frac{1}{2}\right)$. Plot several more points: for example, $\left(1, \frac{2}{3}\right)$ and $(-3, 2)$. Refer to the graph in the back of the textbook.

21. The vertical asymptotes are $x = -2$ and $x = 2$ (where the denominator is zero). The degree of the denominator exceeds that of the numerator, so $y = 0$ is the horizontal asymptote. The x and y-intercept is $(0, 0)$. Plot additional points, such as $\left(1, -\frac{2}{3}\right)$ and $\left(-1, \frac{2}{3}\right)$. Refer to the graph in the back of the textbook.

23. $x^2 + 5x + 4 = (x + 4)(x + 1)$, so the vertical asymptotes are $x = -4$ and $x = -1$. The degree of the denominator exceeds that of the numerator, so $y = 0$ is the horizontal asymptote. The numerator is zero for $x = 2$, so $(2, 0)$ is the x-intercept. The y-intercept is $\left(0, -\frac{1}{2}\right)$. Plot additional points, such as $(-2, 2)$ and $\left(-5, -\frac{7}{4}\right)$. Refer to the graph in the back of the textbook.

25. The vertical asymptotes are $x = 2$ and $x = -2$ (where the denominator is zero). The numerator's and denominator's degrees are equal, with a leading coefficient ratio of 1, so $y = 1$ is the horizontal asymptote. The numerator is zero for $x = -1$ and $x = 1$, so the x-intercepts are $(-1, 0)$ and $(1, 0)$. The y-intercept is $\left(0, \frac{1}{4}\right)$.

Plot additional points, such as $\left(3, \frac{8}{5}\right)$ and $\left(-3, \frac{8}{5}\right)$. Refer to the graph in the back of the textbook.

27. The vertical asymptote is $x = 1$ (where the denominator is zero). The denominator's degree exceeds the numerator's so $y = 0$ is the horizontal asymptote. The numerator is equal to zero at $x = -1$, so $(-1, 0)$ is the x-intercept. The y-intercept is at $(0, 1)$. Plot additional points, such as $(2, 3)$ and $(3, 1)$. Refer to the graph in the back of the textbook.

29. $x^2 + 3$ is never equal to zero so there are no vertical asymptotes. The denominator's degree exceeds the numerator's, so $y = 0$ is the horizontal asymptote. The x and y-intercept is $(0, 0)$. Plot additional points, such as $\left(-1, -\frac{1}{4}\right)$ and $\left(1, \frac{1}{4}\right)$. Refer to the graph in the back of the textbook.

31. $x^2 + 1$ is never equal to zero so there are no vertical asymptotes. The denominator's degree exceeds the numerator's, so $y = 0$ is the horizontal asymptote. The x and y-intercept is $(0, 0)$. Plot additional points, such as $(1, 2)$ and $(-1, -2)$. Refer to the graph in the back of the textbook.

33. a. $\dfrac{1}{y} = \dfrac{1}{k} + \dfrac{1}{x} = \dfrac{x}{kx} + \dfrac{k}{kx} = \dfrac{x+k}{kx}$.

But if $\dfrac{1}{y} = \dfrac{x+k}{kx}$, then their reciprocals are also equal:

$$y = \dfrac{kx}{x+k}.$$

b. Refer to the graph in the back of the textbook. The graphs increase from the origin and approach a horizontal asymptote at $y = k$.

35. a. $y = 12$ is the horizontal asymptote, so $a = 12$. The graph passes through the point $(100, 10)$, so solve the following for k:

$$10 = \dfrac{12(100)}{100 + k}$$
$$1000 + 10k = 1200$$
$$10k = 200$$
$$k = 20$$

The equation is $y = \dfrac{12x}{x+20}$.

b. Plot using Xmin $= 0$, Xmax $= 100$, Ymin $= 0$, and Ymax $= 14$. Make sure to write parentheses around the denominator $x + 20$. The calculator's graph agrees with the graph given in the problem.

37. a. As s increases, v approaches V.

b. $v = \dfrac{VK}{K+K} = \dfrac{VK}{2K} = \dfrac{V}{2}$

c. Refer to the graph in the back of the textbook. $V \approx 0.7$ since it appears that the values of v are approaching 0.7 (see part (a)). Using this value of V and a point from the table (for example (1.00, 0.20)), we can find K:

$$0.2 = \frac{0.7(1)}{1+K}$$

$$0.2 + 0.2K = 0.7$$

$$0.2K = 0.5$$

$$K = 2.5$$

These values of V and K are only approximations, so many answers are possible.

d. Refer to the graph in the back of the textbook.

Homework 8.3

1. Factoring $7c^2d$ out of the numerator and denominator yields $-\dfrac{2}{d^2}$.

3. Factoring 2 out of the numerator and denominator yields $\dfrac{2x+3}{3}$.

5. $\dfrac{6a^3-4a^2}{4a}=\dfrac{2a^2(3a-2)}{4a}=\dfrac{a(3a-2)}{2}$

7. $\dfrac{6-6t^2}{(t-1)^2}=\dfrac{6\left(1-t^2\right)}{(t-1)^2}=\dfrac{6(1-t)(1+t)}{(t-1)^2}=-\dfrac{6(1+t)}{t-1}$ or $\dfrac{6(1+t)}{1-t}$

9. $\dfrac{2y^2-8}{2y+4}=\dfrac{2\left(y^2-4\right)}{2(y+2)}=\dfrac{2(y+2)(y-2)}{2(y+2)}=y-2$

11. $\dfrac{6-2v}{v^3-27}=\dfrac{2(3-v)}{(v-3)(v^2+3v+9)}=\dfrac{-2}{v^2+3v+9}$

13. $\dfrac{4x^3-36x}{6x^2+18x}=\dfrac{4x(x^2-9)}{6x(x+3)}=\dfrac{4x(x-3)(x+3)}{6x(x+3)}=\dfrac{2(x-3)}{3}$

15. $\dfrac{y^2-9x^2}{(3x-y)^2}=\dfrac{(y-3x)(y+3x)}{(y-3x)^2}=\dfrac{y+3x}{y-3x}$

17. $\dfrac{2x^2+x-6}{x^2+x-2}=\dfrac{(2x-3)(x+2)}{(x-1)(x+2)}=\dfrac{2x-3}{x-1}$

19. $\dfrac{8z^3-27}{4z^2-9}=\dfrac{(2z-3)(4z^2+6z+9)}{(2z+3)(2z-3)}=\dfrac{4z^2+6z+9}{2z+3}$

21. (b), since $\dfrac{4a^2-2a}{2a-1}=\dfrac{2a(2a-1)}{2a-1}=2a$

23. None

25. Combining and factoring out $24np$ yields $-\dfrac{np^2}{2}$.

27. Combining and factoring out a^2b^2 yields $\dfrac{5}{ab}$.

29. $\dfrac{5x+25}{2x}\cdot\dfrac{4x}{2x+10}=\dfrac{5(x+5)\cdot 2(2x)}{2x\cdot 2(x+5)}=5$

31. $\dfrac{4a^2-1}{a^2-16}\cdot\dfrac{a^2-4a}{2a+1}=\dfrac{(2a+1)(2a-1)\cdot a(a-4)}{(a-4)(a+4)\cdot(2a+1)}=\dfrac{a(2a-1)}{a+4}$

33. $\dfrac{2x^2-x-6}{3x^2+4x+1}\cdot\dfrac{3x^2+7x+2}{2x^2+7x+6}=\dfrac{(2x+3)(x-2)\cdot(3x+1)(x+2)}{(3x+1)(x+1)\cdot(2x+3)(x+2)}=\dfrac{x-2}{x+1}$

35. $\dfrac{3x^4-48}{x^4-4x^2-32}\cdot\dfrac{4x^4-8x^3+4x^2}{2x^4+16x}=\dfrac{3\left(x^4-16\right)}{\left(x^2-8\right)\left(x^2+4\right)}\cdot\dfrac{4x^2\left(x^2-2x+1\right)}{2x\left(x^3+8\right)}$

$$=\dfrac{3\left(x^2-4\right)\left(x^2+4\right)\cdot 4x^2(x-1)(x-1)}{\left(x^2-8\right)\left(x^2+4\right)\cdot 2x(x+2)\left(x^2-2x+4\right)}$$

$$=\dfrac{3(x-2)(x+2)\left(x^2+4\right)\cdot 4x^2(x-1)^2}{\left(x^2-8\right)\left(x^2+4\right)\cdot 2x(x+2)\left(x^2-2x+4\right)}$$

$$=\dfrac{6x(x-2)(x-1)^2}{\left(x^2-8\right)\left(x^2-2x+4\right)}$$

37. $\dfrac{4x-8}{3y}\div\dfrac{6x-12}{y}=\dfrac{4x-8}{3y}\cdot\dfrac{y}{6x-12}=\dfrac{4(x-2)\cdot y}{3y\cdot 6(x-2)}=\dfrac{2}{9}$

39. $\dfrac{a^2-a-6}{a^2+2a-15}\div\dfrac{a^2-4}{a^2+6a+5}=\dfrac{a^2-a-6}{a^2+2a-15}\cdot\dfrac{a^2+6a+5}{a^2-4}$

$=\dfrac{(a-3)(a+2)\cdot(a+1)(a+5)}{(a+5)(a-3)\cdot(a+2)(a-2)}=\dfrac{a+1}{a-2}$

41. $\dfrac{x^3+y^3}{x}\div\dfrac{x+y}{3x}=\dfrac{x^3+y^3}{x}\cdot\dfrac{3x}{x+y}=\dfrac{(x+y)(x^2-xy+y^2)(3x)}{x(x+y)}=3(x^2-xy+y^2)$

43. $1\div\dfrac{x^2-1}{x+2}=1\cdot\dfrac{x+2}{x^2-1}=\dfrac{x+2}{x^2-1}$

45. $(x^2-5x+4)\div\dfrac{x^2-1}{x^2}=(x^2-5x+4)\cdot\dfrac{x^2}{x^2-1}=(x-1)(x-4)\cdot\dfrac{x^2}{(x+1)(x-1)}$

$=\dfrac{x^2(x-4)}{x+1}$

47. $\dfrac{x^2+3x}{2y}\div 3x=\dfrac{x^2+3x}{2y}\cdot\dfrac{1}{3x}=\dfrac{x(x+3)}{6xy}=\dfrac{x+3}{6y}$

49. Write the expression as a sum of fractions and then simplify each fraction:

$$\frac{18r^2s^2 - 15rs + 6}{3rs} = \frac{18r^2s^2}{3rs} - \frac{15rs}{3rs} + \frac{6}{3rs} = 6rs - 5 + \frac{2}{rs}$$

51. Write the expression as a sum of fractions and then simplify each fraction:

$$\frac{15s^{10} - 21s^5 + 6}{-3s^2} = -\frac{15s^{10}}{3s^2} + \frac{21s^5}{3s^2} - \frac{6}{3s^2} = -5s^8 + 7s^3 - \frac{2}{s^2}$$

53. Since there is more than one term in the denominator, use long division:

$$2y + 1 \overline{)4y^2 + 12y + 7} \quad\Rightarrow\quad 2y + 5 + \frac{2}{2y+1}$$

$$\underline{-(4y^2 + 2y)}$$
$$10y + 7$$
$$\underline{-(10y + 5)}$$
$$2$$

55. Use long division:

$$x - 2 \overline{)x^3 + 2x^2 + x + 1} \quad\Rightarrow\quad x^2 + 4x + 9 + \frac{19}{x-2}$$

$$\underline{-(x^3 - 2x^2)}$$
$$4x^2 + x$$
$$\underline{-(4x^2 - 8x)}$$
$$9x + 1$$
$$\underline{-(9x - 18)}$$
$$19$$

57. Use long division:

$$2z + 1 \overline{)8z^4 + 0z^3 + 4z^2 + 5z + 3} \quad\Rightarrow\quad 4z^3 - 2z^2 + 3z + 1 + \frac{2}{2z+1}$$

$$\underline{-(8z^4 + 4z^3)}$$
$$-4z^3 + 4z^2$$
$$\underline{-(-4z^3 - 2z^2)}$$
$$6z^2 + 5z$$
$$\underline{-(6z^2 + 3z)}$$
$$2z + 3$$
$$\underline{-(2z + 1)}$$
$$2$$

59. Use long division:

$$x - 2 \overline{)x^4 + 0x^3 + 0x^2 + 0x - 1} \quad\Rightarrow\quad x^3 + 2x^2 + 4x + 8 + \frac{15}{x-2}$$

$$\underline{-(x^4 - 2x^3)}$$
$$2x^3 + 0x^2$$
$$\underline{-(2x^3 - 4x^2)}$$
$$4x^2 + 0x$$
$$\underline{-(4x^2 - 8x)}$$
$$8x - 1$$
$$\underline{-(8x - 16)}$$
$$15$$

1. $(t+4)\left(t^2-t-1\right)=t\left(t^2-t-1\right)+4\left(t^2-t-1\right)=t^3-t^2-t+4t^2-4t-4$

$=t^3+3t^2-5t-4$

3. $(v-10)^3=v^3-3v^2(10)+3v(10)^2-10^3=v^3-30v^2+300v-1000$

5. $y^3+27x^3=y^3+(3x)^3=(y+3x)\left(y^2-3xy+9x^2\right)$

7. a. The vertical asymptotes occur where the denominator equals zero. The denominator factors to $x(x+2)(x-2)$, so the vertical asymptotes are the lines $x=0$, $x=-2$, and $x=2$.

  b. The x-intercepts occur where $h(x)$ is equal to zero. This occurs where the numerator equals zero. The numerator factors to $(x-3)(x+3)$, so the x-intercepts are the points $(3, 0)$ and $(-3, 0)$.

9. The vertical asymptotes are $x=-1$ and $x=1$ (where the denominator is zero). The degree of the denominator exceeds the degree of the numerator so the horizontal asymptote is $y=0$. The numerator is equal to zero for $x=0$, so $(0, 0)$ is the

x-intercept. This point is also the y-intercept. Plot additional points, such as $\left(\frac{1}{2},-\frac{4}{3}\right)$

and $\left(2,\frac{4}{3}\right)$. Refer to the graph in the back of the textbook.

11. $\dfrac{x-12+6x^2}{17x-12-6x^2}=\dfrac{6x^2+x-12}{-(6x^2-17x+12)}=\dfrac{(3x-4)(2x+3)}{-(3x-4)(2x-3)}=\dfrac{2x+3}{-(2x-3)}=-\dfrac{2x+3}{2x-3}$

13. $\dfrac{2x^2+6x}{2(x+3)^2}=\dfrac{2x(x+3)}{2(x+3)^2}=\dfrac{x}{x+3}$

15. $\dfrac{15n^2}{3p}\cdot\dfrac{5p^2}{n^3}=\dfrac{3\cdot5\cdot5n^2p^2}{3pn^3}=\dfrac{25p}{n}$

17. $\dfrac{x^2-6x+5}{x^2+2x-3}\cdot\dfrac{x^2-4x-21}{x^2-10x+25}=\dfrac{(x-5)(x-1)}{(x+3)(x-1)}\cdot\dfrac{(x-7)(x+3)}{(x-5)(x-5)}=\dfrac{x-7}{x-5}$

19. $\dfrac{7a+14}{14a-28}\cdot\dfrac{4-2a}{a+2}=\dfrac{7(a+2)}{14(a-2)}\cdot\dfrac{-2(a-2)}{(a+2)}=-\dfrac{14}{14}=-1$

21. $\dfrac{2y-6}{y+2x}\div\dfrac{4y-12}{2y+4x}=\dfrac{2y-6}{y+2x}\cdot\dfrac{2y+4x}{4y-12}=\dfrac{2(y-3)}{(y+2x)}\cdot\dfrac{2(y+2x)}{4(y-3)}=1$

23. $\dfrac{x^3-1}{x^2+x+1} \div \dfrac{(x-1)^3}{x^2-1} = \dfrac{x^3-1}{x^2+x+1} \cdot \dfrac{x^2-1}{(x-1)^3} = \dfrac{(x-1)(x^2+x+1)}{(x^2+x+1)} \cdot \dfrac{(x-1)(x+1)}{(x-1)^3}$

$= \dfrac{x+1}{x-1}$

25. Use long division as below. The answer is y^2+2y-4.

$$
\begin{array}{r}
y^2+2y-4 \\
y+1\overline{)\,y^3+3y^2-2y-4} \\
\underline{-(y^3+y^2)} \\
2y^2-2y \\
\underline{-(2y^2+2y)} \\
-4y-4 \\
\underline{-(4y-4)} \\
0
\end{array}
$$

27. a.

X	Y₁
-74	763.1
-64	785.56
-54	806.35
-44	824.73
-34	840.08
-24	851.91
-14	859.85

X= -74

X	Y₁
-4	863.66
6	863.23
16	858.58
26	849.84
36	837.28
46	821.28
56	802.37

X= -4

X	Y₁
66	781.18
76	758.49
86	735.18
96	712.29
106	690.95
116	672.45
126	658.17

X=66

b. Refer to the graph in the back of the textbook.

c. $H(0) = 864$ min.

d. $H(-14) = 859.8$ min.

e. $(14)(60) = 840$ minutes, and the graph is above 840 for $-34 \le t \le 34$.

f. $(13)(60) = 780$ minutes, and the graph is below 780 for $t > 66$ or $t < -66$.

Homework 8.4

1. $\dfrac{x}{2} - \dfrac{3}{2} = \dfrac{x-3}{2}$

3. $\dfrac{1}{6}a + \dfrac{1}{6}b - \dfrac{5}{6}c = \dfrac{a}{6} + \dfrac{b}{6} - \dfrac{5c}{6} = \dfrac{a+b-5c}{6}$

5. $\dfrac{x-1}{2y} + \dfrac{x}{2y} = \dfrac{x-1+x}{2y} = \dfrac{2x-1}{2y}$

7. $\dfrac{3}{x+2y} - \dfrac{x-3}{x+2y} - \dfrac{x-1}{x+2y} = \dfrac{3-(x-3)-(x-1)}{x+2y} = \dfrac{3-x+3-x+1}{x+2y} = \dfrac{-2x+7}{x+2y}$

9. $6(x+y)^2 = 2\cdot 3\cdot(x+y)^2;\ 4xy^2 = 2^2xy^2;\ \text{LCD} = 2^2\cdot 3xy^2(x+y)^2 = 12xy^2(x+y)^2$

11. $a^2 + 5a + 4 = (a+1)(a+4);\ \text{LCD} = (a+4)(a+1)^2$

13. $x^2 - x = x(x-1);\ \text{LCD} = x(x-1)^3$

15. $\dfrac{x}{2} + \dfrac{2x}{3} = \dfrac{x}{2}\cdot\dfrac{3}{3} + \dfrac{2x}{3}\cdot\dfrac{2}{2} = \dfrac{3x}{6} + \dfrac{4x}{6} = \dfrac{7x}{6}$

17. $\dfrac{5y}{6} - \dfrac{3y}{4} = \dfrac{5y}{6}\cdot\dfrac{2}{2} - \dfrac{3y}{4}\cdot\dfrac{3}{3} = \dfrac{10y}{12} - \dfrac{9y}{12} = \dfrac{y}{12}$

19. $\dfrac{x+1}{2x} + \dfrac{2x-1}{3x} = \dfrac{x+1}{2x}\cdot\dfrac{3}{3} + \dfrac{2x-1}{3x}\cdot\dfrac{2}{2} = \dfrac{3x+3+4x-2}{6x} = \dfrac{7x+1}{6x}$

21. $\dfrac{5}{x} + \dfrac{3}{x-1} = \dfrac{5\cdot(x-1)}{x\cdot(x-1)} + \dfrac{3\cdot x}{(x-1)\cdot x} = \dfrac{5x-5+3x}{x(x-1)} = \dfrac{8x-5}{x(x-1)}$

23. $\dfrac{y}{2y-1} - \dfrac{2y}{y+1} = \dfrac{y\cdot(y+1)}{(2y-1)\cdot(y+1)} - \dfrac{2y\cdot(2y-1)}{(y+1)\cdot(2y-1)} = \dfrac{y^2+y-4y^2+2y}{(2y-1)(y+1)}$

$= \dfrac{-3y^2+3y}{(2y-1)(y+1)} = \dfrac{-3y(y-1)}{(2y-1)(y+1)}$

25. $\dfrac{y-1}{y+1} - \dfrac{y-2}{2y-3} = \dfrac{(y-1)\cdot(2y-3)}{(y+1)\cdot(2y-3)} - \dfrac{(y-2)\cdot(y+1)}{(2y-3)\cdot(y+1)} = \dfrac{2y^2-5y+3-\left(y^2-y-2\right)}{(y+1)(2y-3)}$

$= \dfrac{y^2-4y+5}{(y+1)(2y-3)}$

27. $\dfrac{7}{5x-10} - \dfrac{5}{3x-6} = \dfrac{7}{5(x-2)}\cdot\dfrac{3}{3} - \dfrac{5}{3(x-2)}\cdot\dfrac{5}{5} = \dfrac{-4}{15(x-2)}$

29. $\dfrac{y-1}{y^2-3y} - \dfrac{y+1}{y^2+2y} = \dfrac{y-1}{y(y-3)} - \dfrac{y+1}{y(y+2)} = \dfrac{(y-1)\cdot(y+2)}{y(y-3)\cdot(y+2)} - \dfrac{(y+1)\cdot(y-3)}{y(y+2)\cdot(y-3)}$

$= \dfrac{y^2+y-2-\left(y^2-2y-3\right)}{y(y-3)(y+2)} = \dfrac{3y+1}{y(y-3)(y+2)}$

31. $x - \dfrac{1}{x} = \dfrac{x}{1}\cdot\dfrac{x}{x} - \dfrac{1}{x} = \dfrac{x^2-1}{x}$

33. $x + \dfrac{1}{x-1} - \dfrac{1}{(x-1)^2} = \dfrac{x\cdot(x-1)^2}{(x-1)^2} + \dfrac{1\cdot(x-1)}{(x-1)\cdot(x-1)} - \dfrac{1}{(x-1)^2}$

$= \dfrac{x\cdot\left(x^2-2x+1\right)+x-1-1}{(x-1)^2} = \dfrac{x^3-2x^2+2x-2}{(x-1)^2}$

35. a. Let s represent the speed of the paddle wheel in still water. The speed with the current is $s+8$, the distance is 25, and the time is $t_1 = \dfrac{25}{s+8}$.

b. The speed against the current is $s-8$, the distance is 25, and the time is $t_2 = \dfrac{25}{s-8}$.

c. $t_1 + t_2 = \dfrac{25\cdot(s-8)}{(s+8)\cdot(s-8)} + \dfrac{25\cdot(s+8)}{(s-8)\cdot(s+8)} = \dfrac{50s}{(s-8)(s+8)}$

37. a. Let w represent the speed of the wind. Orville's ground speed is $400+w$, the distance is 900, and the time is $t_1 = \dfrac{900}{400+w}$.

b. Wilbur's ground speed is $400-w$, the distance is 900, and the time is $t_2 = \dfrac{900}{400-w}$.

c. Orville has a shorter time and arrives sooner, by

$t_1 - t_2 = \dfrac{900\cdot(400+w)}{(400-w)\cdot(400+w)} - \dfrac{900\cdot(400-w)}{(400+w)\cdot(400-w)} = \dfrac{1800w}{(400+w)(400-w)}$

39. $\dfrac{\frac{2}{a}+\frac{3}{2a}}{5+\frac{1}{a}} = \dfrac{2a\left(\frac{2}{a}+\frac{3}{2a}\right)}{2a\left(5+\frac{1}{a}\right)} = \dfrac{4+3}{10a+2} = \dfrac{7}{10a+2}$

41. $\dfrac{1+\frac{2}{a}}{1-\frac{4}{a^2}} = \dfrac{a^2\left(1+\frac{2}{a}\right)}{a^2\left(1-\frac{4}{a^2}\right)} = \dfrac{a^2+2a}{a^2-4} = \dfrac{a(a+2)}{(a+2)(a-2)} = \dfrac{a}{a-2}$

43. $\dfrac{h+\frac{h}{m}}{1+\frac{1}{m}} = \dfrac{m\left(h+\frac{h}{m}\right)}{m\left(1+\frac{1}{m}\right)} = \dfrac{mh+h}{m+1} = \dfrac{h(m+1)}{m+1} = h$

45. $\dfrac{1}{1-\frac{1}{q}} = \dfrac{q}{q\left(1-\frac{1}{q}\right)} = \dfrac{q}{q-1}$

47. $\dfrac{L+C}{\frac{1}{L}+\frac{1}{C}} = \dfrac{LC(L+C)}{LC\left(\frac{1}{L}+\frac{1}{C}\right)} = \dfrac{LC(L+C)}{C+L} = LC$

49. $\dfrac{\frac{4}{x^2}-\frac{4}{z^2}}{\frac{2}{z}-\frac{2}{x}} = \dfrac{x^2z^2\left(\frac{4}{x^2}-\frac{4}{z^2}\right)}{x^2z^2\left(\frac{2}{z}-\frac{2}{x}\right)} = \dfrac{4z^2-4x^2}{2x^2z-2xz^2} = \dfrac{4(z-x)(z+x)}{2xz(x-z)} = -\dfrac{2(z+x)}{xz}$

51. a. $p = q + 60$, so $\dfrac{1}{f} = \dfrac{1}{q+60} + \dfrac{1}{q}$.

  b. $\dfrac{1}{f} = \dfrac{1}{q+60} + \dfrac{1}{q} = \dfrac{q}{q(q+60)} + \dfrac{q+60}{q(q+60)} = \dfrac{2q+60}{q(q+60)} = \dfrac{2q+60}{q^2+60q}$

  Take the reciprocal of both sides: $f = \dfrac{q^2+60q}{2q+60}$

53. a. First leg: $t = \dfrac{d}{r_1}$; second leg: $t = \dfrac{d}{r_2}$

  b. Total distance $= 2d$; total time $= \dfrac{d}{r_1} + \dfrac{d}{r_2}$

  c. Average speed $= \dfrac{\text{total distance}}{\text{total time}} = \dfrac{2d}{\frac{d}{r_1}+\frac{d}{r_2}}$

  d. $\dfrac{2d}{\frac{d}{r_1}+\frac{d}{r_2}} \cdot \dfrac{r_1 r_2}{r_1 r_2} = \dfrac{2dr_1 r_2}{dr_2+dr_1} = \dfrac{2r_1 r_2}{r_1+r_2}$

  e. Average speed $= \dfrac{2(70 \cdot 50)}{70+50} = 58\frac{1}{3}\text{mi/h}$

55. $x^{-2}+y^{-2} = \dfrac{1}{x^2} + \dfrac{1}{y^2} = \dfrac{1}{x^2}\cdot\dfrac{y^2}{y^2} + \dfrac{1}{y^2}\cdot\dfrac{x^2}{x^2} = \dfrac{x^2+y^2}{x^2 y^2}$

57. $2w^{-1}-(2w)^{-2} = \dfrac{2}{w} - \dfrac{1}{(2w)^2} = \dfrac{2}{w} - \dfrac{1}{4w^2} = \dfrac{2}{w}\cdot\dfrac{4w}{4w} - \dfrac{1}{4w^2} = \dfrac{8w-1}{4w^2}$

59. $a^{-1}b - ab^{-1} = \dfrac{b}{a} - \dfrac{a}{b} = \dfrac{b^2}{ab} - \dfrac{a^2}{ab} = \dfrac{b^2 - a^2}{ab}$

61. $\left(x^{-1} + y^{-1}\right)^{-1} = \dfrac{1}{x^{-1} + y^{-1}} = \dfrac{1}{\frac{1}{x} + \frac{1}{y}} = \dfrac{xy}{xy\left(\frac{1}{x} + \frac{1}{y}\right)} = \dfrac{xy}{y + x}$

63. $\dfrac{x + x^{-2}}{x} = \dfrac{x + \frac{1}{x^2}}{x} = \dfrac{x^2\left(x + \frac{1}{x^2}\right)}{x^2 \cdot x} = \dfrac{x^3 + 1}{x^3}$

65. $\dfrac{a^{-1} + b^{-1}}{(ab)^{-1}} = \dfrac{\frac{1}{a} + \frac{1}{b}}{\frac{1}{ab}} = \dfrac{ab\left(\frac{1}{a} + \frac{1}{b}\right)}{ab \cdot \frac{1}{ab}} = \dfrac{b + a}{1} = b + a$

67. $\dfrac{\frac{2}{\sqrt{7}}}{1 - \frac{\sqrt{3}}{\sqrt{7}}} = \dfrac{\sqrt{7} \cdot \frac{2}{\sqrt{7}}}{\sqrt{7}\left(1 - \frac{\sqrt{3}}{\sqrt{7}}\right)} = \dfrac{2}{\sqrt{7} - \sqrt{3}} = \dfrac{2\left(\sqrt{7} + \sqrt{3}\right)}{\left(\sqrt{7} - \sqrt{3}\right)\left(\sqrt{7} + \sqrt{3}\right)} = \dfrac{2\left(\sqrt{7} + \sqrt{3}\right)}{7 - 3}$

$= \dfrac{2\left(\sqrt{7} + \sqrt{3}\right)}{4} = \dfrac{\sqrt{7} + \sqrt{3}}{2}$

69. $\dfrac{\frac{\sqrt{3}}{2} + \frac{1}{\sqrt{2}}}{1 - \frac{\sqrt{3}}{2} \cdot \frac{1}{\sqrt{2}}} = \dfrac{2\sqrt{2}\left(\frac{\sqrt{3}}{2} + \frac{1}{\sqrt{2}}\right)}{2\sqrt{2}\left(1 - \frac{\sqrt{3}}{2\sqrt{2}}\right)} = \dfrac{\sqrt{6} + 2}{2\sqrt{2} - \sqrt{3}} = \dfrac{\left(\sqrt{6} + 2\right)\left(2\sqrt{2} + \sqrt{3}\right)}{\left(2\sqrt{2} - \sqrt{3}\right)\left(2\sqrt{2} + \sqrt{3}\right)}$

$= \dfrac{2\sqrt{12} + \sqrt{18} + 4\sqrt{2} + 2\sqrt{3}}{4 \cdot 2 - 3} = \dfrac{4\sqrt{3} + 3\sqrt{2} + 4\sqrt{2} + 2\sqrt{3}}{5} = \dfrac{6\sqrt{3} + 7\sqrt{2}}{5}$

Homework 8.5

1.
$$\frac{6}{w+2} = 4$$
$$(w+2)\frac{6}{w+2} = 4(w+2)$$
$$6 = 4w + 8$$
$$-2 = 4w$$
$$-\frac{1}{2} = w$$
This value checks in the original equation.

3.
$$9 = \frac{h-5}{h-2}$$
$$9(h-2) = \frac{h-5}{h-2}(h-2)$$
$$9h - 18 = h - 5$$
$$8h = 13$$
$$h = \frac{13}{8}$$
This value checks in the original equation.

5.
$$\frac{15}{s^2} = 8$$
$$s^2 \cdot \frac{15}{s^2} = 8s^2$$
$$15 = 8s^2$$
$$\frac{15}{8} = s^2$$
$$\pm\sqrt{\frac{15}{8}} = s$$
$$s = \pm\frac{\sqrt{15}}{2\sqrt{2}} = \pm\frac{\sqrt{30}}{4}$$
These values check in the original equation.

7.
$$4.3 = \sqrt{\frac{18}{y}}$$
$$(4.3)^2 = \left(\sqrt{\frac{18}{y}}\right)^2$$
$$18.49 = \frac{18}{y}$$
$$18.49y = 18$$
$$y = \frac{18}{18.49} \approx 0.97$$
This value checks in the original equation.

9. $S = 1600$ lbs, $w = 4$ in, $h = 6$ in, and we need to solve for l:
$$1600 = \frac{182.6(4)(9)^2}{l}$$
$$1600 = \frac{59,162.4}{l}$$
$$1600l = 59,162.4$$
$$l = \frac{59,162.4}{1600} \approx 37 \text{ ft}$$

11. **a.** The ducks' speed against the wind is $50 - v$ mph, so $t = \frac{150}{50-v}$. Refer to the graph in the back of the textbook.

   b. Let $t = 4$ and solve for v:
$$4 = \frac{150}{50-v}$$
$$4(50-v) = 150$$
$$200 - 4v = 150$$
$$-4v = -50$$
$$v = 12.5 \text{ mph}$$
This corresponds to the point (12.5, 4) on the graph.

13. Let $C = 168$ and solve for p:

$$168 = \frac{72p}{100 - p}$$

$$168(100 - p) = 72p$$

$$16,800 - 168p = 72p$$

$$16,800 = 240p$$

$$70 = p$$

70% of the population can be immunized for $168,000.

15. a. The functions intersect for $x = 1$, so $x = 1$ is the solution. Refer to the graph in the back of the textbook.

b.

$$\frac{2x}{x+1} = \frac{x+1}{2}$$

$$2(x+1) \cdot \frac{2x}{x+1} = \frac{x+1}{2} \cdot 2(x+1)$$

$$4x = x^2 + 2x + 1$$

$$0 = x^2 - 2x + 1$$

$$0 = (x-1)(x-1)$$

Setting each factor equal to zero gives the solution $x = 1$.

17. a. The solution is the x-value of the intersection point: $x = \frac{1}{2}$. Refer to the graph in the back of the textbook.

b.

$$\frac{2}{x+1} = \frac{x}{x+1} + 1$$

$$(x+1)\left(\frac{2}{x+1}\right) = \left(\frac{x}{x+1} + 1\right)(x+1)$$

$$2 = x + x + 1$$

$$1 = 2x$$

$$\frac{1}{2} = x$$

19. a. Refer to the graph in the back of the textbook. The graphs cross at $x = 2\frac{1}{2} = \$2.50$.

b. $\dfrac{160}{x} = 6x + 49$; $160 = 6x^2 + 49x$;

$$6x^2 + 49x - 160 = 0$$

$$x = \frac{-49 \pm \sqrt{49^2 + 4(6)(160)}}{12}$$

$$= \frac{-49 \pm 79}{12}$$

$$x = \frac{5}{2} = 2\frac{1}{2}$$

(The negative solution is extraneous since the price of a hamburger cannot be negative.)

21. a. $3200 = lw$, so $l = \frac{3200}{w}$.

b. $P = 2w + 2l = 2w + \frac{6400}{w}$.

c. Refer to the graph in the back of the textbook. The lowest point is at $w = 56.57$, $P = 226.27$. Choosing $w = 56.57$ feet minimizes P for the given area, at $P = 226.27$ feet.

d. $240 = 2w + \frac{6400}{w}$

e. $240 = 2w + \frac{6400}{w}$

$2w^2 - 240w + 6400 = 0$

Divide both sides by 2.

$w^2 - 120w + 3200 = 0$

$w = \frac{120 \pm \sqrt{120^2 - 4(1)(3200)}}{2}$

$= \frac{120 \pm 40}{2} = 60 \pm 20$

$w = 60 - 20 = 40$ or $w = 60 + 20 = 80$. (Note: This equation could have also be solved by factoring.) If $w = 40$, then $l = 80$ and if $w = 80$, then $l = 40$. Thus, the dimensions of the garden are 40 ft by 80 ft.

23. Multiply each side of $\frac{a}{b} = \frac{c}{d}$ by the least common denominator bd:

$bd \cdot \frac{a}{b} = \frac{c}{d} \cdot bd$

$ad = cb$

25. $\frac{3}{4} = \frac{y+2}{12-y}$

$3(12 - y) = 4(y + 2)$

$36 - 3y = 4y + 8$

$28 = 7y$

$4 = y$

This value checks in the original equation.

27. $\frac{50}{r} = \frac{75}{r+20}$

$50(r + 20) = 75r$

$50r + 1000 = 75r$

$1000 = 25r$

$40 = r$

This value checks in the original equation.

29. Let t represent the amount of taxes on the $275,000 home.

$\frac{t}{275,000} = \frac{2700}{120,000}$

$120,000t = 742,500,000$

$t = \$6187.50$

31. Let l represent the actual length of the island.

$\frac{\frac{3}{8}}{10} = \frac{\frac{27}{16}}{l}$

$\frac{3}{8}l = \frac{270}{16}$

$l = \frac{270}{16} \cdot \frac{8}{3} = 45$ miles

33. Let p represent the original perch population.

$\frac{18}{80} = \frac{200}{p}$

$18p = 16,000$

$p = 889$ perch

35. a. height variation
$$= 8848 - (-11{,}034) = 19{,}882 \text{ m}$$

b. $\dfrac{19.882 \text{ km}}{6400 \text{ km}} \approx 0.003 = 0.3\%$

c. Let h represent the height of Mount Everest on the basketball-sized world.
$$\frac{h}{4.75 \text{ in}} = \frac{8848 \text{ m}}{6{,}400{,}000 \text{ m}}$$
$$6{,}400{,}000h = 42{,}028$$
$$h = 0.00657 \text{ inches}$$

37. a. *AEFB* is a square, so $AE = AB = 1$. *ABCD* is a rectangle, so $CD = AB = 1$. $ED = AD - AE = x - 1$.

b. $\dfrac{AB}{AD} = \dfrac{ED}{CD}$ or $\dfrac{1}{x} = \dfrac{x-1}{1}$

c. $1 = x(x-1); \quad 0 = x^2 - x - 1;$
$$x = \frac{-(-1) \pm \sqrt{(-1)^2 - 4(1)(-1)}}{2(1)}$$
$$= \frac{1 \pm \sqrt{5}}{2}$$
Since x must be positive, the golden ratio is $\dfrac{1+\sqrt{5}}{2}$.

39. Solve $S = \dfrac{a}{1-r}$ for r:
$$S(1-r) = a$$
$$S - Sr = a$$
$$S - a = Sr$$
$$\frac{S-a}{S} = r \ \text{ or } \ r = 1 - \frac{a}{S}$$

41. Solve $H = \dfrac{2xy}{x+y}$ for x:
$$H(x+y) = 2xy$$
$$Hx + Hy = 2xy$$
$$Hy = 2xy - Hx$$
$$Hy = x(2y - H)$$
$$\frac{Hy}{2y - H} = x$$

43. Solve $F = \dfrac{Gm_1 m_2}{d^2}$ for d:
$$Fd^2 = Gm_1 m_2$$
$$d^2 = \frac{Gm_1 m_2}{F}$$
$$d = \pm\sqrt{\frac{Gm_1 m_2}{F}}$$

45. Solve $\dfrac{1}{Q} + \dfrac{1}{I} = \dfrac{2}{r}$ for r:
$$QIr\left(\frac{1}{Q} + \frac{1}{I}\right) = \left(\frac{2}{r}\right)QIr$$
$$Ir + Qr = 2QI$$
$$r(I + Q) = 2QI$$
$$r = \frac{2QI}{I + Q}$$

47. a.
$$T = \frac{1}{v}\left(\frac{rD}{R} - d\right)$$

$$v \cdot T = v \cdot \frac{1}{v}\left(\frac{rD}{R} - d\right)$$

$$vT = \frac{rD}{R} - d$$

$$\frac{1}{T} \cdot vT = \frac{1}{T} \cdot \left(\frac{rD}{R} - d\right)$$

$$v = \frac{1}{T}\left(\frac{rD}{R} - d\right)$$

b. Substitute the given values and then compute on your calculator as shown in the following screen:

```
(1/2.68)*(3.82E5
*1.41E6/1.48E8-3
.48E3)
         59.44937475
```

The speed of the moon is about 59.45 km/min.

49.
$$\frac{3}{x-2} = \frac{1}{2} + \frac{2x-7}{2x-4}$$

$$\frac{3}{x-2} = \frac{1}{2} + \frac{2x-7}{2(x-2)}$$

$$2(x-2) \cdot \left(\frac{3}{x-2}\right) = \left(\frac{1}{2} + \frac{2x-7}{2(x-2)}\right) \cdot 2(x-2)$$

$$6 = x - 2 + 2x - 7$$

$$15 = 3x$$

$$5 = x$$

This value checks in the original equation.

51.
$$\frac{4}{x+2} - \frac{1}{x} = \frac{2x-1}{x(x+2)}$$

$$x(x+2) \cdot \left(\frac{4}{x+2} - \frac{1}{x}\right) = \left(\frac{2x-1}{x(x+2)}\right) \cdot x(x+2)$$

$$4x - (x+2) = 2x - 1$$

$$4x - x - 2 = 2x - 1$$

$$x = 1$$

This value checks in the original equation.

53.

$$\frac{x}{x+2} - \frac{3}{x-2} = \frac{x^2+8}{(x+2)(x-2)}$$

$$(x+2)(x-2) \cdot \left(\frac{x}{x+2} - \frac{3}{x-2} \right) = \left(\frac{x^2+8}{(x+2)(x-2)} \right) \cdot (x+2)(x-2)$$

$$x(x-2) - 3(x+2) = x^2+8$$

$$x^2 - 2x - 3x - 6 = x^2 + 8$$

$$-5x = 14$$

$$x = -\frac{14}{5}$$

This value checks in the original equation.

55.

$$\frac{4}{3x} + \frac{3}{3x+1} + 2 = 0$$

$$3x(3x+1) \cdot \left(\frac{4}{3x} + \frac{3}{3x+1} + 2 \right) = 0 \cdot 3x(3x+1)$$

$$4(3x+1) + 3(3x) + 2(3x)(3x+1) = 0$$

$$12x + 4 + 9x + 18x^2 + 6x = 0$$

$$18x^2 + 27x + 4 = 0$$

$$(6x+1)(3x+4) = 0$$

Setting each factor equal to zero gives $x = -\frac{1}{6}$ or $x = -\frac{4}{3}$, both of which check in the original equation.

57. a. Let s represent the airspeed of the plane. Then the ground speed is $s - 20$ and the time for the outward journey is $t_1 = \dfrac{144}{s - 20}$.

b. For the return journey, the ground speed is $s + 20$ and the time is $t_2 = \dfrac{144}{s + 20}$.

c. Refer to the graph in the back of the textbook. The point with y-coordinate 3 is $(100, 3)$. (Use TRACE or graph the line $y = 100$ and use the *intersect* feature.) If the airspeed of the plane is 100 mph, the round trip will take 3 hours.

d. $\dfrac{144}{s - 20} + \dfrac{144}{s + 20} = 3$

e.
$$\frac{144}{s - 20} + \frac{144}{s + 20} = 3$$
$$(s - 20)(s + 20) \cdot \left(\frac{144}{s - 20} + \frac{144}{s + 20} \right) = 3 \cdot (s - 20)(s + 20)$$
$$144(s + 20) + 144(s - 20) = 3\left(s^2 - 400\right)$$
$$144s + 2880 + 144s - 2880 = 3s^2 - 1200$$
$$0 = 3s^2 - 288s - 1200$$
$$0 = 3\left(s^2 - 96s - 400\right)$$
$$0 = 3(s - 100)(s + 4)$$

Setting each factor equal to zero gives $s = 100$ and $s = -4$, both of which check in the original equation. However, since airspeed is not negative, we disregard the negative value. The airspeed to maintain is 100 miles per hour.

Chapter 8 Review

1. $(2x - 5)(x^2 - 3x + 2) = 2x(x^2 - 3x + 2) - 5(x^2 - 3x + 2)$

    $= 2x^3 - 6x^2 + 4x - 5x^2 + 15x - 10 = 2x^3 - 11x^2 + 19x - 10$

3. $8x^3 - 27z^3 = (2x)^3 - (3z)^3 = (2x - 3z)((2x)^2 + 6xz + (3z)^2)$

    $= (2x - 3z)(4x^2 + 6xz + 9z^2)$

5. a. $\dfrac{n(n-1)(n-2)}{6} = \dfrac{n^3 - 3n^2 + 2n}{6} = \dfrac{n^3}{6} - \dfrac{3n^2}{6} + \dfrac{2n}{6} = \dfrac{n^3}{6} - \dfrac{n^2}{2} + \dfrac{n}{3}$

    b. $\dfrac{12}{6}(12 - 1)(12 - 2) = 220$ combinations of pizza

    c. Graph $y_1 = \dfrac{x}{6}(x - 1)(x - 2)$ and $y_2 = 1000$ for Xmin = 0, Xmax = 30, Ymin = 0, Ymax = 1200. Use the *intersect* feature.

    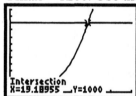

    Intersection
    X=19.18955   Y=1000

    $n = 20$ are needed. (Since 19 toppings give less than 1000 combinations, we round up to the nearest integer.)

7. a. $V = \pi r^2 h = \pi \left(\dfrac{h}{2}\right)^2 h = \dfrac{\pi h^3}{4}$

    b. $\dfrac{\pi(2)^3}{4} = 2\pi \approx 6.28 \text{ cm}^3;\ \dfrac{\pi(4)^3}{4} = 16\pi = 50.27 \text{ cm}^3$

    c. Refer to the graph in the back of the textbook. The answers to part (b) are verified by the points (2, 6.28) and (4, 50.27) on the graph. The point (5.03, 100) shows that a 100 cubic centimeter can will have a height of 5.03 cm. Algebraically:

    $100 = \dfrac{\pi h^3}{4};\ h = \sqrt[3]{\dfrac{400}{\pi}} \approx 5.03 \text{ cm}$

9. a. Refer to the graph in the back of the textbook (or below in part (d)).

b. $N = \dfrac{44(8)}{40 + (8)^2} \approx 3.38$, so the club had about 338 active members.

c. $2 = \dfrac{44t}{40 + t^2}$

$2\left(40 + t^2\right) = 44t$

$80 + 2t^2 = 44t$

$2\left(t^2 - 22t + 40\right) = 0$

$2(t - 2)(t - 20) = 0$

The club had 200 active members after 2 months and after 20 months.

d. The highest point on the graph is for $x \approx 6.32$, so during the sixth month, the club had the largest number of active members. The number of members eventually decreases to zero.

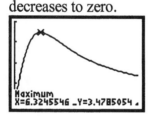

11. a. $x - 4 = 0$ when $x = 4$, so that is a vertical asymptote. The denominator's degree exceeds the numerator's, so $y = 0$ is a horizontal asymptote. $\dfrac{1}{0 - 4} = -\dfrac{1}{4}$ is a y-intercept.

b. Refer to the graph in the back of the textbook.

13. a. $x + 3 = 0$ when $x = -3$, so that is a vertical asymptote. The numerator's and denominator's degrees are equal with a leading coefficient ratio of 1, so $y = 1$ is a horizontal asymptote. $\dfrac{0 - 2}{0 + 3} = -\dfrac{2}{3}$ is a y-intercept. The numerator is zero for $x = 2$, so that is an x-intercept.

b. Refer to the graph in the back of the textbook.

15. a. $x^2 - 4 = 0$ when $x = \pm 2$, so those are two vertical asymptotes. The numerator's and denominator's degrees are equal with a leading coefficient ratio of 3, so $y = 3$ is a horizontal asymptote. $\dfrac{3(0)}{0 - 4} = 0$, so $(0, 0)$ is both a y- and an x-intercept.

b. Refer to the graph in the back of the textbook.

17. a. Let s represent their paddling speed in still water. Traveling upstream, their paddling speed is $s - 2$ and their time is $t = \dfrac{90}{s-2}$.

b. Traveling downstream, their paddling speed is $s + 2$ and their time is $t = \dfrac{90}{s+2}$.

c. Refer to the graph in the back of the textbook. The point is (8, 24). If the club paddles at 8 miles per hour, they will complete the trip after 24 hours of paddling.

d. If they paddle 6 hours per day for 4 days, their total paddling time will be 24 hours. Thus, $\dfrac{90}{s+2} + \dfrac{90}{s-2} = 24$.

e. $90(s-2)+90(s+2)=24(s-2)(s+2)$

$90s - 180 + 90s + 180 = 24\left(s^2 - 4\right)$

$180s = 24s^2 - 96$

$0 = 12\left(2s^2 - 15s - 8\right)$

$0 = 12(2s + 1)(s - 8)$

So $s = -\dfrac{1}{2}$ or $s = 8$. The speed cannot be negative, so the answer is 8 miles per hour.

19. Factoring out $2a(a - 1)^2$ yields $\dfrac{a}{2(a-1)}$.

21. Factoring out $2x^2y$ yields $\dfrac{y^2 - 2x}{2}$.

23. Factoring out $(a - 3)$ yields $\dfrac{a - 3}{2(a+3)}$.

25. Combining and factoring out $3ab$ yields $10ab$.

27. Combining and factoring out $2x(2x + 3)$ yields $\dfrac{6x}{2x+3}$.

29. $\dfrac{a^2 - a - 2}{a^2 - 4} \div \dfrac{a^2 + 2a + 1}{a^2 - 2a} = \dfrac{a^2 - a - 2}{a^2 - 4} \cdot \dfrac{a^2 - 2a}{a^2 + 2a + 1} = \dfrac{(a+1)(a-2)}{(a+2)(a-2)} \cdot \dfrac{a(a-2)}{(a+1)^2}$

$= \dfrac{a(a-2)}{(a+2)(a+1)}$

31. $1 \div \dfrac{4x^2 - 1}{2x+1} = 1 \cdot \dfrac{2x+1}{4x^2 - 1} = \dfrac{2x+1}{(2x+1)(2x-1)} = \dfrac{1}{2x-1}$

33. $\dfrac{36x^6 - 28x^4 + 16x^2 - 4}{4x^4} = \dfrac{36x^6}{4x^4} - \dfrac{28x^4}{4x^4} + \dfrac{16x^2}{4x^4} - \dfrac{4}{4x^4} = 9x^2 - 7 + \dfrac{4}{x^2} - \dfrac{1}{x^4}$

35.

$$\begin{array}{r} x^2 - 2x - 2 - \frac{1}{x-2} \\ x-2\overline{)x^3 - 4x^2 + 2x + 3} \\ \underline{-(x^3 - 2x^2)} \\ -2x^2 + 2x \\ \underline{-(-2x^2 + 4x)} \\ -2x + 3 \\ \underline{-(-2x + 4)} \\ -1 \end{array}$$

37. $\dfrac{x+2-(x-4)}{3x} = \dfrac{6}{3x} = \dfrac{2}{x}$

39. $\dfrac{3}{2(x-3)} \cdot \dfrac{x+3}{x+3} - \dfrac{4}{(x+3)(x-3)} \cdot \dfrac{2}{2} = \dfrac{3x+1}{2(x+3)(x-3)}$

41. $\dfrac{2a+1}{a-3} \cdot \dfrac{a-1}{a-1} - \dfrac{-2}{(a-3)(a-1)} = \dfrac{2a^2 - a + 1}{(a-3)(a-1)}$

43. $\dfrac{\frac{3}{4} - \frac{1}{2}}{\frac{3}{4} + \frac{1}{2}} \cdot \dfrac{4}{4} = \dfrac{3-2}{3+2} = \dfrac{1}{5}$

45. $\dfrac{x-4}{x - \frac{16}{x}} \cdot \dfrac{x}{x} = \dfrac{x^2 - 4x}{x^2 - 16} = \dfrac{x(x-4)}{(x+4)(x-4)} = \dfrac{x}{x+4}$

47. $\dfrac{y+3}{y+5} = \dfrac{1}{3}$

$3(y+3) = y+5$

$3y + 9 = y + 5$

$2y = -4$

$y = -2$

This value checks in the original equation.

49. $\dfrac{x}{x-2} = \dfrac{2}{x-2} + 7$

$x = 2 + 7(x-2)$

$x = 2 + 7x - 14$

$12 = 6x$

$2 = x$

This value of x makes the fractions in the equation undefined, so there is no solution.

51. Solve $V = C\left(1 - \dfrac{t}{n}\right)$ for n:

$V = C\left(1 - \dfrac{t}{n}\right)$

$V = C - \dfrac{Ct}{n}$

$nV = nC - Ct$

$nV - nC = -Ct$

$n(V - C) = -Ct$

$n = \dfrac{-Ct}{V - C} = \dfrac{Ct}{C - V}$

53. Solve $\dfrac{p}{q} = \dfrac{r}{q+r}$ for q:

$p(q + r) = rq$

$pq + pr = rq$

$pq - rq = -pr$

$q(p - r) = -pr$

$q = \dfrac{-pr}{p - r} = \dfrac{pr}{r - p}$

55. $\dfrac{1}{x^3} + \dfrac{1}{y} = \dfrac{y}{x^3 y} + \dfrac{x^3}{x^3 y} = \dfrac{x^3 + y}{x^3 y}$

57. $\dfrac{x^{-1} - y}{x - y^{-1}} = \dfrac{\frac{1}{x} - y}{x - \frac{1}{y}} = \dfrac{\frac{1}{x} - y}{x - \frac{1}{y}} \cdot \dfrac{xy}{xy} = \dfrac{y - xy^2}{x^2 y - x} = \dfrac{y(1 - xy)}{x(xy - 1)} = -\dfrac{y}{x}$

59. $\dfrac{x^{-1} - y^{-1}}{(x - y)^{-1}} = \dfrac{\frac{1}{x} - \frac{1}{y}}{\frac{1}{x-y}} = \dfrac{\frac{1}{x} - \frac{1}{y}}{\frac{1}{x-y}} \cdot \dfrac{xy(x - y)}{xy(x - y)} = \dfrac{y(x - y) - x(x - y)}{xy} = \dfrac{-(x - y)^2}{xy}$

Chapter Nine: Sequences and Series

Homework 9.1

1. $a_n = n - 5$
$a_1 = 1 - 5 = -4$
$a_2 = 2 - 5 = -3$
$a_3 = 3 - 5 = -2$
$a_4 = 4 - 5 = -1$

3. $c_n = \dfrac{n^2 - 2}{2}$
$c_1 = \dfrac{1^2 - 2}{2} = -\dfrac{1}{2}$
$c_2 = \dfrac{2^2 - 2}{2} = 1$
$c_3 = \dfrac{3^2 - 2}{2} = \dfrac{7}{2}$
$c_4 = \dfrac{4^2 - 2}{2} = 7$

5. $s_n = 1 + \dfrac{1}{n}$
$s_1 = 1 + \dfrac{1}{1} = 2$
$s_2 = 1 + \dfrac{1}{2} = \dfrac{3}{2}$
$s_3 = 1 + \dfrac{1}{3} = \dfrac{4}{3}$
$s_4 = 1 + \dfrac{1}{4} = \dfrac{5}{4}$

7. $u_n = \dfrac{n(n-1)}{2}$
$u_1 = \dfrac{1(1-1)}{2} = \dfrac{0}{2} = 0$
$u_2 = \dfrac{2(2-1)}{2} = \dfrac{2}{2} = 1$
$u_3 = \dfrac{3(3-1)}{2} = \dfrac{6}{2} = 3$
$u_4 = \dfrac{4(4-1)}{2} = \dfrac{12}{2} = 6$

9. $w_n = (-1)^n$
$w_1 = (-1)^1 = -1$
$w_2 = (-1)^2 = 1$
$w_3 = (-1)^3 = -1$
$w_4 = (-1)^4 = 1$

11. $B_n = \dfrac{(-1)^n (n-2)}{n}$
$B_1 = \dfrac{(-1)^1 (1-2)}{1} = 1$
$B_2 = \dfrac{(-1)^2 (2-2)}{2} = 0$
$B_3 = \dfrac{(-1)^3 (3-2)}{3} = -\dfrac{1}{3}$
$B_4 = \dfrac{(-1)^4 (4-2)}{4} = \dfrac{1}{2}$

13. $D_n = 1$
$D_1 = 1$
$D_2 = 1$
$D_3 = 1$
$D_4 = 1$

15. $a_1 = \dfrac{4}{3}$

17. $a_n = 3a_{n-1}$

19. $a_{n+1} = \dfrac{1}{3} a_{n+2}$

21. $D_6 = 2^6 - 6 = 58$

23. $x_{26} = \log 26 \approx 1.415$

25. $z_{20} = 2\sqrt{20} = 4\sqrt{5} \approx 8.944$

27. $s_1 = 3, \; s_n = s_{n-1} + 2$

$s_2 = s_1 + 2 = 3 + 2 = 5$

$s_3 = s_2 + 2 = 5 + 2 = 7$

$s_4 = s_3 + 2 = 7 + 2 = 9$

$s_5 = s_4 + 2 = 9 + 2 = 11$

n	1	2	3	4	5
s_n	3	5	7	9	11

29. $d_1 = 24, \; d_{n+1} = \frac{-1}{2} d_n$

$d_2 = \frac{-1}{2} d_1 = \frac{-1}{2}(24) = -12$

$d_3 = \frac{-1}{2} d_2 = \frac{-1}{2}(-12) = 6$

$d_4 = \frac{-1}{2} d_3 = \frac{-1}{2}(6) = -3$

$d_5 = \frac{-1}{2} d_4 = \frac{-1}{2}(-3) = \frac{3}{2}$

n	1	2	3	4	5
d_n	24	-12	6	-3	3/2

31. $t_1 = 1, \; t_{n+1} = (n+1)t_n$

$t_2 = (1+1)t_1 = 2(1) = 2$

$t_3 = (2+1)t_2 = 3(2) = 6$

$t_4 = (3+1)t_3 = 4(6) = 24$

$t_5 = (4+1)t_4 = 5(24) = 120$

n	1	2	3	4	5
t_n	1	2	6	24	120

33. $w_1 = 100, \; w_n = 1.10w_{n-1} + 100$

$w_2 = 1.10w_1 + 100$

$\quad = 1.10(100) + 100 = 210$

$w_3 = 1.10w_2 + 100$

$\quad = 1.10(210) + 100 = 331$

$w_4 = 1.10w_3 + 100$

$\quad = 1.10(331) + 100 = 464.1$

$w_5 = 1.10w_4 + 100$

$\quad = 1.10(464.1) + 100 = 610.51$

n	1	2	3	4	5
w_n	100	210	331	464.1	610.51

35. a. During the second year, the car is worth $14,000 - (0.15)(\$14,000)$ = \$11,900$.
During the third year, the car is worth $\$11,900 - (0.15)(\$11,900)$ = \$10,115$.
During the fourth year, the car is worth $\$10,115 - (0.15)(\$10,115)$ =\$8597.75$

n	1	2	3	4
a_n	\$14,000	\$11,900	\$10,115	\$8597.75

b. Each term of the sequence can be found by multiplying the previous term by 0.85.
$a_1 = 14,000; \; a_{n+1} = 0.85a_n$

37. a.

n	1	2	3	4
a_n	\$1.55	\$2.00	\$2.45	\$2.90

b. Each term of the sequence can be found by adding 0.45 to the previous term.
$a_1 = 1.55; \; a_{n+1} = a_n + 0.45$

39. a. 12% is the annual interest rate, so the monthly interest rate is $12\%/12 = 1\%$. Therefore, the interest earned after the first month is $\$50,000(0.01) = \$500$. But since that is the amount that Geraldo withdraws, he still has $\$50,000$ in his account during the second month.

n	1	2	3	4
a_n	\$50,000	\$50,000	\$50,000	\$50,000

b. The account balance never changes:
$a_1 = 50,000; \; a_{n+1} = a_n$

41. a. After the first dose there are 10 ml present. After the second dose there are $10 + 10(0.8) = 18$ ml present. After the third dose there are $10 + 18(0.8) = 24.4$ ml present. After the fourth dose there are $10 + 24.4(0.8) = 29.52$ ml present.

n	1	2	3	4
a_n	10	18	24.4	29.52

b. Each term of the sequence can be found by multiplying the previous term by 0.8 and adding 10.
$a_1 = 10; \quad a_{n+1} = 0.8a_n + 10$

43. a. No matter where the points are drawn or how the points are connected, only 3 lines can be drawn. (The three lines form a triangle.)

b. Six lines can be drawn. There are four lines which form the "outer perimeter" and two other lines which form the "diagonals."

c.

n	3	4	5	6	7
L_n	3	6	10	15	21

d. $L_3 = 3; \quad L_{n+1} = n + L_n$

45. a. Each term, after the first two, is found by adding the previous two terms.

n	1	2	3	4	5	6	7	8	9
f_n	1	1	2	3	5	8	13	21	34

n	10	11	12	13	14	15	16
f_n	55	89	144	233	377	610	987

b.
$$\frac{f_2}{f_1} = \frac{1}{1} = 1$$
$$\frac{f_3}{f_2} = \frac{2}{1} = 2$$
$$\frac{f_4}{f_3} = \frac{3}{2} = 1.5$$
$$\frac{f_5}{f_4} = \frac{5}{3} = 1.\overline{6}$$
$$\frac{f_6}{f_5} = \frac{8}{5} = 1.6$$
$$\frac{f_7}{f_6} = \frac{13}{8} = 1.625$$
$$\frac{f_8}{f_7} = \frac{21}{13} = 1.615385$$
$$\frac{f_9}{f_8} = \frac{34}{21} = 1.619048$$
$$\frac{f_{10}}{f_9} = \frac{55}{34} = 1.617647$$
$$\frac{f_{11}}{f_{10}} = \frac{89}{55} = 1.61818\overline{18}$$
$$\frac{f_{12}}{f_{11}} = \frac{144}{89} = 1.617978$$
$$\frac{f_{13}}{f_{12}} = \frac{233}{144} = 1.61805\overline{5}$$
$$\frac{f_{14}}{f_{13}} = \frac{377}{233} = 1.618026$$
$$\frac{f_{15}}{f_{14}} = \frac{610}{377} = 1.618037$$
$$\frac{f_{16}}{f_{15}} = \frac{987}{610} = 1.618033$$

$\dfrac{1+\sqrt{5}}{2} \approx 1.61803$, so the limit of $\dfrac{f_{n+1}}{f_n}$ as n approaches ∞ is equal to the "golden ratio."

47. In your calculator, clear the screen, type 1 and hit ENTER. Then type
$1/(1 + \text{Ans}) + 1$. ("Ans" can be found on the TI-83 by using "2nd" then "(−)".)
Now hit ENTER over and over again to see the terms of the sequence.
a_n approaches 1.414214, or $\sqrt{2}$.

49. In your calculator, clear the screen, type 3 and hit ENTER. Then type
$\sqrt{(1 + \text{Ans})/2}$. Now hit ENTER over and over again to see the terms of the

sequence. c_n approaches 0.640388. (It turns out that this is equal to $\frac{1+\sqrt{17}}{8}$.)

51. In your calculator, clear the screen, type 1 and hit ENTER. Then type
$0.5*(\text{Ans} + 4/\text{Ans})$. Now hit ENTER over and over again to see the terms of the
sequence. s_n approaches 2.

Homework 9.2

1. This sequence is geometric. Each term is obtained from the previous term by multiplying by 3.

3. This sequence is arithmetic. Each term is obtained from the previous term by adding –8.

5. This sequence is geometric. Each term is obtained from the previous term by multiplying by –1.

7. This sequence is neither arithmetic or geometric.

9. This sequence is geometric. Each term is obtained from the previous term by multiplying by $\frac{1}{3}$

11. This sequence is geometric. Each term is obtained from the previous term by multiplying by $-\frac{3}{2}$.

13. Start with $a_1 = 2$ and add 4 to get from one term to the next. The first four terms are 2, 6, 10, 14.

15. Start with $a_1 = \frac{1}{2}$ and add $\frac{1}{4}$ to get from one term to the next. The first four terms are $\frac{1}{2}, \frac{3}{4}, 1, \frac{5}{4}$.

17. Start with $a_1 = 2.7$ and add –0.8 to get from one term to the next. The first four terms are 2.7, 1.9, 1.1, 0.3.

19. Start with $a_1 = 5$ and multiply by –2 to get from one term to the next. The first four terms are 5, –10, 20, –40.

21. Start with $a_1 = 9$ and multiply by $\frac{2}{3}$ to get from one term to the next. The first four terms are 9, 6, 4, $\frac{8}{3}$.

23. Start with $a_1 = 60$ and multiply by 0.4 to get from one term to the next. The first four terms are 60, 24, 9.6, 3.84.

25. Add 4 to each successive term. The next three terms are 15, 19, 23.
$a = 3$ and $d = 4$:
$a_n = 3 + (n - 1)4$

27. Add –4 to each successive term: The next three terms are –13, –17, –21.
$a = -1$ and $d = -4$:
$a_n = -1 + (n - 1)(-4)$

29. Multiply each successive term by 2. The next three terms are
$\frac{16}{3}, \frac{32}{3}, \frac{64}{3}$.
$a = \frac{2}{3}$ and $r = 2$
$a_n = \frac{2}{3}(2)^{n-1}$

31. Multiply each successive term by $-\frac{1}{2}$. The next three terms are
$-\frac{1}{2}, \frac{1}{4}, -\frac{1}{8}$.
$a = 4$ and $r = -\frac{1}{2}$
$a_n = 4\left(-\frac{1}{2}\right)^{n-1}$

33. $a = 2$ and $d = \frac{5}{2} - 2 = \frac{1}{2}$
$a_n = 2 + (n - 1)\frac{1}{2}$
$a_{12} = 2 + (12 - 1)\frac{1}{2} = \frac{15}{2}$

35. $a = -3$ and $r = -\frac{1}{2}$

$$a_n = -3\left(-\frac{1}{2}\right)^{n-1}$$

$$a_8 = -3\left(-\frac{1}{2}\right)^{8-1} = \frac{3}{128}$$

37. $a_5 = 48$ and $r = 2$

$$a_5 = ar^4$$

$$48 = a\,2^4$$

$$a = \frac{48}{2^4} = 3$$

39. $a = \frac{1}{8}$ and $r = 2$

$$a_n = ar^{n-1}$$

$$512 = \frac{1}{8} \cdot 2^{n-1}$$

$$4096 = 2^{n-1}$$

$$2^{12} = 2^{n-1}$$

$$12 = n - 1$$

$$13 = n$$

There are 13 terms in this sequence.

41. $s_1 = 3$ and $d = 2$

$$s_n = 3 + 2(n - 1)$$

43. $x_1 = 0$ and $d = -3$

$$x_n = 0 + (n - 1)(-3)$$

$$x_n = -3(n - 1)$$

45. $d_1 = 24$ and $r = -\frac{1}{2}$

$$d_n = 24\left(-\frac{1}{2}\right)^{n-1}$$

47. $w_1 = 1$ and $r = 2$

$$w_n = 1(2)^{n-1}$$

$$w_n = 2^{n-1}$$

49. a.

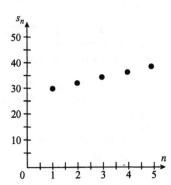

b. $s_1 = 30$ and $d = 2$

$s_{50} = 30 + (50 - 1)2 = 128$

There are 128 seats in the fiftieth row.

51. a.

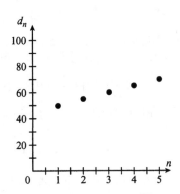

b. $d_1 = 50$, $d = 5$, $n = \frac{70}{5} = 14$

$d_{14} = 50 + (14 - 1)5 = 115$

The charge is $115.

53. a.

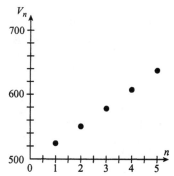

b. $V_1 = 500(1.05) = 525$
and $r = 1.05$

$V_{18} = 525(1.05)^{18-1} = 1203.31$
The deposit will be worth
$1203.31 on Valerie's eighteenth
birthday.

55. a.

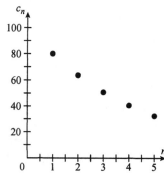

b. $c_1 = 80$ and $r = 0.8$
$c_{20} = 80(0.8)^{20-1}$
$c_{20} = 1.15$
There will be 1.15 kg remaining.

Homework 9.3

1. $a_1 = -4 + 3(1) = -1$
$a_9 = -4 + 3(9) = 23$
$S_9 = \frac{9}{2}(-1 + 23) = 99$
The sum is 99.

3. $a_1 = 18 - \frac{4}{3}(1) = \frac{50}{3}$
$a_{16} = 18 - \frac{4}{3}(16) = -\frac{10}{3}$
$S_{16} = \frac{16}{2}\left(\frac{50}{3} - \frac{10}{3}\right) = \frac{320}{3} = 106\frac{2}{3}$
The sum is $106\frac{2}{3}$.

5. $a_1 = 1.6 + 0.2(1) = 1.8$
$a_{30} = 1.6 + 0.2(30) = 7.6$
$S_{30} = \frac{30}{2}(1.8 + 7.6) = 141$
The sum is 141.

7. $a_1 = 2(-4)^{1-1} = 2$
$a_6 = 2(-4)^{6-1} = -2048$
$S_5 = \frac{-2048 - 2}{-4 - 1} = 410$
The sum is 410.

9. $a_1 = -48\left(\frac{1}{2}\right)^{1-1} = -48$
$a_{10} = -48\left(\frac{1}{2}\right)^{10-1} = -\frac{3}{32}$
$S_9 = \frac{-\frac{3}{32} - (-48)}{\frac{1}{2} - 1} = -\frac{1533}{16}$
The sum is $= -\frac{1533}{16} = -95.8125$.

11. $a_1 = 18(1.15)^{1-1} = 18$
$a_5 = 18(1.15)^{5-1} = 18(1.15)^4$
$S_4 = \frac{18(1.15)^4 - 18}{1.15 - 1} \approx 89.881$
The sum is 89.881.

13. Arithmetic: Each term is found from the previous term by adding 2.
$a = 2$ and $d = 2$
$a_n = 2 + (n - 1)2 = 2n$
$a_1 = 2(1) = 2$
$a_{50} = 2(50) = 100$
100 is the 50th term.
$S_{50} = \frac{50}{2}(2 + 100) = 2550$
The sum is 2550.

15. Geometric: Each term is obtained from previous term by multiplying by 2.
$a = 2$ and $r = 2$
$a_n = 2(2)^{n-1} = 2^n$
$a_1 = 2^1 = 2$
$a_{10} = 2^{10} = 1024$
1024 is the 10th term
$a_{11} = 2^{11} = 2048$
$S_{10} = \frac{2048 - 2}{2 - 1} = 2046$
The sum is 2046.

17. Neither: each term cannot be obtained from adding or multiplying the previous term by a constant.
$1 + 8 + 27 + 64 + 125 + 216 + 343$
$= 784$
The sum is 784.

19. Arithmetic: each term can be obtained from the previous term by adding -3.
$a = 87$ and $d = -3$
$a_n = 87 - 3(n - 1) = 90 - 3n$
$a_1 = 90 - 3(1) = 87$
$a_{17} = 90 - 3(17) = 39$
39 is the 17th term.
$S_{17} = \frac{17}{2}(87 + 39) = 1071$
The sum is 1071.

21. Geometric: each term is obtained from the previous term by multiplying by $\frac{1}{3}$.

$a = 6$ and $r = \frac{1}{3}$

$a_n = 6\left(\frac{1}{3}\right)^{n-1}$

$a_1 = 6\left(\frac{1}{3}\right)^{1-1} = 6$

$a_7 = 6\left(\frac{1}{3}\right)^{7-1} = \frac{2}{243}$

$\frac{2}{243}$ is the 7th term.

$a_8 = 6\left(\frac{1}{3}\right)^{8-1} = \frac{2}{729}$

$S_7 = \frac{\frac{2}{729} - 6}{\frac{1}{3} - 1} = \frac{2186}{243}$

The sum is $\frac{2186}{243} \approx 8.996$.

23. $14 + 16 + 18 + \ldots + 88$
arithmetic: $a = 14$ and $d = 2$
$a_n = 14 + (n-1)2 = 12 + 2n$
$a_1 = 14$
$a_{38} = 12 + 2(38) = 88$
88 is the 38th term.

$S_{38} = \frac{38}{2}(14 + 88) = 1938$
The sum is 1938.

25. $1 + 2 + 3 + \ldots + 12$
arithmetic: $a = 1$ and $d = 1$
$a_n = 1 + (n-1)(1) = n$
$a_1 = 1$
$a_{12} = 12$
There are 12 terms in the series.

$S_{12} = \frac{12}{2}(1 + 12) = 78$
The clock will strike 78 times.

27. a. Drops first time 24 ft;

Rises first time $24\left(\frac{3}{4}\right) = 18$ ft;

Drops second time 18 feet;
Rises second time

$18\left(\frac{3}{4}\right) = 13.5$ ft;

Drops third time 13.5 ft;
Rises third time

$13.5\left(\frac{3}{4}\right) = 10.125$ ft.

It will bounce 10.125 feet after hitting the floor the third time.

b. First round down and up:
24 + 18
Second round down and up:
18 + 13.5
Third round down and up:
13.5 + 10.125
Fourth time down only:
10.125
The ball will travel 24 + 18 + 18 + 13.5 + 13.5 + 10.125 + 10.125 = 107.25 feet.

29. arithmetic:
$a = 920,000$ and $d = -40,000$
$a_n = 920,000 - 40,000(n-1)$
$a_1 = 920,000$
$a_{10} = 920,000 - 40,000(10 - 1)$
$= 560,000$
The year 2000 corresponds to the 10th term.

$S_{10} = \frac{10}{2}(920,000 + 560,000)$
$= 7,400,000$
The total revenue should be $7,400,000.

31. arithmetic: $a = 0.10$ and $d = 0.05$

$a_n = 0.10 + (n-1)(0.05)$

$\quad = 0.05 + 0.05n$

$a_1 = 0.10$

$a_{50} = 0.05 + 0.05(50) = 2.55$

$S_{50} = \frac{50}{2}(0.10 + 2.55) = 66.25$

It will take 66.25 seconds.

33. geometric: $a = 920,000$ and $r = 0.92$

$a_n = 920,000(0.92)^{n-1}$

$a_1 = 920,000$

$a_{10} = 920,000(0.92)^{10-1}$

$\quad = 434,388.45$

The year 2000 corresponds to the 10th term.

$a_{11} = 920,000(0.92)^{11-1}$

$\quad = 399,637.38$

$S_{10} = \frac{399,637.38 - 920,000}{0.92 - 1}$

$\quad = 6,504,532.78$

The total revenue is \$6,504,532.78.

35. geometric: $a = 0.1$ and $r = 1.2$

$a_n = 0.1(1.2)^{n-1}$

$a_1 = 0.1$

$a_{51} = 0.1(1.2)^{51-1} = 910.0438$

$S_{50} = \frac{910.0438 - 0.1}{1.2 - 1} = 4549.7$

The performance time is 4549.7 seconds, which is about 75.8 minutes.

37. geometric: $n = 1$ corresponds to the day David was born, $n = 2$ to David's 1st birthday, $n = 3$ to David's 2nd birthday, and so on until David's 18th birthday when $n = 19$.

$a = 500$ and $r = 1.05$

$a_n = 500(1.05)^{n-1}$

$a_{19} = 500(1.05)^{19-1} = 1203.31$

$a_{20} = 500(1.05)^{20-1} = 1263.475$

$S_{19} = \frac{1263.475 - 500}{1.05 - 1} = 15,269.50$

There will be \$15,269.50 in the account.

39. geometric: $a = 0.01$ and $r = 2$

$a_n = 0.01(2)^{n-1}$

$a_1 = 0.01$

$a_{31} = 0.01(2)^{31-1} = 10,737,418.24$

$S_{30} = \frac{10,737,418.24 - 0.01}{2 - 1}$

$\quad = 10,737,418.23$

The total income is \$10,737,418.23.

Chapter 9 Review

1. $B_n = 1.0025 B_{n-1} - 200$

3. $c_n = \dfrac{n(n+1)(2n+1)}{6}$

$c_1 = \dfrac{1(2)(3)}{6} = 1$

$c_2 = \dfrac{2(3)(5)}{6} = 5$

$c_3 = \dfrac{3(4)(7)}{6} = 14$

$c_4 = \dfrac{4(5)(9)}{6} = 30$

5. $r_i = 3$
$r_1 = 3$
$r_2 = 3$
$r_3 = 3$
$r_4 = 3$

7. $c_{n+1} = c_n - 3$
$c_1 = 5$
$c_2 = c_1 - 3 = 5 - 3 = 2$
$c_3 = c_2 - 3 = 2 - 3 = -1$
$c_4 = c_3 - 3 = -1 - 3 = -4$
$c_5 = c_4 - 3 = -4 - 3 = -7$
The first five terms are
$5, 2, -1, -4, -7$.

9. a. $c_1 = 100$ mg
$c_2 = 100(0.93) = 93$ mg
$c_3 = 93(0.93) = 86.49$ mg
$c_4 = 86.49(0.93) = 80.4357$ mg

b. $c_1 = 100, \ c_n = 0.93 c_{n-1}$

c. $c_n = 100(0.93)^{n-1}$

11. a. $a_1 = -7$
$a_2 = -7 + (-6) = -13$
$a_3 = -13 + (-6) = -19$
$a_4 = -19 + (-6) = -25$

b. $a_1 = -7, \ a_n = a_{n-1} - 6$

c. $a_n = -7 + (n-1)(-6) = -1 - 6n$

13. arithmetic: $d = -5.3$

15. neither

17. $a_1 = 72, \ r = \dfrac{1}{2}$

$a_n = 72\left(\dfrac{1}{2}\right)^{n-1}$

$a_8 = 72\left(\dfrac{1}{2}\right)^{8-1} = \dfrac{9}{16}$

19. $a = \dfrac{16}{27}$ and $r = -\dfrac{3}{2}$

$a_8 = \dfrac{16}{27}\left(-\dfrac{3}{2}\right)^{8-1}$

$a_8 = -\dfrac{81}{8}$

The eighth term is $-\dfrac{81}{8}$.

21. $a_1 = 8$
$a_{28} = 89 = 8 + (28 - 1)d$
$d = 3$
$a_{21} = 8 + (21 - 1)3$
$a_{21} = 68$
The twenty-first term is 68.

23. $p_n = 1000(1.1)^n$ (geometric)
$p_1 = 1000(1.1)^1 = 1100$
$p_6 = 1000(1.1)^6 = 1771.561$
$S_5 = \dfrac{1771.561 - 1100}{1.1 - 1} = 6715.61$

25. $q_n = 3.42 + 2.16n$ (arithmetic)

$q_1 = 3.42 + 2.16(1) = 5.58$

$q_{15} = 3.42 + 2.16(15) = 35.82$

$S_{15} = \frac{15}{2}(5.58 + 35.82) = 310.5$

27. After hitting the floor once, the ball reaches a height of $12\left(\frac{2}{3}\right) = 8$ feet. After hitting the floor twice, the ball reaches a height of $12\left(\frac{2}{3}\right)^2 = \frac{16}{3}$ feet. Let h be the height the ball reaches after each bounce. Then $h_n = 12\left(\frac{2}{3}\right)^n$. After the fourth bounce, the ball reaches a height of $h_4 = 12\left(\frac{2}{3}\right)^4 = \frac{64}{27}$ feet.

29. $12 + 18 + 24 + \cdots + 96$

$= \frac{n}{2}(a_1 + a_n)$

$= \frac{15}{2}(12 + 96)$

$= 810$

Chapter Ten: Conic Sections

Homework 10.1

1. $4x^2 = 16 - 4y^2$

   $4x^2 + 4y^2 = 16$

   $x^2 + y^2 = 4$

   $x^2 + y^2 = 2^2$

   This is the equation for a circle of radius 2 centered at (0, 0). Refer to the graph in the back of the textbook.

3. $\dfrac{x^2}{16} + \dfrac{y^2}{4} = 1$

   This is the equation for an ellipse centered at (0, 0). The vertices are at (4, 0) and (–4, 0), and the covertices are (0, 2) and (0, –2). Refer to the graph in the back of the textbook.

5. $\dfrac{x^2}{10} + \dfrac{y^2}{25} = 1$

   This is the equation for an ellipse centered at (0, 0). The vertices are at (0, 5) and (0, –5), and the covertices are at $\left(\sqrt{10},\ 0\right)$ and $\left(-\sqrt{10},\ 0\right)$ or approximately at (3.2, 0) and (–3.2, 0). Refer to the graph in the back of the textbook.

7. $x^2 + \dfrac{y^2}{14} = 1$

   $\dfrac{x^2}{1} + \dfrac{y^2}{14} = 1$

   This is the equation for an ellipse centered at (0, 0). The vertices are at $\left(0,\ \sqrt{14}\right)$ and $\left(0,\ -\sqrt{14}\right)$ or approximately at (0, 3.7) and (0, –3.7), and the covertices are at (1, 0) and (–1, 0). Refer to the graph in the back of the textbook.

9. $3y^2 = 30 - 2x^2$

   $2x^2 + 3y^2 = 30$

   $\dfrac{x^2}{15} + \dfrac{y^2}{10} = 1$

   This is the equation for an ellipse centered at (0, 0). The vertices are at $\left(\sqrt{15},\ 0\right)$ and $\left(-\sqrt{15},\ 0\right)$ or approximately at (3.9, 0) and (–3.9, 0), and the covertices are at $\left(0,\ \sqrt{10}\right)$ and $\left(0,\ -\sqrt{10}\right)$ or approximately at (0, 3.2) and (0, –3.2). Refer to the graph in the back of the textbook.

11. $\dfrac{x^2}{25} - \dfrac{y^2}{9} = 1$

This is the equation for a hyperbola centered at $(0, 0)$. The branches open to the left and the right. The vertices are at $(5, 0)$ and $(-5, 0)$. To draw the "central rectangle," move 3 in each vertical direction and 5 in each horizontal direction from the origin.

The asymptotes are $y = \pm\dfrac{3}{5}x$. Refer to the graph in the back of the textbook.

13. $\dfrac{y^2}{12} - \dfrac{x^2}{8} = 1$

This is the equation for a hyperbola centered at $(0, 0)$. The branches open upward and downward. The vertices are at $\left(0,\ \sqrt{12}\right) = \left(0,\ 2\sqrt{3}\right)$ and $\left(0,\ -\sqrt{12}\right) = \left(0,\ -2\sqrt{3}\right)$ or approximately at $(0, 3.5)$ and $(0, -3.5)$. To draw the "central rectangle," move $\sqrt{12} \approx 3.5$ in each vertical direction and $\sqrt{8} \approx 2.8$ in each horizontal direction from the origin. The asymptotes are

$y = \pm\dfrac{\sqrt{12}}{\sqrt{8}}x \approx \pm1.2x$. Refer to the graph in the back of the textbook.

15. $9x^2 - 4y^2 = 36$

$\dfrac{x^2}{4} - \dfrac{y^2}{9} = 1$

This is the equation for a hyperbola centered at $(0, 0)$. The branches open to the left and the right. The vertices are at $(2, 0)$ and $(-2, 0)$. To draw the "central rectangle," move 3 in each vertical direction and 2 in each horizontal direction from the origin.

The asymptotes are $y = \pm\dfrac{3}{2}x$. Refer to the graph in the back of the textbook.

17. $\dfrac{1}{2}x^2 = y^2 - 12$

$y^2 - \dfrac{1}{2}x^2 = 12$

$\dfrac{y^2}{12} - \dfrac{x^2}{24} = 1$

This is the equation for a hyperbola centered at $(0, 0)$. The branches open upward and downward. The vertices are at $\left(0,\ \sqrt{12}\right) = \left(0,\ 2\sqrt{3}\right)$ and $\left(0,\ -\sqrt{12}\right) = \left(0,\ -2\sqrt{3}\right)$ or approximately at $(0, 3.5)$ and $(0, -3.5)$. To draw the "central rectangle," move $\sqrt{12} \approx 3.5$ in each vertical direction and $\sqrt{24} \approx 4.9$ in each horizontal direction from the origin. The asymptotes are

$y = \pm\dfrac{\sqrt{12}}{\sqrt{24}}x \approx \pm0.7x$. Refer to the graph in the back of the textbook.

19. $x^2 = 2y$

$y = \dfrac{x^2}{2}$

This is the equation for a parabola with vertex at (0, 0) and opening upward. $4p = 2$ so $p = \dfrac{1}{2}$ and the focus is at $\left(0, \dfrac{1}{2}\right)$. The directrix is the horizontal line $y = -\dfrac{1}{2}$. Draw a horizontal line segment of length 2 centered at the focus. Draw the parabola through the endpoints of this line segment. Refer to the graph in the back of the textbook.

21. $x = -\dfrac{1}{16}y^2$

$x = -\dfrac{y^2}{16}$

This is the equation for a parabola with vertex at (0, 0) and opening to the left. $4p = 16$ so $p = 4$ and the focus is at (–4, 0), The directrix is the vertical line $x = 4$. Draw a vertical line segment of length 16 centered at the focus. Draw the parabola through the endpoints of this line segment. Refer to the graph in the back of the textbook.

23. $4x^2 = 3y$

$y = \dfrac{4x^2}{3} = \dfrac{x^2}{3/4}$

This is the equation for a parabola with vertex at (0, 0) and opening upward. $4p = \dfrac{3}{4}$ so $p = \dfrac{3}{16}$ and the focus is at $\left(0, \dfrac{3}{16}\right)$. The directrix is the horizontal line $y = -\dfrac{3}{16}$. Draw a horizontal line segment of length $\dfrac{3}{4}$ centered at the focus. Draw the parabola through the endpoints of this line segment. For this problem, it may also be helpful to plot a few additional points: for example, (3, 12) and (–3, 12). Refer to the graph in the back of the textbook.

25. $2y^2 - 3x = 0$

$2y^2 = 3x$

$x = \dfrac{2y^2}{3} = \dfrac{y^2}{3/2}$

This is the equation for a parabola with vertex at (0, 0) and opening to the right. $4p = \dfrac{3}{2}$ so $p = \dfrac{3}{8}$ and the focus is at $\left(\dfrac{3}{8}, 0\right)$. The directrix is the vertical line $x = -\dfrac{3}{8}$. Draw a vertical line segment of length $\dfrac{3}{2}$ centered at the focus. Draw the parabola through the endpoints of this line segment. For this problem, it may also be helpful to plot a few additional points: for example, (6, 3) and (6, –3). Refer to the graph in the back of the textbook.

Homework 10.1

27. a. $y^2 = 4 - x^2$

$x^2 + y^2 = 4$

This is a circle of radius 2 centered at $(0, 0)$.

b. Let $x = -1$:

$(-1)^2 + y^2 = 4$

$1 + y^2 = 4$

$y^2 = 3$

$y = \pm\sqrt{3}$

The points are $\left(-1, \pm\sqrt{3}\right)$.

29. a. $4y^2 = x^2 - 8$

$x^2 - 4y^2 = 8$

$\dfrac{x^2}{8} - \dfrac{y^2}{2} = 1$

This is a hyperbola centered at $(0, 0)$. The branches open to the left and right. The vertices are at $\left(\sqrt{8},\ 0\right) = \left(2\sqrt{2},\ 0\right)$ and $\left(-\sqrt{8},\ 0\right) = \left(-2\sqrt{2},\ 0\right)$.

b. Let $y = 2$:

$\dfrac{x^2}{8} - \dfrac{2^2}{2} = 1$

$\dfrac{x^2}{8} - 2 = 1$

$\dfrac{x^2}{8} = 3$

$x^2 = 24$

$x = \pm\sqrt{24} = \pm 2\sqrt{6}$

The points are $(\pm 2\sqrt{6},\ 2)$.

31. a. $4x^2 = 12 - 2y^2$

$4x^2 + 2y^2 = 12$

$\dfrac{x^2}{3} + \dfrac{y^2}{6} = 1$

This is an ellipse with x-intercepts $\left(\pm\sqrt{3},\ 0\right)$ and y-intercepts $\left(0,\ \pm\sqrt{6}\right)$.

b. Let $x = 4$:

$4(4)^2 = 12 - 2y^2$

$2y^2 = -52$

$y^2 = -26$

But this gives $y = \pm\sqrt{-26}$, which are not real numbers, so there are no points on the ellipse which have an x-value of 4.

33. a. $12x = y^2$ so $x = \dfrac{y^2}{12}$.

$4p = 12$ and thus $p = 3$.

This is a parabola opening to the right from vertex $(0, 0)$. It has focus $(3, 0)$.

b. Let $y = -2$:

$12x = (-2)^2$

$12x = 4$

$x = \dfrac{1}{3}$

The point is $\left(\dfrac{1}{3}, -2\right)$.

35. a. $6 + \dfrac{x^2}{4} = y^2$

$24 + x^2 = 4y^2$

$4y^2 - x^2 = 24$

$\dfrac{y^2}{6} - \dfrac{x^2}{24} = 1$

This is a hyperbola with branches that open upward and downward. It has vertices at $(0, \pm\sqrt{6}\,)$.

b. Let $y = -\sqrt{5}$:

$6 + \dfrac{x^2}{4} = \left(-\sqrt{5}\right)^2$

$6 + \dfrac{x^2}{4} = 5$

$\dfrac{x^2}{4} = -1$

$x^2 = -4$

But this gives $x = \pm\sqrt{-4}$, which are not real numbers, so there are no points on the hyperbola which have a y-value of $-\sqrt{5}$.

37. a. $\dfrac{x^2}{9} + \dfrac{y^2}{4} = 1$

b. *Values rounded to two decimals:*

x	± 3	0	-2	± 2.6
y	0	± 2	± 1.49	1

39. a. $y^2 - \dfrac{x^2}{4} = 1$

b. *Values rounded to two decimals:*

x	0	None	4	± 3.46
y	± 1	0	± 2.24	-2

41. a. Let the center of the ellipse be at the origin. Then the ellipse can be described by the equation

$\dfrac{x^2}{a^2} + \dfrac{y^2}{b^2} = 1$ where $2a = 20$ and $b = 7$. Therefore, the equation is

$\dfrac{x^2}{100} + \dfrac{y^2}{49} = 1$.

b. To find the height of the arch at a distance of 8 feet from the peak, substitute $x = 8$ into the equation and solve for y.

$\dfrac{(8)^2}{100} + \dfrac{y^2}{49} = 1$

$\dfrac{64}{100} + \dfrac{y^2}{49} = 1$

$\dfrac{y^2}{49} = \dfrac{36}{100}$

$y^2 = \dfrac{1764}{100}$

$y = \pm\dfrac{42}{10} = \pm\dfrac{21}{5}$

Therefore the height at 8 feet from the peak is $\dfrac{21}{5}$ ft.

43. a. Let the center of the ellipse be at the origin. Then the ellipse can be described by the equation

$\frac{x^2}{a^2} + \frac{y^2}{b^2} = 1$ where $a = 50$ and $2b = 360$. Therefore, the equation

is $\frac{x^2}{2500} + \frac{y^2}{32,400} = 1$.

b. The cut edge is 330 cm long. Let y equal half of this value, so $y = 165$, and solve for x:

$\frac{x^2}{2500} + \frac{(165)^2}{32,400} = 1$

$\frac{x^2}{2500} + \frac{27,225}{32,400} = 1$

$\frac{x^2}{2500} + \frac{121}{144} = 1$

$\frac{x^2}{2500} = \frac{23}{144}$

$x^2 = \frac{57,500}{144}$

$x \approx \pm 19.98$

Therefore the width of the keel at its widest point is approximately $50 + 19.98 = 69.98$ cm.

45. a. Let the vertex of the parabola be at the origin. Then the parabola can be described by an equation of the form $y = \frac{x^2}{4p}$. Substitute $x = 36$ and $y = 3$ into the equation and solve for p.

$3 = \frac{36^2}{4p}$

$12p = 1296$

$p = 108$

$y = \frac{x^2}{4(108)} = \frac{x^2}{432}$ or $432y = x^2$

b. Since $p = 108$, the focus is 108 in. from the vertex.

47. a. Let the vertex of the parabola be at the origin. Then the parabola can be described by an equation of the form $y = \frac{x^2}{4p}$. Substitute $x = 30$ and $y = 18$ into the equation and solve for p.

$18 = \frac{30^2}{4p}$

$72p = 900$

$p = 12.5$

Therefore, $y = \frac{x^2}{4(12.5)} = \frac{x^2}{50}$ or $50y = x^2$.

b. Since $p = 12.5$, the receiver is 12.5 cm from the vertex.

49. Substitute $y = -360$ into the equation and solve for x.

$$\frac{x^2}{100^2} - \frac{(-360)^2}{150^2} = 1$$

$$\frac{x^2}{10,000} - \left(\frac{-360}{150}\right)^2 = 1$$

$$\frac{x^2}{10,000} - \left(\frac{-12}{5}\right)^2 = 1$$

$$\frac{x^2}{10,000} = \frac{169}{25}$$

$$x^2 = 67,600$$

$$x = \pm 260$$

Therefore, the radius of the base is 260 feet and the diameter of the base is $260(2) = 520$ feet.

51. Substitute $x = 125$ into the equation and solve for y.

$$\frac{(125)^2}{100^2} - \frac{y^2}{150^2} = 1$$

$$\left(\frac{125}{100}\right)^2 - \frac{y^2}{22,500} = 1$$

$$\left(\frac{5}{4}\right)^2 - \frac{y^2}{22,500} = 1$$

$$-\frac{y^2}{22,500} = -\frac{9}{16}$$

$$y^2 = \frac{50,625}{4}$$

$$y = \pm\frac{225}{2} = \pm 112.5$$

Therefore, the greater of the two heights (measured from the base) is $360 + 112.5 = 472.5$ feet. (Note that the 360 was given in problem 49.)

53. $x^2 - y^2 = 0$

$(x - y)(x + y) = 0$

$y = x$ or $y = -x$

These are two lines that pass through the origin, one with slope -1 and the other with slope 1. Refer to the graph in the back of the textbook.

55. $x^2 - y^2 = 4$

$$\frac{x^2}{4} - \frac{y^2}{4} = 1$$

This graph is a hyperbola with vertices $(2, 0)$ and $(-2, 0)$.

$x^2 - y^2 = 1$

$$\frac{x^2}{1} - \frac{y^2}{1} = 1$$

This graph is a hyperbola with vertices $(1, 0)$ and $(-1, 0)$.

$x^2 - y^2 = 0$

This graph is two lines, $y = x$ and $y = -x$. (See Exercise 53.) Note the asymptotes of the two hyperbolas are $y = x$ and $y = -x$. Refer to the graph in the back of the textbook.

Homework 10.2

1. a. $\dfrac{(x-3)^2}{16} + \dfrac{(y-4)^2}{9} = 1$

$\dfrac{(x-3)^2}{4^2} + \dfrac{(y-4)^2}{3^2} = 1$

This is the equation for an ellipse centered at (3, 4). The vertices lie four units to the left and right of the center and the covertices lie three units above and below the center. Refer to the graph in the back of the textbook.

b. $(-1, 4), (7, 4), (3, 7), (3, 1)$

3. a. $\dfrac{(x+2)^2}{6} + \dfrac{(y-5)^2}{12} = 1$

This is the equation for an ellipse centered at (–2, 5). The vertices lie $\sqrt{12} = 2\sqrt{3} \approx 3.5$ units above and below the center and the covertices lie $\sqrt{6} \approx 2.4$ units to the left and right of the center. Refer to the graph in the back of the textbook.

b. $\left(-2, 5+2\sqrt{3}\right), \left(-2, 5-2\sqrt{3}\right),$ $\left(-2-\sqrt{6}, 5\right), \left(-2+\sqrt{6}, 5\right)$

5. a. Complete the square in y:

$9x^2 + 4y^2 - 16y = 20$

$9x^2 + 4(y^2 - 4y + \underline{\ }) = 20 + 4(\underline{\ })$

$9x^2 + 4(y^2 - 4y + 4) = 20 + 4(4)$

$9x^2 + 4(y-2)^2 = 36$

$\dfrac{x^2}{4} + \dfrac{(y-2)^2}{9} = 1$

This is the equation for an ellipse centered at (0, 2). The vertices lie three units above and below the center and the covertices lie two units to the left and right of the center. Refer to the graph in the back of the textbook.

b. $(0, 5), (0, -1), (-2, 2), (2, 2)$

7. a. Complete the square in x and y:

$9x^2 + 16y^2 - 18x + 96y + 9 = 0$

$9(x^2 - 2x + \underline{\ }) + 16(y^2 + 6y + \underline{\ })$
$= -9 + 9(\underline{\ }) + 16(\underline{\ })$

$9(x^2 - 2x + 1) + 16(y^2 + 6y + 9)$
$= -9 + 9(1) + 16(9)$

$9(x-1)^2 + 16(y+3)^2 = 144$

$\dfrac{(x-1)^2}{16} + \dfrac{(y+3)^2}{9} = 1$

This is the equation for an ellipse centered at (1, –3). The vertices lie four units to the left and right of the center and the covertices lie three units above and below the center. Refer to the graph in the back of the textbook.

b. $(-3, -3), (5, -3), (1, 0),$ and $(1, -6)$

9. a. Complete the square in x and y:

$$8x^2 + y^2 - 48x + 4y + 68 = 0$$

$$8(x^2 - 6x + \_\_) + (y^2 + 4y + \_\_)$$
$$= -68 + 8(\_\_) + (\_\_)$$

$$8(x^2 - 6x + 9) + (y^2 + 4y + 4)$$
$$= -68 + 8(9) + 4$$

$$8(x - 3)^2 + (y + 2)^2 = 8$$

$$\frac{(x-3)^2}{1} + \frac{(y+2)^2}{8} = 1$$

This is the equation of an ellipse centered at $(3, -2)$. The vertices lie $\sqrt{8} = 2\sqrt{2} \approx 2.8$ units above and below the center and the covertices lie one unit left and right of the center. Refer to the graph in the back of the textbook.

b. $(3, -2 + 2\sqrt{2}), (3, -2 - 2\sqrt{2}),$
$(2, -2), (4, -2)$

11. $\dfrac{(x-1)^2}{3^2} + \dfrac{(y-6)^2}{2^2} = 1$

$$\frac{(x-1)^2}{9} + \frac{(y-6)^2}{4} = 1$$

$$4(x-1)^2 + 9(y-6)^2 = 36$$

$$4(x^2 - 2x + 1) + 9(y^2 - 12y + 36)$$
$$= 36$$

$$4x^2 + 9y^2 - 8x - 108y + 292 = 0$$

13. The center is between the vertices at $(-2, 2)$. The distance between the vertices is 10, and this is a horizontal distance, so $a = 5$. Also, 6 is the length of the minor axis, which in this case is vertical, so $b = 3$.

$$\frac{(x+2)^2}{5^2} + \frac{(y-2)^2}{3^2} = 1$$

$$\frac{(x+2)^2}{25} + \frac{(y-2)^2}{9} = 1$$

$$9(x+2)^2 + 25(y-2)^2 = 225$$

$$9(x^2 + 4x + 4) + 25(y^2 - 4y + 4)$$
$$= 225$$

$$9x^2 + 25y^2 + 36x - 100y - 89 = 0$$

15. The center is between the vertices at $(-4, 3)$. The distance between the vertices is 12 and this is a vertical distance so $b = 6$. The distance between the covertices, is 6 and this is a horizontal distance so $a = 3$.

$$\frac{(x+4)^2}{3^2} + \frac{(y-3)^2}{6^2} = 1$$

$$\frac{(x+4)^2}{9} + \frac{(y-3)^2}{36} = 1$$

$$4(x+4)^2 + (y-3)^2 = 36$$

$$4(x^2 + 8x + 16) + (y^2 - 6y + 9) = 36$$

$$4x^2 + y^2 + 32x - 6y - 37 = 0$$

Homework 10.2

17. a. $\dfrac{(x-4)^2}{9} - \dfrac{(y+2)^2}{16} = 1$

This is the equation for a hyperbola opening left and right and centered at $(4, -2)$. The vertices are three units to the left and right of the center. To draw the central rectangle, move 3 units in each horizontal direction and 4 units in each vertical direction from $(4, -2)$. Draw the asymptotes through the corners of the central rectangle. Refer to the graph in the back of the textbook.

b. Vertices: $(1, -2)$ and $(7, -2)$;

To find two more exact points on the hyperbola, choose a value of x or y that has a point on the graph. For example, choose $y = 2$. Then

$$\dfrac{(x-4)^2}{9} - \dfrac{(2+2)^2}{16} = 1$$

$$\dfrac{(x-4)^2}{9} - 1 = 1$$

$$(x-4)^2 = 18$$

$$x - 4 = \pm\sqrt{18} = \pm 3\sqrt{2}$$

$$x = 4 \pm 3\sqrt{2}$$

So two additional points are $(4 - 3\sqrt{2}, 2)$ and $(4 + 3\sqrt{2}, 2)$.

19. a. $\dfrac{x^2}{4} - \dfrac{(y-3)^2}{8} = 1$

This is the equation for a hyperbola opening to the left and right and centered at $(0, 3)$. The vertices are two units to the left and right of the center. To draw the central rectangle, move 2 units in each horizontal direction and $\sqrt{8} \approx 2.8$ units in each vertical direction from $(0, 3)$. Draw the asymptotes through the corners of the central rectangle. Refer to the graph in the back of the textbook.

b. Vertices: $(-2, 3)$ and $(2, 3)$;

To find two more exact points on the hyperbola, choose a value of x or y that has a point on the graph. For example, choose $y = 7$. Then

$$\dfrac{x^2}{4} - \dfrac{(7-3)^2}{8} = 1$$

$$\dfrac{x^2}{4} - 2 = 1$$

$$x^2 = 12$$

$$x = \pm\sqrt{12} = \pm 2\sqrt{3}$$

So two additional points are $(-2\sqrt{3}, 7)$ and $(2\sqrt{3}, 7)$.

21. a. Complete the square in x:

$16y^2 - 4x^2 + 32x - 128 = 0$

$16y^2 - 4(x^2 - 8x + \underline{})$
$\qquad\qquad = 128 - 4(\underline{})$

$16y^2 - 4(x^2 - 8x + 16) = 128 - 64$

$16y^2 - 4(x-4)^2 = 64$

$\dfrac{y^2}{4} - \dfrac{(x-4)^2}{16} = 1$

This is the equation for a hyperbola opening up and down and centered at (4, 0). The vertices are 2 units above and below the center. To draw the central rectangle, move 2 units in each vertical direction and 4 units in each horizontal direction from (4, 0). Draw the asymptotes through the corners of the central rectangle. Refer to the graph in the back of the textbook.

b. Vertices: (4, –2) and (4, 2);
To find two more exact points on the hyperbola, choose a value of x or y that has a point on the graph. For example, choose $x = 0$. Then

$\dfrac{y^2}{4} - \dfrac{(0-4)^2}{16} = 1$

$\dfrac{y^2}{4} - 1 = 1$

$y^2 = 8$

$y = \pm\sqrt{8} = \pm 2\sqrt{2}$

So two additional points are
(0, $-2\sqrt{2}$) and (0, $2\sqrt{2}$).

23. a. Complete the square in x and y:

$4x^2 - 6y^2 - 32x - 24y + 16 = 0$

$4(x^2 - 8x + \underline{}) - 6(y^2 + 4y + \underline{})$
$\qquad\qquad = -16 + 4(\underline{}) - 6(\underline{})$

$4(x^2 - 8x + 16) - 6(y^2 + 4y + 4)$
$\qquad\qquad = -16 + 64 - 24$

$4(x-4)^2 - 6(y+2)^2 = 24$

$\dfrac{(x-4)^2}{6} - \dfrac{(y+2)^2}{4} = 1$

This is the equation for a hyperbola opening left and right and centered at (4, –2). The vertices are $\sqrt{6} \approx 2.4$ units to the left and right of the center. To draw the central rectangle, move $\sqrt{6}$ units in each horizontal direction and 2 units in each vertical direction from (4, –2). Draw the asymptotes through the corners of the central rectangle. Refer to the graph in the back of the textbook.

b. Vertices: $(4 - \sqrt{6}, -2)$ and $(4 + \sqrt{6}, -2)$;
To find two more exact points on the hyperbola, choose a value of x or y that has a point on the graph. For example, choose $y = 0$. Then

$\dfrac{(x-4)^2}{6} - \dfrac{(0+2)^2}{4} = 1$

$\dfrac{(x-4)^2}{6} - 1 = 1$

$(x-4)^2 = 12$

$x - 4 = \pm\sqrt{12} = \pm 2\sqrt{3}$

$x = 4 \pm 2\sqrt{3}$

So two additional points are
$(4 - 2\sqrt{3}, 0)$ and $(4 + 2\sqrt{3}, 0)$.

25. a. Complete the square in y:
$$12x^2 - 3y^2 + 24y - 84 = 0$$
$$12x^2 - 3(y^2 - 8y + \underline{\ \ }) = 84 - 3(\underline{\ \ })$$
$$12x^2 - 3(y^2 - 8y + 16) = 84 - 3(16)$$
$$12x^2 - 3(y-4)^2 = 36$$
$$\frac{x^2}{3} - \frac{(y-4)^2}{12} = 1$$

This is the equation for a hyperbola opening left and right and centered at (0, 4). The vertices are $\sqrt{3} \approx 1.7$ units to the left and right of the center. To draw the central rectangle, move $\sqrt{3}$ units in each horizontal direction and $\sqrt{12} = 2\sqrt{3} \approx 3.5$ units in each vertical direction from (0, 4). Draw the asymptotes through the corners of the central rectangle. Refer to the graph in the back of the textbook.

b. Vertices: $(-\sqrt{3}, 4)$ and $(\sqrt{3}, 4)$;
To find two more exact points on the hyperbola, choose a value of x or y that has a point on the graph. For example, choose $y = -2$. Then
$$\frac{x^2}{3} - \frac{(-2-4)^2}{12} = 1$$
$$\frac{x^2}{3} - 3 = 1$$
$$x^2 = 12$$
$$x = \pm\sqrt{12} = \pm 2\sqrt{3}$$
So two additional points are $(-2\sqrt{3}, -2)$ and $(2\sqrt{3}, -2)$.

27. $\dfrac{(y-5)^2}{6^2} - \dfrac{(x+1)^2}{8^2} = 1$

$$\frac{(y-5)^2}{36} - \frac{(x+1)^2}{64} = 1$$
$$16(y^2 - 10y + 25) - 9(x^2 + 2x + 1) = 576$$
$$16y^2 - 9x^2 - 160y - 18x - 185 = 0$$

29. Since the conjugate axis is horizontal, it lies on the line $y = 1$ and the hyperbola opens up and down. Therefore, the transverse axis is vertical and lies on the line $x = -1$. The center is the intersection of the two axes at $(-1, 1)$. The distance from the center to a vertex is 2, so $b = 2$. The distance from the center to one end of the conjugate axis is 4, so $a = 4$.
$$\frac{(y-1)^2}{2^2} - \frac{(x+1)^2}{4^2} = 1$$
$$\frac{(y-1)^2}{4} - \frac{(x+1)^2}{16} = 1$$
$$4(y^2 - 2y + 1) - (x^2 + 2x + 1) = 16$$
$$4y^2 - x^2 - 8y - 2x - 13 = 0$$

31. a. Rewrite the equation as

$$(x-0)=\tfrac{1}{2}(y+3)^2$$

This is the equation for a parabola with vertex at $(0, -3)$ and opening to the right. Refer to the graph in the back of the textbook.

b. To find other points on the parabola, substitute values of y into the equation. For example:

$$y=0 \Rightarrow x=\tfrac{1}{2}(0+3)^2=\tfrac{9}{2}$$

$$y=1 \Rightarrow x=\tfrac{1}{2}(1+3)^2=8$$

$$y=-1 \Rightarrow x=\tfrac{1}{2}(-1+3)^2=2$$

So, including the vertex, four points on the parabola are $(0, -3)$, $(9/2, 0)$, $(8, 1)$, and $(2, -1)$.

33. a. Rewrite the equation as

$$(y+4)=-\tfrac{1}{6}(x-3)^2$$

This is the equation for a parabola with vertex at $(3, -4)$ and opening downward. Refer to the graph in the back of the textbook.

b. To find other points on the parabola, substitute values of x into the equation. For example:

$x=0$

$$\Rightarrow y=-\tfrac{1}{6}(0-3)^2-4=-\tfrac{11}{2}$$

$x=-3$

$$\Rightarrow y=-\tfrac{1}{6}(-3-3)^2-4=-10$$

$$x=9 \Rightarrow y=-\tfrac{1}{6}(9-3)^2-4=-10$$

So, including the vertex, four points on the parabola are $(3, -4)$, $(0, -11/2)$, $(-3, -10)$, and $(9, -10)$.

35. a. Complete the square in y:

$$y^2-4y+8x+6=0$$

$$(y^2-4y+\underline{})=-8x-6+\underline{}$$

$$(y^2-4y+4)=-8x-6+4$$

$$(y-2)^2=-8x-2$$

$$(y-2)^2=-8(x+\tfrac{1}{4})$$

$$(x+\tfrac{1}{4})=-\tfrac{1}{8}(y-2)^2$$

This is the equation of a parabola opening to the left with vertex at $(-1/4, 2)$. Refer to the graph in the back of the textbook.

b. Substitute values of y into the equation. For example:

$$y=0:\ x=-\tfrac{1}{8}(0-2)^2-\tfrac{1}{4}=-\tfrac{3}{4}$$

$$y=-2:\ x=-\tfrac{1}{8}(-2-2)^2-\tfrac{1}{4}=-\tfrac{9}{4}$$

$$y=4:\ x=-\tfrac{1}{8}(4-2)^2-\tfrac{1}{4}=-\tfrac{3}{4}$$

So, including the vertex, four points are $(-1/4, 2)$, $(-3/4, 0)$, $(-9/4, -2)$, and $(-3/4, 4)$.

37. a. Complete the square in y:

$$9y^2 = 6y + 12x - 1$$

$$9(y^2 - \tfrac{2}{3}y + \underline{}) = 12x - 1 + 9(\underline{})$$

$$9(y^2 - \tfrac{2}{3}y + \tfrac{1}{9}) = 12x - 1 + 9(\tfrac{1}{9})$$

$$9(y - \tfrac{1}{3})^2 = 12x$$

$$x = \tfrac{3}{4}(y - \tfrac{1}{3})^2$$

This is the equation of a parabola opening to the right with vertex $(0, \tfrac{1}{3})$. Refer to the graph in the back of the textbook.

b. Substitute values of y into the equation. For example:

$$y = 0: \ x = \tfrac{3}{4}(0 - \tfrac{1}{3})^2 = \tfrac{1}{12}$$

$$y = -2: \ x = \tfrac{3}{4}(-2 - \tfrac{1}{3})^2 = 4\tfrac{1}{12}$$

$$y = 3: \ x = \tfrac{3}{4}(3 - \tfrac{1}{3})^2 = 5\tfrac{1}{3}$$

So four points on the graph, including the vertex, are $(0, \tfrac{1}{3})$, $(\tfrac{1}{12}, 0)$, $(4\tfrac{1}{12}, -2)$, and $(5\tfrac{1}{3}, 3)$.

39. $y^2 = 6 - 4x^2$

$$4x^2 + y^2 = 6$$

$$\frac{x^2}{3/2} + \frac{y^2}{6} = 1$$

This is the equation of an ellipse that has center at $(0, 0)$ and major axis vertical. $a = \sqrt{\tfrac{3}{2}} \approx 1.2$ and $b = \sqrt{6} \approx 2.4$, so it has vertices at approximately $(0, -2.4)$, $(0, 2.4)$ and covertices at approximately $(-1.2, 0)$ and $(1.2, 0)$.

41. $x^2 + 2y - 4 = 0$

$$2(y - 2) = -x^2$$

$$(y - 2) = -\tfrac{1}{2}x^2$$

This is the equation of a parabola which opens downward and has vertex $(0, 2)$; $a = -\tfrac{1}{2}$.

43. $6 + \dfrac{x^2}{4} = y^2$

$$y^2 - \frac{x^2}{4} = 6$$

$$\frac{y^2}{6} - \frac{x^2}{24} = 1$$

This is the equation for a hyperbola centered at $(0, 0)$. The transverse axis lies on the y-axis, $a = \sqrt{24} = 2\sqrt{6} \approx 4.9$ and $b = \sqrt{6} \approx 2.4$, The vertices are $\sqrt{6}$ units above and below the center at approximately $(0, 2.4)$ and $(0, -2.4)$.

45. $\tfrac{1}{2}y^2 - x = 4$

$$(x + 4) = \tfrac{1}{2}y^2$$

This is the equation of a parabola which opens to the right and has vertex $(-4, 0)$; $a = \tfrac{1}{2}$.

47. $\dfrac{(x + 3)^2}{5} + \dfrac{y^2}{12} = 1$

This is the equation of an ellipse with center at $(-3, 0)$ and major axis vertical. $a = \sqrt{5} \approx 2.2$ and $b = \sqrt{12} = 2\sqrt{3} \approx 3.5$ so the vertices are at approximately $(-3, -3.5)$ and $(-3, 3.5)$ and the covertices are at approximately $(-5.2, 0)$ and $(-0.8, 0)$.

49. Complete the square in x:

$$2x^2 + y^2 + 4x = 2$$

$$2(x^2 + 2x + \underline{\quad}) + y^2 = 2 + 2(\underline{\quad})$$

$$2(x^2 + 2x + 1) + y^2 = 2 + 2(1)$$

$$2(x + 1)^2 + y^2 = 4$$

$$\frac{(x+1)^2}{2} + \frac{y^2}{4} = 1$$

This is the equation for an ellipse centered at $(-1, 0)$. The major axis is vertical, $a = \sqrt{2} \approx 1.4$ and $b = 2$ so the vertices are at $(-1, 2)$ and $(-1, -2)$ and the covertices are at approximately $(-2.4, 0)$ and $(0.4, 0)$.

Homework 10.3

1. Solve the first equation for y: $y = \dfrac{4}{x}$.
Substitute this expression into the second equation and solve for x:

$$x^2 + \left(\frac{4}{x}\right)^2 = 8$$

$$x^2 + \frac{16}{x^2} = 8$$

$$x^4 + 16 = 8x^2$$

$$x^4 - 8x^2 + 16 = 0$$

$$(x^2 - 4)(x^2 - 4) = 0$$

$$(x + 2)(x - 2)(x + 2)(x - 2) = 0$$

So $x = -2$ or $x = 2$.

When $x = -2$, $y = \dfrac{4}{-2} = -2$.

When $x = 2$, $y = \dfrac{4}{2} = 2$. Thus, the solutions are $(-2, -2)$ and $(2, 2)$.

To check these answers graphically on your calculator, solve the initial equations for y and enter the resulting expressions into your calculator. Then use the intersection feature of your calculator.

$$y_1 = \frac{4}{x}$$

$$y_2 = \sqrt{8 - x^2}$$

$$y_3 = -\sqrt{8 - x^2}$$

Xmin $= -6$, Xmax $= 6$,
Ymin $= -5$, Ymax $= 6$

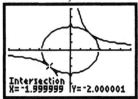

3. From the second equation, $y = \dfrac{6}{x}$.
Substitute this expression into the first equation and solve for x:

$$x^2 - \left(\frac{6}{x}\right)^2 = 16$$

$$x^2 - \frac{36}{x^2} = 16$$

$$x^4 - 36 = 16x^2$$

$$x^4 - 16x^2 - 36 = 0$$

$$(x^2 + 2)(x^2 - 18) = 0$$

This gives $x^2 = -2$, which has no real solutions, and $x^2 = 18$, which has solutions

$$x = \pm\sqrt{18} = \pm 3\sqrt{2} \approx \pm 4.24$$

When $x = 3\sqrt{2}$,

$$y = \frac{6}{3\sqrt{2}} = \frac{6\sqrt{2}}{6} = \sqrt{2} \approx 1.41.$$

When $x = -3\sqrt{2}$,

$$y = \frac{6}{-3\sqrt{2}} = -\frac{6\sqrt{2}}{6} = -\sqrt{2} \approx -1.41.$$

Thus, the solutions are $(3\sqrt{2}, \sqrt{2})$ and $(-3\sqrt{2}, -\sqrt{2})$.

To check these answers graphically on your calculator, use the intersection feature with equations

$$y_1 = \frac{6}{x}, \ y_2 = \sqrt{x^2 - 16}, \text{ and}$$

$$y_3 = -\sqrt{x^2 - 16}.$$

Standard window:

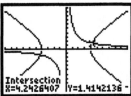

5. Solve by elimination:

$$x^2 + y^2 = 10$$
$$\underline{x^2 - y^2 = 8}$$
$$2x^2 = 18$$
$$x^2 = 9$$
$$x = \pm 3$$

Substitute $x^2 = 9$ into the first equation:

$$x^2 + y^2 = 10$$
$$9 + y^2 = 10$$
$$y^2 = 1$$
$$y = \pm 1$$

Thus, the solutions are $(-3, -1)$, $(3, -1)$, $(-3, 1)$, and $(3, 1)$.

To confirm this result graphically, solve each of the original equations for y and graph $y_1 = \sqrt{10 - x^2}$, $y_2 = -\sqrt{10 - x^2}$, $y_3 = \sqrt{x^2 - 8}$, and $y_4 = -\sqrt{x^2 - 8}$. Use the intersection feature of the calculator.

Xmin = –6, Xmax = 6,
Ymin = –6, Ymax = 6

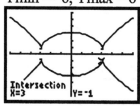

7. To clear fractions, multiply the first equation by 70:

$$70\left(\frac{x^2}{35}\right) - 70\left(\frac{y^2}{10}\right) = 70(1)$$
$$2x^2 - 7y^2 = 70$$

Now add this result to the second equation:

$$2x^2 - 7y^2 = 70$$
$$\underline{x^2 + 7y^2 = 77}$$
$$3x^2 = 147$$
$$x^2 = 49$$
$$x = \pm 7$$

Substitute $x^2 = 49$ into the second equation:

$$x^2 + 7y^2 = 77$$
$$49 + 7y^2 = 77$$
$$7y^2 = 28$$
$$y^2 = 4$$
$$y = \pm 2$$

Thus, the solutions are $(-7, -2)$, $(7, -2)$, $(-7, 2)$, and $(7, 2)$.

To confirm this result graphically, solve each of the original equations for y and graph $y_1 = \sqrt{\frac{1}{7}(2x^2 - 70)}$, $y_2 = -\sqrt{\frac{1}{7}(2x^2 - 70)}$, $y_3 = \sqrt{\frac{1}{7}(77 - x^2)}$, and $y_4 = -\sqrt{\frac{1}{7}(77 - x^2)}$.

Standard window:

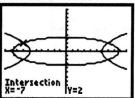

9. To clear fractions, multiply the first equation by 16:

$$16\left(\frac{x^2}{2}\right) - 16\left(\frac{y^2}{16}\right) = 16(1)$$

$$8x^2 - y^2 = 16$$

Write the second equation in standard form and add the result to the above equation:

$$8x^2 - y^2 = 16$$
$$\underline{-x^2 + y^2 = 12}$$
$$7x^2 = 28$$
$$x^2 = 4$$
$$x = \pm 2$$

Substitute $x^2 = 4$ into the second equation:

$$-x^2 + y^2 = 12$$
$$-(4) + y^2 = 12$$
$$y^2 = 16$$
$$y = \pm 4$$

Thus, the solutions are $(-2, -4)$, $(2, -4)$, $(-2, 4)$, and $(2, 4)$.

To confirm this result graphically, solve each of the original equations for y and graph $y_1 = \sqrt{8x^2 - 16}$, $y_2 = -\sqrt{8x^2 - 16}$, $y_3 = \sqrt{12 + x^2}$, and $y_4 = -\sqrt{12 + x^2}$.
Standard window:

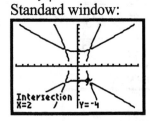

11. Write the second equation in standard form and solve by elimination:

$$x^2 - 6y^2 = 10$$
$$\underline{-x^2 + 28y^2 = 12}$$
$$22y^2 = 22$$
$$y^2 = 1$$
$$y = \pm 1$$

Substitute $y^2 = 1$ into the first equation:

$$x^2 - 6y^2 = 10$$
$$x^2 - 6(1) = 10$$
$$x^2 = 16$$
$$x = \pm 4$$

Thus, the solutions are $(-4, -1)$, $(4, -1)$, $(-4, 1)$, and $(4, 1)$.

To confirm this result graphically, solve each of the original equations for y and graph $y_1 = \sqrt{\frac{1}{6}(x^2 - 10)}$, $y_2 = -\sqrt{\frac{1}{6}(x^2 - 10)}$, $y_3 = \sqrt{\frac{1}{28}(12 + x^2)}$, and $y_4 = -\sqrt{\frac{1}{28}(12 + x^2)}$.
Xmin $= -10$, Xmax $= 10$, Ymin $= -5$, Ymax $= 5$:

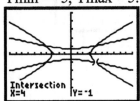

13. Multiply the second equation by -1 and add the result to the first equation:

$$x^2 \qquad + y^2 + 2y = 19$$
$$-x^2 + 2x - y^2 - 8y = -33$$
$$\overline{ 2x - 6y = -14}$$

Solve this equation for x:

$$2x = 6y - 14$$
$$x = 3y - 7$$

Substitute this expression for x in the first equation:

$$x^2 + y^2 + 2y = 19$$
$$(3y - 7)^2 + y^2 + 2y = 19$$
$$9y^2 - 42y + 49 + y^2 + 2y = 19$$
$$10y^2 - 40y + 30 = 0$$
$$10(y^2 - 4y + 3) = 0$$
$$10(y - 3)(y - 1) = 0$$

So $y = 3$ or $y = 1$. When $y = 1$, $x = 3(1) - 7 = -4$. When $y = 3$, $x = 3(3) - 7 = 2$. Thus, the solutions are $(-4, 1)$ and $(2, 3)$.

To confirm this result graphically, first complete the square in y in the first equation, then solve this equation for y:

$$x^2 + (y^2 + 2y + \underline{}) = 19 + \underline{}$$
$$x^2 + (y^2 + 2y + 1) = 19 + 1$$
$$x^2 + (y + 1)^2 = 20$$
$$(y + 1)^2 = 20 - x^2$$
$$y + 1 = \pm\sqrt{20 - x^2}$$
$$y = -1 \pm \sqrt{20 - x^2}$$

Next, complete the square in x and y in the second equation, and then solve this equation for y:

$$(x^2 - 2x + \underline{}) + (y^2 + 8y + \underline{})$$
$$= 33 + \underline{} + \underline{}$$
$$(x^2 - 2x + 1) + (y^2 + 8y + 16)$$
$$= 33 + 1 + 16$$
$$(x - 1)^2 + (y + 4)^2 = 50$$
$$(y + 4)^2 = 50 - (x - 1)^2$$
$$y + 4 = \pm\sqrt{50 - (x - 1)^2}$$
$$y = -4 \pm \sqrt{50 - (x - 1)^2}$$

Now we can graph the four resulting equations on the calculator:

$$y_1 = -1 + \sqrt{20 - x^2}$$
$$y_2 = -1 - \sqrt{20 - x^2}$$
$$y_3 = -4 + \sqrt{50 - (x - 1)^2}$$
$$y_4 = -4 - \sqrt{50 - (x - 1)^2}$$

Xmin $= -7$, Xmax $= 10$, Ymin $= -12$, Ymax $= 7$

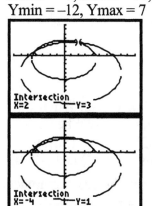

15. FOIL the binomials in the first equation and collect like terms:

$$(x+1)^2 + (y-2)^2 = 5$$
$$x^2 + 2x + 1 + y^2 - 4y + 4 = 5$$
$$x^2 + 2x + y^2 - 4y = 0$$

FOIL the binomials in the second equation and collect like terms:

$$(x-5)^2 + (y+1)^2 = 50$$
$$x^2 - 10x + 25 + y^2 + 2y + 1 = 50$$
$$x^2 - 10x + y^2 + 2y = 24$$

Add the result of the first equation to the opposite of the result of the second equation:

$$
\begin{aligned}
x^2 + 2x + y^2 - 4y &= \ \ \ 0 \\
-x^2 + 10x - y^2 - 2y &= -24 \\
\hline
12x - 6y &= -24
\end{aligned}
$$

Solve this last equation for y:

$$-6y = -12x - 24$$
$$y = 2x + 4$$

Substitute this in for y in one of the equations. Here we take the first equation:

$$x^2 + 2x + (2x+4)^2 - 4(2x+4) = 0$$
$$x^2 + 2x + 4x^2 + 16x + 16 - 8x - 16 = 0$$
$$5x^2 + 10x = 0$$
$$5x(x+2) = 0$$

So $x = 0$ or $x = -2$. When $x = 0$, $y = 2(0) + 4 = 4$. When $x = -2$, $y = 2(-2) + 4 = 0$. Thus, the solutions are $(0, 4)$ and $(-2, 0)$.

To confirm this result graphically using the graphing calculator, solve each equation for y.

Solve the first equation for y:

$$(x+1)^2 + (y-2)^2 = 5$$
$$(y-2)^2 = 5 - (x+1)^2$$
$$y - 2 = \pm\sqrt{5 - (x+1)^2}$$
$$y = 2 \pm \sqrt{5 - (x+1)^2}$$

Solve the second equation for y:

$$(x-5)^2 + (y+1)^2 = 50$$
$$(y+1)^2 = 50 - (x-5)^2$$
$$y + 1 = \pm\sqrt{50 - (x-5)^2}$$
$$y = -1 \pm \sqrt{50 - (x-5)^2}$$

Now we can graph the four resulting equations on the calculator:

$$y_1 = 2 + \sqrt{5 - (x+1)^2}$$
$$y_2 = 2 - \sqrt{5 - (x+1)^2}$$
$$y_3 = -1 + \sqrt{50 - (x-5)^2}$$
$$y_4 = -1 - \sqrt{50 - (x-5)^2}$$

Xmin = –8, Xmax = 8, Ymin = –8, Ymax = 8

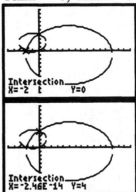

(Note that the above x-value is the calculator's best guess at $x = 0$.)

17. FOIL the binomial in the first equation and collect like terms:

$$(x-100)^2 + y^2 = 90^2$$

$$x^2 - 200x + 10,000 + y^2 = 8100$$

$$x^2 - 200x + y^2 = -1900$$

FOIL the binomials in the second equation and collect like terms:

$$(x-28)^2 + (y-96)^2 = 42^2$$

$$x^2 - 56x + 784 + y^2 - 192y + 9216$$
$$= 1764$$

$$x^2 - 56x + y^2 - 192y = -8236$$

Add the result of the first equation to the opposite of the result of the second equation:

$$\begin{array}{r} x^2 - 200x + y^2 \qquad\quad = -1900 \\ -x^2 + 56x - y^2 + 192y = \;\; 8236 \\ \hline -144x + 192y = 6336 \end{array}$$

Solve this last equation for x:

$$-144x = -192y + 6336$$

$$x = \frac{4}{3}y - 44$$

Substitute this in for x in the first equation:

$$\left(\frac{4}{3}y - 44\right)^2 - 200\left(\frac{4}{3}y - 44\right) + y^2$$
$$= -1900$$

$$\frac{16}{9}y^2 - \frac{352}{3}y + 1936 - \frac{800}{3}y$$
$$+ 8800 + y^2 = -1900$$

$$\frac{25}{9}y^2 - 384y + 12,636 = 0$$

$$25y^2 - 3456y + 113,724 = 0$$

Using the quadratic formula, we find that $y = 84.24$ or $y = 54$. When $y = 84.24$, $x = \frac{4}{3}(84.24) - 44 = 68.32$.

When $y = 54$, $x = \frac{4}{3}(54) - 44 = 28$.

Thus, the solutions are $(68.32, 84.24)$ and $(28, 54)$.

19. It takes 7 seconds for the signal to reach you and, from Investigation 15, we assume that the transmission travels at 5 meters per second. Therefore, you are $(5)(7) = 35$ meters away from the satellite. Say your position is at the point (x, y). Then the square of the distance between you and $(60, 80)$ is

$(x-60)^2 + (y-80)^2 = 35^2$ and the square of the distance between you and $(93.6, 35.2)$ is

$(x-93.6)^2 + (y-35.2)^2 = 35^2$.

Foiling and combining like terms in each of the equations gives:

$$x^2 - 120x + y^2 - 160y = -8775$$

$$x^2 - 187.2x + y^2 - 70.4y = -8775$$

Subtract the bottom equation from the top equation and solve for x:

$$67.2x - 89.6y = 0$$

$$x = \frac{89.6}{67.2}y = \frac{4}{3}y$$

Substitute this into the first equation:

$$\left(\frac{4}{3}y\right)^2 - 120\left(\frac{4}{3}y\right) + y^2 - 160y$$
$$= -8775$$

$$\frac{25}{9}y^2 - 320y + 8775 = 0$$

$$25y^2 - 2880y + 78,975 = 0$$

$$25(y^2 - 115.2y + 3159) = 0$$

Using the quadratic formula with $a = 1$, $b = -115.2$, and $c = 3159$, we find that $y = 70.2$ or $y = 45$. When $y = 70.2$, $x = \frac{4}{3}(70.2) = 93.6$. When $y = 45$, $x = \frac{4}{3}(45) = 60$. Therefore, your position is at the point $(93.6, 70.2)$ or at the point $(60, 45)$.

21. Let w be the width of the rectangle and let l be the length of the rectangle. Using formulas for perimeter and area, we form the following system of equations:

$lw = 216$

$2l + 2w = 60$

Solving the second equation for l gives $l = 30 - w$. Substitute this into the first equation and solve for w:

$(30 - w)w = 216$

$30w - w^2 = 216$

$w^2 - 30w + 216 = 0$

$(w - 12)(w - 18) = 0$

So $w = 12$ or $w = 18$. (One could also use the quadratic formula to solve the above equation.)
When $w = 12$, then $l = 30 - 12 = 18$.
When $w = 18$, then $l = 30 - 18 = 12$.
Thus, the rectangle has dimensions 12 feet by 18 feet.

23. We are given that $PV = K$ and that $K = 30$ inch-pounds, so $PV = 30$. Let P_1 and V_1 be the original pressure and volume and let P_2 and V_2 be the new pressure and volume. So $P_2 = P_1 + 4$ and $V_2 = V_1 - 2$. Since $PV = 30$ both for the original and old values of P and V, we form the following system of two equations:

$P_1 V_1 = 30$

$(P_1 + 4)(V_1 - 2) = 30$

From the first equation, $V_1 = \dfrac{30}{P_1}$.

Substitute this into the second equation:

$(P_1 + 4)(\dfrac{30}{P_1} - 2) = 30$

$30 - 2P_1 + \dfrac{120}{P_1} - 8 = 30$

$-2P_1 + \dfrac{120}{P_1} - 8 = 0$

$-2P_1^2 + 120 - 8P_1 = 0$

$-2(P_1^2 + 4P_1 - 60) = 0$

$2(P_1 + 10)(P_1 - 6) = 0$

So $P_1 = -10$ or $P_1 = 6$. Since pressure is positive, disregard the negative result. Then

$V_1 = \dfrac{30}{P_1} = \dfrac{30}{6} = 5$. Therefore, the

original pressure was 6 pounds per square inch and the original volume was 5 cubic inches.

25. a. The first hyperbola has center (0, 0) and branches opening upward and downward from the vertices (0, –12) and (0, 12). The second hyperbola has center (0, 22) and branches opening upward and downward from the vertices (0, 21) and (0, 23). Refer to the graph in part (b).

b. Solve the first equation for y:

$$\frac{y^2}{144} - \frac{x^2}{81} = 1$$

$$y^2 = 144\left(\frac{x^2}{81} + 1\right)$$

$$y = \pm\sqrt{144\left(\frac{x^2}{81} + 1\right)}$$

Solve the second equation for y:

$$(y-22)^2 - \frac{x^2}{48} = 1$$

$$(y-22)^2 = \frac{x^2}{48} + 1$$

$$y - 22 = \pm\sqrt{\frac{x^2}{48} + 1}$$

$$y = 22 \pm \sqrt{\frac{x^2}{48} + 1}$$

Graph the four resulting equations as $y_1, y_2, y_3,$ and y_4.

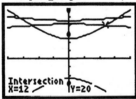

The intersection shown above is at the point P. The location of the transmitters are indicated by the small "box" points, with ordering lowest to highest: B, A, and C.

c. By part (b), your ship is at the location $P(12, 20)$. Calculate the following distances:

$$\overline{AP} = \sqrt{(0-12)^2 + (15-20)^2} = 13$$

$$\overline{BP} = \sqrt{(0-12)^2 + (-15-20)^2} = 37$$

$$\overline{CP} = \sqrt{(0-12)^2 + (29-20)^2} = 15$$

It follows that $\overline{BP} = 24 + \overline{AP}$ and $\overline{CP} = 2 + \overline{AP}$.

27. a. The first hyperbola has center $(0, 0)$ and branches opening to the left and right from the vertices $(-7, 0)$ and $(7, 0)$. The second hyperbola has center $(-28, -45)$ and branches opening upward and downward from the vertices $(-28, -9)$ and $(-28, -81)$. Refer to the graph in part (b).

   b. Solve the first equation for y:

$$\frac{x^2}{49} - \frac{y^2}{735} = 1$$

$$\frac{y^2}{735} = \frac{x^2}{49} - 1$$

$$y^2 = 735\left(\frac{x^2}{49} - 1\right)$$

$$y = \pm\sqrt{735\left(\frac{x^2}{49} - 1\right)}$$

Solve the second equation for y:

$$\frac{(y+45)^2}{36^2} - \frac{(x+28)^2}{27^2} = 1$$

$$(y+45)^2 = 36^2 \cdot \left(\frac{(x+28)^2}{27^2} + 1\right)$$

$$y + 45 = \pm 36\sqrt{\frac{(x+28)^2}{729} + 1}$$

$$y = -45 \pm 36\sqrt{\frac{(x+28)^2}{729} + 1}$$

Graph the four resulting equations as y_1, y_2, y_3, and y_4.

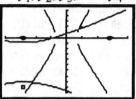

The location of the transmitters are indicated by the small "box" points, with transmitter A on the x-axis to the right of the origin, transmitter B on the x-axis to the left of the origin, and transmitter C in quadrant three.

Find the intersection point that is closer to point A than to point B and closer to point B than to point C. This intersection, point P, is shown below.

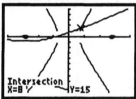

   c. By part (b), your ship is at the location $P(8, 15)$. Calculate the following distances:

$$\overline{AP} = \sqrt{(8-28)^2 + (15-0)^2}$$
$$= 25$$

$$\overline{BP} = \sqrt{(8-(-28))^2 + (15-0)^2}$$
$$= 39$$

$$\overline{CP} = \sqrt{(8-(-28))^2 + (15-(-90))^2}$$
$$= 111$$

It follows that $\overline{BP} = 14 + \overline{AP}$ and $\overline{CP} = 72 + \overline{BP}$.

Chapter 10 Review

1. $4x^2 + y^2 = 25$

$$\frac{x^2}{25/4} + \frac{y^2}{25} = 1$$

So $a = \frac{5}{2}$ and $b = 5$. The ellipse has center at $(0, 0)$, vertices at $(0, 5)$ and $(0, -5)$, and covertices at $(-\frac{5}{2}, 0)$ and $(\frac{5}{2}, 0)$. Refer to the graph in the back of the textbook.

3. Let $y = -4$ and solve for x:

$$4x^2 + (-4)^2 = 25$$
$$4x^2 = 9$$
$$x^2 = \frac{9}{4}$$
$$x = \pm\frac{3}{2}$$

The points are $(-\frac{3}{2}, -4)$ and $(\frac{3}{2}, -4)$.

5. $4y^2 = 9 + 36x^2$

$$4y^2 - 36x^2 = 9$$

$$\frac{y^2}{9/4} - \frac{x^2}{1/4} = 1$$

So $a = \frac{1}{2}$ and $b = \frac{3}{2}$. This is a hyperbola opening upward and downward with center at $(0, 0)$. It has vertices at $(0, -\frac{3}{2})$ and $(0, \frac{3}{2})$. To draw the "central rectangle," move $\frac{3}{2}$ units in each vertical direction and $\frac{1}{2}$ unit in each horizontal direction from the origin. Draw the asymptotes through the corners of the central rectangle. The slopes of the asymptotes are $m = \pm\frac{\frac{3}{2}}{\frac{1}{2}} = \pm 3$. Since the asymptotes pass through the origin, their equations are $y = -3x$ and $y = 3x$. Refer to the graph in the back of the textbook.

7. $x = 4y^2$ is a parabola opening to the right with vertex at the origin. It passes through points $(4, 1)$ and $(4, -1)$. Refer to the graph in the back of the textbook.

9. a. $\frac{1}{2}x^2 - y = 4$

$y + 4 = \frac{1}{2}x^2$

This is a parabola opening upward with vertex at $(0, -4)$.

b. Let $y = -\frac{5}{2}$ and solve for x:

$\frac{1}{2}x^2 - \left(-\frac{5}{2}\right) = 4$

$\frac{1}{2}x^2 + \frac{5}{2} = 4$

$x^2 + 5 = 8$

$x^2 = 3$

$x = \pm\sqrt{3}$

The points are $\left(\pm\sqrt{3}, -\frac{5}{2}\right)$.

11. $\frac{(x-5)^2}{15} + \frac{y^2}{25} = 1$

This is an ellipse with center at $(5, 0)$. The vertices are 5 units above and below the center, at $(5, -5)$ and $(5, 5)$. The covertices are $\sqrt{15} \approx 3.9$ units left and right of the center, at $(5 - \sqrt{15}, 0)$ and $(5 + \sqrt{15}, 0)$. Refer to the graph in the back of the textbook.

13. Write the equation as:

$(y - 2) = \frac{1}{4}(x + 3)^2$

This is the equation for a parabola with vertex at $(-3, 2)$ and opening upward. $a = \frac{1}{4}$. Refer to the graph in the back of the textbook.

15. This is the equation of a hyperbola with center at $(2, -3)$ and branches opening left and right. It has vertices two units to the left and right of the center, at $(0, -3)$ and $(4, -3)$. To draw the "central rectangle", move 3 units in each vertical direction and 2 units in each horizontal direction from $(2, -3)$. Draw the asymptotes through the corners of the rectangle. Refer to the graph in the back of the textbook.

17. a. Complete the square in x and y:

$4y^2 - 3x^2 - 24y - 24x - 24 = 0$

$4(y^2 - 6y + \underline{}) - 3(x^2 + 8x + \underline{})$
$= 24 + 4(\underline{}) - 3(\underline{})$

$4(y^2 - 6y + 9) - 3(x^2 + 8x + 16)$
$= 24 + 4(9) - 3(16)$

$4(y - 3)^2 - 3(x + 4)^2 = 12$

$\frac{(y-3)^2}{3} - \frac{(x+4)^2}{4} = 1$

b. This is a hyperbola opening upward and downward with center $(-4, 3)$. The vertices are $(-4, 3 - \sqrt{3})$ and $(-4, 3 + \sqrt{3})$. To draw the "central rectangle," move $\sqrt{3} \approx 1.7$ units in each vertical direction and 2 units in each horizontal direction. Draw the asymptotes through the corners of the rectangle. Refer to the graph in the back of the textbook.

19. a. Complete the square in y:

$$y^2 + 6y + 4x + 1 = 0$$

$$(y^2 + 6y + \underline{9}) = -4x - 1 + \underline{9}$$

$$(y + 3)^2 = -4x + 8$$

$$(x - 2) = -\frac{1}{4}(y + 3)^2$$

b. This is the equation for a parabola with vertex at $(2, -3)$ and opening to the left. $a = -\frac{1}{4}$.

21. a. Complete the square in x and y:

$$4x^2 + y^2 - 16x + 4y + 4 = 0$$

$$4(x^2 - 4x + \underline{4}) + (y^2 + 4y + \underline{4})$$

$$= -4 + 4(\underline{4}) + \underline{4}$$

$$4(x - 2)^2 + (y + 2)^2 = 16$$

$$\frac{(x - 2)^2}{4} + \frac{(y + 2)^2}{16} = 1$$

b. This is the equation for an ellipse centered at $(2, -2)$. The major axis is vertical. The vertices lie four units above and below the center at $(2, 2)$ and $(2, -6)$, and the covertices lie two units to the left and right of the center at $(0, -2)$ and $(4, -2)$.

23.
$$\frac{(x + 1)^2}{4^2} + \frac{(y - 4)^2}{2^2} = 1$$

$$\frac{(x + 1)^2}{16} + \frac{(y - 4)^2}{4} = 1$$

25. Since one end of the vertical conjugate axis is at $(-3, 4)$, it lies on the line $x = -3$. The horizontal line $y = 1$ passes through the vertex $(-4, 1)$. The intersection of the these two lines is the center of the hyperbola, at $(-3, 1)$. The hyperbola opens to the left and right, and $(-4, 1)$ is 1 unit left of the center, so $a = 1$. The point $(-3, 4)$ is 3 units above the center, so $b = 3$.

$$\frac{(x + 3)^2}{1^2} - \frac{(y - 4)^2}{3^2} = 1$$

$$(x + 3)^2 - \frac{(y - 4)^2}{9} = 1$$

27. The ellipse must be tangent to the y-axis at one of its vertices. So one of its vertices is $(0, -2)$, which is 5 units away from the center. Thus, $a = 5$. The ellipse must be tangent to the x-axis at one of its covertices. So one of its covertices is $(-5, 0)$, which is 2 units away from the center. Thus, $b = 2$. The equation is then

$$\frac{(x + 5)^2}{5^2} + \frac{(y + 2)^2}{2^2} = 1 \text{ or}$$

$$\frac{(x + 5)^2}{25} + \frac{(y + 2)^2}{4} = 1.$$

29. From the second equation,
$y = \dfrac{16}{2x} = \dfrac{8}{x}$. Substitute this into the
first equation:

$$x^2 + \left(\frac{8}{x}\right)^2 = 20$$

$$x^2 + \frac{64}{x^2} = 20$$

$$x^4 + 64 = 20x^2$$

$$x^4 - 20x^2 + 64 = 0$$

Using the quadratic formula,

$$x^2 = \frac{-(-20) \pm \sqrt{(-20)^2 - 4(1)(64)}}{2(1)}$$

$$= \frac{20 \pm \sqrt{144}}{2} = \frac{20 \pm 12}{2}$$

So $x^2 = \dfrac{20 + 12}{2} = 16$, which gives

$x = \pm 4$, or $x^2 = \dfrac{20 - 12}{2} = 4$, which

gives $x = \pm 2$. Substitute these values
of x into the equation $y = \dfrac{8}{x}$ to find
the ordered pairs (4, 2), (–4, –2),
(2, 4), and (2, –4). Note that these
ordered pairs satisfy both of the
original equations. Refer to the
graph in the back of the textbook.

31. Write the first equation in standard
form and then add the two equations:

$$-x^2 + y^2 = 24$$
$$\underline{2x^2 - y^2 = 1}$$
$$x^2 = 25$$
$$x = \pm 5$$

Put $x^2 = 25$ into the first equation:

$$-(25) + y^2 = 24$$
$$y^2 = 49$$
$$y = \pm 7$$

The solutions are (–5, –7), (–5, 7),
(5, –7), and (5, 7). Refer to the graph
in the back of the textbook.

33. In the second equation, subtract 2
from both sides and then multiply
each term by –1. Add the result to
the first equation:

$$x^2 + 4x + y^2 - 4y = 26$$
$$\underline{-x^2 + 12x - y^2 = 2}$$
$$16x - 4y = 28$$

Solve this last equation for y:

$$-4y = -16x + 28$$
$$y = 4x - 7$$

Substitute this expression into the
second equation:

$$x^2 - 12x + y^2 + 2 = 0$$
$$x^2 - 12x + (4x - 7)^2 + 2 = 0$$
$$x^2 - 12x + 16x^2 - 56x + 49 + 2 = 0$$
$$17x^2 - 68x + 51 = 0$$
$$17(x^2 - 4x + 3) = 0$$
$$17(x - 3)(x - 1) = 0$$

So $x = 3$ or $x = 1$. When $x = 3$,
$y = 4(3) - 7 = 5$. When $x = 1$,
$y = 4(1) - 7 = -3$. Thus, the
solutions are (3, 5) and (1, –3).

In order to graph the first equation
on your calculator, first complete the
square in y and then solve for y:

$$x^2 + 4x + (y^2 - 4y + \underline{4}) = 26 + \underline{4}$$
$$x^2 + 4x + (y - 2)^2 = 30$$
$$(y - 2)^2 = 30 - 4x - x^2$$
$$y - 2 = \pm\sqrt{30 - 4x - x^2}$$
$$y = 2 \pm \sqrt{30 - 4x - x^2}$$

In order to graph the second
equation, solve for y:

$$y^2 = -x^2 + 12x - 2$$

$$y = \pm\sqrt{-x^2 + 12x - 2}$$

Graph the resulting four equations as
$y_1, y_2, y_3,$ and y_4. Refer to the graph
in the back of the textbook.

35. Let x represent Moia's speed and let y represent Fran's speed. Then $y = x + 5$ since Fran drives 5 mph faster than Moia. Using $t = \dfrac{d}{r}$, Moia's time is $\dfrac{180}{x}$ and Fran's time is $\dfrac{200}{y}$. Since their times are equal, $\dfrac{180}{x} = \dfrac{200}{y}$. Substituting $y = x + 5$ into this last equation, we have

$$\frac{180}{x} = \frac{200}{x+5}$$
$$180(x + 5) = 200x$$
$$180x + 900 = 200x$$
$$900 = 20x$$
$$45 = x$$

Thus, Moia's speed is 45 mph and Fran's speed is 50 mph.

37. Let x represent the length and let y represent the width of the rectangle. Then from the formulas for perimeter and area, our two equations are $2x + 2y = 34$ and $xy = 70$. The first equation is equivalent to $x + y = 17$, or $y = 17 - x$. Substitute this into the second equation:

$$xy = 70$$
$$x(17 - x) = 70$$
$$17x - x^2 = 70$$
$$0 = x^2 - 17x + 70$$
$$0 = (x - 7)(x - 10)$$

So $x = 7$ or $x = 10$. When $x = 7$, $y = 17 - 7 = 10$. When $x = 10$, $y = 17 - 10 = 7$. Thus, the rectangle measures 7 cm by 10 cm.

39. Let x represent the rate of the morning train and let y represent the rate of the evening train. The evening train is 10 mph faster than the morning train, so $y = 10 + x$. Using the formula $t = \dfrac{d}{r}$ and the fact that Norm's distance to work is 10 miles, Norm's time in the morning is $\dfrac{10}{x}$ and Norm's time in the evening is $\dfrac{10}{y}$. The sum of these times is 50 minutes, or $\dfrac{5}{6}$ hour, so $\dfrac{10}{x} + \dfrac{10}{y} = \dfrac{5}{6}$. Substituting $y = 10 + x$ into this last equation gives the following:

$$\frac{10}{x} + \frac{10}{10+x} = \frac{5}{6}$$

Multiply each term by $6x(10+x)$:
$$60(10 + x) + 60x = 5x(10 + x)$$
$$600 + 60x + 60x = 50x + 5x^2$$
$$0 = 5x^2 - 70x - 600$$
$$0 = 5(x^2 - 14x - 120)$$
$$0 = 5(x - 20)(x + 6)$$

So, $x = 20$ or $x = -6$. Since the rate cannot be negative, we disregard this last result. When $x = 20$, $y = 10 + 20 = 30$. So the rate of the morning train is 20 mph and the rate of the evening train is 30 mph.

41. a. The first hyperbola has center (0, 256) and branches opening upward and downward. The vertices are 60 units above and below the center, at (0, 316) and (0, 196). The second hyperbola has center (0, 0) and branches opening upward and downward. The vertices are 60 units above and below the center, at (0, 60) and (0, –60). Refer to the graph in part (b).

b. Solve the first equation for y:

$$\frac{(y-256)^2}{60^2} = 1 + \frac{x^2}{20736}$$

$$(y-256)^2 = 3600\left(1 + \frac{x^2}{20736}\right)$$

$$y - 256 = \pm 60\sqrt{1 + \frac{x^2}{20736}}$$

$$y = 256 \pm 60\sqrt{1 + \frac{x^2}{20736}}$$

Solve the second equation for y:

$$\frac{y^2}{3600} - \frac{x^2}{6400} = 1$$

$$y^2 = 3600\left(\frac{x^2}{6400} + 1\right)$$

$$y = \pm 60\sqrt{\frac{x^2}{6400} + 1}$$

Graph the four resulting equations as $y_1, y_2, y_3,$ and y_4. First, use the window Xmin = –240, Xmax = 240, Ymin = –150, and Ymax = 450. The location of the transmitters are indicated by the small "box" points, with ordering lowest to highest: B, A, and C:

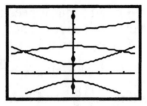

Next, graph using the window suggested in the problem. The point P is the intersection point shown in the graph below.

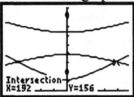

c. By part (b), your ship is at the location $P(192, 156)$. Calculate the following distances:

$$\overline{AP} = \sqrt{(0-192)^2 + (100-156)^2}$$
$$= 200$$

$$\overline{BP} = \sqrt{(0-192)^2 + (412-156)^2}$$
$$= 320$$

$$\overline{CP} = \sqrt{(0-192)^2 + (-100-156)^2}$$
$$= 320$$

It follows that $\overline{BP} = 120 + \overline{AP}$ and $\overline{CP} = 120 + \overline{AP}$.

Appendix A: Review Topics

Exercise A.1

1. a. Let j represent Jim's age and let a represent Ana's age. Jim is 27 years older than Ana, so $j = 27 + a$.

  b. Evaluate the above expression for $a = 22$: $j = 27 + 22 = 49$. So Jim is 49 years old when Ana is 22.

3. a. Let r represent Helen's rate (or speed) and let t represent her time. Rate multiplied by time equals distance. Helen must travel a distance of 1260 miles so $rt = 1260$. We solve for t by dividing both sides by r. This gives $t = \dfrac{1260}{r}$.

  b. Evaluating the equation above at $r = 45$, gives $t = \dfrac{1260}{45} = 28$ hours.

5. a. Let y represent the number of yards of fabric and let f represent the number of feet of fabric. The number of feet of fabric is three times the number of yards of fabric, so $f = 3y$. We solve for y by dividing both sides by 3. Therefore, $y = \dfrac{f}{3}$.

  b. Fabric costs \$5.79 per yard so the cost of y yards is $5.79y$ dollars. Substituting $\dfrac{f}{3}$ in for y gives the cost of f feet of fabric:
  $$\text{Cost} = 5.79\left(\frac{f}{3}\right) = 1.93f \text{ dollars.}$$

  c. Evaluate the expression from (b) at $f = 10$. This gives:
  $$\text{Cost} = 5.79\left(\frac{10}{3}\right) = \$19.30.$$

7. a. Let A represent the area and let r represent the radius. Area is π times radius squared, so $A = \pi r^2$.

  b. Evaluate at $r = 5$:
  $$A = \pi(5)^2 = 25\pi = 78.54 \text{ cm}^2.$$

9. a. Let n represent the total number of assignments and let m represent the number of assignments that Farshid missed. The total number of assignments equals the number he turned in (20) plus the number he missed. So $n = 20 + m$.

b. His average homework score, a, is the total number of points (198) divided by the total number of assignments. So, $a = \dfrac{198}{n}$. Now substitute $20 + m$ in for n to get:
$$a = \frac{198}{20+m}.$$

c. Evaluate the equation above at $m = 5$; $a = \dfrac{198}{20+5} = 7.92$.

11. a. Let t represent the amount of sales tax and let p represent the price of an item. The amount of sales tax is equal to the price of an item times the sales tax rate. So $t = p(7.9\%) = p(0.079)$.

b. The total bill is the price plus the amount of sales tax:
$b = p + p(0.079) = 1.079p$.

c. Evaluate at $p = 490$:
$b = 1.079(490) = \$528.71$.

13. a. Let C be the total cost of talking for m minutes. Then C equals $\$1.97$ plus $\$0.39$ times each minute. So $C = 1.97 + 0.39m$.

b. Evaluate at $m = 27$;
$C = 1.97 + 0.39(27) = \$12.50$.

15. a. Let c represent the amount of rice Juan consumes in w weeks. Since Juan consumes 0.4 lbs per week, the amount he consumes in w weeks is 0.4 times w. So $c = 0.4w$.

b. Let r represent the amount of rice left in the bag after w weeks. Since the bag originally held 50 lbs, if Juan has consumed c lbs after w weeks, then $r = 50 - c$. However, we can substitute $0.4w$ in for c, therefore the amount left is given by the formula:
$r = 50 - 0.4w$.

c. Evaluate the formula in (b) at $w = 6$; $r = 50 - 0.4(6) = 47.6$ lbs.

17. a. If Leon's truck uses u gallons of gas to drive m miles then the gas mileage that Leon's truck gets is $\dfrac{m}{u}$ miles per gallon. We know that Leon's truck gets 20 miles per gallon. So $\dfrac{m}{u} = 20$. To solve for u, multiply both sides by u to clear fractions: $m = 20u$. Then divide by 20. So $u = \dfrac{m}{20}$.

b. The truck holds 14.6 gallons, so if Leon has used u gallons of gas then there are $g = 14.6 - u$ gallons remaining in the tank. Substituting in $\dfrac{m}{20}$ for u (from the expression in (a)), we get $g = 14.6 - \dfrac{m}{20}$.

c. Evaluate the expression in (b) at $m = 110$. Then:
$$g = 14.6 - \frac{110}{20} = 9.1 \text{ gallons}$$

19. $\dfrac{3(6-8)}{-2} - \dfrac{6}{-2} = \dfrac{3(-2)}{-2} - \dfrac{6}{-2} = \dfrac{-6}{-2} - \dfrac{6}{-2} = 3 - (-3) = 3 + 3 = 6$

21. $6[3 - 2(4+1)] = 6[3 - 2(5)] = 6[3 - 10] = 6(-7) = -42$

23. $(4-3)[2 + 3(2-1)] = (1)[2 + 3(1)] = 2 + 3 = 5$

25. $64 \div \big(8[4 - 2(3+1)]\big) = 64 \div \big(8[4 - 2(4)]\big) = 64 \div \big(8[4 - 8]\big) = 64 \div \big(8[-4]\big)$
$= 64 \div (-32) = -2$

27. $5[3 + (8-1)] \div (-25) = 5[3 + (7)] \div (-25) = 5[10] \div (-25) = 50 \div (-25) = -2$

29. $[-3(8-2) + 3] \cdot [24 \div 6] = [-3(6) + 3] \cdot [24 \div 6] = [-18 + 3] \cdot [24 \div 6] = [-15] \cdot [4] = -60$

31. $-5^2 = -25$ since $-5^2 = -1 \cdot 5^2 = -1 \cdot 25 = -25$

33. $(-3)^4 = 81$ since $(-3)^4 = (-3)(-3)(-3)(-3) = 81$

35. $-4^3 = -64$ since $-4^3 = -1 \cdot 4^3 = -1 \cdot 64 = -64$

37. $(-2)^5 = -32$ since $(-2)^5 = (-2)(-2)(-2)(-2)(-2) = -32$

39. $\dfrac{4 \cdot 2^3}{16} + 3 \cdot 4^2 = \dfrac{4 \cdot 8}{16} + 3 \cdot 16 = \dfrac{32}{16} + 48 = 2 + 48 = 50$

41. $\dfrac{3^2 - 5}{6 - 2^2} - \dfrac{6^2}{3^2} = \dfrac{9 - 5}{6 - 4} - \dfrac{36}{9} = \dfrac{4}{2} - \dfrac{36}{9} = 2 - 4 = -2$

43. $\dfrac{(-5)^2 - 3^2}{4 - 6} + \dfrac{(-3)^2}{2 + 1} = \dfrac{25 - 9}{4 - 6} + \dfrac{9}{2 + 1} = \dfrac{16}{-2} + \dfrac{9}{3} = -8 + 3 = -5$

45. $\dfrac{-8398}{26 \cdot 17} = \dfrac{-8398}{442} = -19$

47. $\dfrac{112.78 + 2599.124}{27.56} = \dfrac{2711.904}{27.56} = 98.4$

49. $\sqrt{24 \cdot 54} = \sqrt{1296} = = 36$

51. $\dfrac{116 - 35}{215 - 242} = \dfrac{81}{-27} = -3$

53. $\sqrt{27^2 + 36^2} = \sqrt{729 + 1296} = \sqrt{2025} = 45$

55. $\dfrac{-27 - \sqrt{27^2 - 4(4)(35)}}{2 \cdot 4} = \dfrac{-27 - \sqrt{729 - 560}}{8} = \dfrac{-27 - \sqrt{169}}{8} = \dfrac{-27 - 13}{8} = \dfrac{-40}{8} = -5$

57. a. $2 + \frac{3}{4}$

   b. $\frac{2+3}{4}$

59. a. -23^2

   b. $(-23)^2$

61. a. $\sqrt{9+16}$

   b. $\sqrt{9} + 16$

63. $\frac{5(F-32)}{9}$; $F = 212$:

$$\frac{5(212-32)}{9} = \frac{5(180)}{9}$$

$$= \frac{900}{9} = 100$$

65. $P + Prt$; $P = 1000, r = 0.04, t = 2$:

$$1000 + 1000 \cdot (0.04) \cdot 2$$
$$= 1000 + 80 = 1080$$

67. $\frac{1}{2}gt^2 - 12t$; $g = 32, t = \frac{3}{4}$:

$$\frac{1}{2} \cdot 32 \cdot \left(\frac{3}{4}\right)^2 - 12\left(\frac{3}{4}\right)$$

$$= \frac{1}{2} \cdot 32 \cdot \frac{9}{16} - \frac{36}{4} = \frac{288}{32} - 9$$

$$= 9 - 9 = 0$$

69. $\frac{32(V-v)^2}{g}$; $V = 12.78, v = 4.26,$ and $g = 32$:

$$\frac{32(12.78 - 4.26)^2}{32} = \frac{32(8.52)^2}{32}$$

$$= (8.52)^2 = 72.5904$$

Exercise A.2

1. $3x + 5 = 26$ Subtract 5 from both sides.

    $3x = 21$ Divide both sides by 3.

    $x = 7$

3. $3(z + 2) = 37$ Distribute.

    $3z + 6 = 37$ Subtract 6 from both sides.

    $3z = 31$ Divide both sides by 3.

    $z = \dfrac{31}{3}$

5. $3y - 2(y - 4) = 12 - 5y$ Distribute.

    $3y - 2y + 8 = 12 - 5y$ Collect like terms.

    $y + 8 = 12 - 5y$ Add $5y$ to both sides.

    $6y + 8 = 12$ Subtract 8 from both sides.

    $6y = 4$ Divide both sides by 6.

    $y = \dfrac{4}{6} = \dfrac{2}{3}$

7. $0.8w - 2.6 = 1.4w + 0.3$ Subtract $0.8w$ from both sides.

    $-2.6 = 0.6w + 0.3$ Subtract 0.3 from both sides.

    $-2.9 = 0.6w$ Divide both sides by 0.6.

    $-4.83 \approx w$

9. $0.25t + 0.10(t - 4) = 11.60$ Distribute.

    $0.25t + 0.10t - 0.40 = 11.60$ Combine like terms.

    $0.35t - 0.40 = 11.60$ Add 0.40 to both sides.

    $0.35t = 12.00$ Divide both sides by 0.35.

    $t \approx 34.29$

11. Celine's demand is given by the expression $200 - 5p$, and their supply is given by the expression $56 + 3p$. Setting supply equal to demand gives the equation:

    $56 + 3p = 200 - 5p$ Add $5p$ to both sides.

    $56 + 8p = 200$ Subtract 56 from both sides.

    $8p = 144$ Divide both sides by 8.

    $p = \$18.00$ Include appropriate units.

13. a. We are asked to find the amount of time that it takes Roger's wife to find him. Let t = the time in hours that has gone by since Roger's wife left the house.

  b. Roger's wife travels at 45 miles per hour so she will have traveled $45t$ miles after t hours.

  c. Roger leaves six hours ahead of his wife and cycles at 16 miles per hour so the distance that he has traveled is $16(t + 6)$.

  d. Roger's wife has caught up to Roger when their distances traveled are equal. So we solve the equation:

$45t = 16(6 + t)$ Distribute.

$45t = 96 + 16t$ Subtract $16t$ from both sides.

$29t = 96$ Divide both sides by 29.

    $t \approx 3.3$ hours   Include appropriate units.

15. a. We are asked to find the number of copies that would need to be made in order for the total cost of both machines to be the same. Let c be the number of copies made.

  b. The total cost of the first machine listed is $20,000 plus $0.02 per copy, or $20,000 + 0.02c$. The total cost of the second machine is $17,500 plus $0.025 per copy or $17,500 + 0.025c$.

  c. We want these total costs to be the same so we set the above expressions equal to one another and solve for c:

$20,000 + 0.02c = 17,500 + 0.025c$ Subtract $0.02c$ from both sides.

$20,000 = 17,500 + 0.005c$ Subtract $17,500$ from both sides.

$2500 = 0.005c$ Divide both sides by 0.005.

$500,000 = c$

Therefore, 500,000 copies must be made to justify the higher price.

17. a. The population now is 135,000 and the rate of increase is 8% per year. Expect that in one year population will increase by 0.08(135,000). So in one year the population will be $135,000 + 0.08(135,000) = 145,800$ people.

  b. Let x represent the population last year. Then, as in (a), an 8% increase would make the population $x + (0.08)x$ this year. We know the population this year is 135,000 so we solve for x in:

$x + (0.08)x = 135,000$ Combine like terms.

$(1.08)x = 135,000$ Divide both sides by 1.08.

$x = 125,000$

19. Virginia's salary was $24,000 per year. A 7% pay cut would reduce her salary to $24,000 - 24,000(0.07) = \$22,320$ per year. If she wants to make $24,000 again, then she must receive a pay increase r so that $22,320 + 22,320r = 24,000$. Solve for r:

$22,320 + 22,320r = 24,000$ Subtract 22,320 from both sides.

$\qquad 22,320r = 1680$ Divide both sides by 22,320.

$\qquad\qquad r \approx 0.07527$ Express r as a percentage.

$\qquad\qquad r \approx 7.527\%$

21. Let $F =$ Delbert's final exam score. Then we have the weighted average: $77 \cdot (0.70) + F \cdot (0.30) =$ term average. Delbert wants a term average of 80, so solve for F in the following equation:

$77 \cdot (0.70) + F \cdot (0.30) = 80$ Multiply $77 \cdot (0.70)$.

$\quad 53.9 + F \cdot (0.30) = 80$ Subtract 53.9 from both sides.

$\qquad\quad F \cdot (0.30) = 26.1$ Divide both sides by 0.30.

$\qquad\qquad\quad F = 87$

23. a. Let A represent the amount in pounds of the first fertilizer (6% potash) that we use. We are asked to find A so that a mixture of A pounds of the first fertilizer and $(10 - A)$ pounds of the second fertilizer (15% potash) has 8% potash.

b.

Pounds of fertilizer	% Potash	Pounds of potash
A	6%	$(0.06)A$
$10 - A$	15%	$0.15(10 - A)$
10	8%	$(0.08)10 = 0.8$

c. According to the table above, there are $(0.06)A$ pounds of potash in A pounds of the first fertilizer, and $0.15(10 - A)$ pounds of potash in $(10 - A)$ pounds of the second fertilizer, and 0.8 pounds of potash in the mixture.

d. The amount of potash in the mixture is $0.06A + 0.15(10 - A)$. We want this amount to be 0.8 pounds so we solve for A in the following equation.

$0.06A + 0.15(10 - A) = 0.8$ $\qquad$ Distribute.

$0.06A + 1.5 - 0.15A = 0.8$ $\qquad$ Combine like terms.

$\quad\; 1.5 - 0.09A = 0.8$ $\qquad$ Add $0.09A$ to both sides.

$\qquad\qquad\quad 1.5 = 0.8 + 0.09A$ $\quad$ Subtract 0.8 from both sides.

$\qquad\qquad\quad 0.7 = 0.09A$ $\qquad$ Divide both sides by 0.09.

$\qquad\quad 7.78 \approx A$

So we should use 7.78 pounds of the first and 2.22 pounds of the second.

25. a. Let s represent the annual salary of Lacy's clerks. We are asked to find s so that the average salary at the store is under $19,000.

   b. Lacy's pays out:

to managers:	$4(\$28,000) = \$112,000$
to department heads:	$12(\$22,000) = \$264,000$
to clerks:	$30(s) = 30s$

   c. The total amount Lacy's pays in salaries is $112,000 + 264,000 + 30s$. If Lacy's has an average salary of under $19,000, then another expression for the total amount Lacy's pays in salaries is $19,000(4 + 12 + 30) = 19,000(46) = \$874,000$.

   d. Solve for s in: $112,000 + 264,000 + 30s = 874,000$.

   $$112,000 + 264,000 + 30s = 874,000 \quad \text{Combine constants on the left.}$$
   $$376,000 + 30s = 874,000 \quad \text{Subtract } 376,000 \text{ from both sides.}$$
   $$30s = 498,000 \quad \text{Divide both sides by 30.}$$
   $$s = 16,600$$

   Lacy's needs to pay clerks less than $16,600 annually to keep the average below $19,000.

27. $v = k + gt$ Subtract k from both sides.

   $v - k = gt$ Divide both sides by g.

   $\dfrac{v - k}{g} = t$

29. $S = 2w(w + 2h)$ Distribute.

   $S = 2w^2 + 4wh$ Subtract $2w^2$ from both sides.

   $S - 2w^2 = 4wh$ Divide both sides by $4w$.

   $\dfrac{S - 2w^2}{4w} = h$

31. $P = a + (n - 1)d$ Distribute.

   $P = a + nd - d$ Subtract a from both sides.

   $P - a = nd - d$ Add d to both sides.

   $P - a + d = nd$ Divide both sides by d.

   $\dfrac{P - a + d}{d} = n$

33. $A = \pi rh + \pi r^2$ Subtract πr^2 from both sides.

   $A - \pi r^2 = \pi rh$ Divide both sides by πr.

   $\dfrac{A - \pi r^2}{\pi r} = h$

Exercise A.3

1. Let the angles of the triangle be A, B, and C. Let angle A be the smallest, $B = 10 + A$, and $C = 29 + A$. We use that the sum of the angles of any triangle is 180°.

$$A + B + C = 180 \quad \text{Substitute the expressions for } B \text{ and } C.$$
$$A + (10 + A) + (29 + A) = 180 \quad \text{Combine like terms.}$$
$$3A + 39 = 180 \quad \text{Subtract 39 from both sides.}$$
$$3A = 141 \quad \text{Divide both sides by 3.}$$
$$A = 47$$

The angles are 47°, 57°, and 76°.

3. Since the triangle is a right triangle, one angle is 90°. Let A be the smallest acute angle. Then the other acute angle is $2A$. Since the sum of the angles of any triangle is 180°, we have:

$$90 + A + 2A = 180 \quad \text{Combine like terms.}$$
$$90 + 3A = 180 \quad \text{Subtract 90 from both sides.}$$
$$3A = 90 \quad \text{Divide both sides by 3.}$$
$$A = 30$$

The angles are 30°, 60°, and 90°.

5. Let A represent the measure of the equal angles. The vertex angle is $2A - 20$. Since the sum of the angles of any triangle is 180°, we have:

$$A + A + (2A - 20) = 180 \quad \text{Combine like terms.}$$
$$4A - 20 = 180 \quad \text{Add 20 to both sides.}$$
$$4A = 200 \quad \text{Divide both sides by 4.}$$
$$A = 50$$

The angles are 50°, 50°, and 80°.

7. Let l be the length of the equal sides. Since the sum of the three sides is 42 cm, we have:

$$l + l + 12 = 42 \quad \text{Combine like terms.}$$
$$2l + 12 = 42 \quad \text{Subtract 12 from both sides.}$$
$$2l = 30 \quad \text{Divide both sides by 2.}$$
$$l = 15$$

The equal sides are 15 cm long.

9. $3x - 2 > 1 + 2x \quad$ Subtract $2x$ from both sides.

$\quad x - 2 > 1 \quad\quad\quad$ Add 2 to both sides.

$\quad\quad x > 3$

11. $\dfrac{-2x-6}{-3} > 2$ Multiply both sides by -3 & change inequality symbol.

$-2x-6 < -6$ Add 6 to both sides.

$\quad\quad -2x < 0$ Divide both sides by -2 and change inequality symbol.

$\quad\quad\quad x > 0$

13. $\dfrac{2x-3}{3} \le \dfrac{3x}{-2}$ Multiply both sides by 3.

$\quad 2x-3 \le \dfrac{9x}{-2}$ Multiply both sides by -2 & change inequality symbol.

$-4x+6 \ge 9x$ Add $4x$ to both sides.

$\quad\quad 6 \ge 13x$ Divide both sides by 13.

$\quad\quad \dfrac{6}{13} \ge x$

15. Let x be the length of the third side. By the triangle inequality, $10+6 > x$ or $16 > x$. Using another pair of sides, $6+x > 10$ or $x > 4$. Therefore, $4 < x < 16$, so the third side is greater than 4 feet but less than 16 feet long.

17. Let x represent the height of the lamppost. By similar triangles, we have:

$\dfrac{x}{12+9} = \dfrac{6}{9}$ Simplify both sides.

$\dfrac{x}{21} = \dfrac{2}{3}$ Multiply both sides by 21.

$x = \dfrac{2 \cdot 21}{3}$ Simplify.

$x = 14$

The lamppost is 14 feet tall.

19. Let r represent the radius of the circle of the exposed surface of water. Form two triangles by "slicing" the cone down the middle so that one triangle has sides of 4 ft and 12 feet and the other triangle has sides of r feet and 7 feet. By similar triangles:

$\dfrac{r}{7} = \dfrac{4}{12}$ Multiply both sides by 7.

$r = \dfrac{4 \cdot 7}{12}$ Simplify.

$r = \dfrac{7}{3}$

The area of this circle is $\pi r^2 = \pi \left(\dfrac{7}{3}\right)^2 \approx 17.10 \ \text{ft}^2$.

21. Let x represent the distance EB. By similar triangles:

$\dfrac{EB}{DE} = \dfrac{AC}{CD}$ Substitute the values given in problem.

$\dfrac{x}{58} = \dfrac{20}{13}$ Multiply both sides by 58.

$x = \dfrac{20 \cdot 58}{13}$ Simplify.

$x = \dfrac{1160}{13} \approx 89.23$

The distance EB is approximately 89.23 feet.

23. a. The volume of a sphere is $\frac{4}{3}\pi r^3$ where r is the radius. So the amount of helium

needed is $\frac{4}{3}\pi(1.2)^3 \approx 7.24$ m^3.

  b. The surface area of a sphere of radius r is $4\pi r^2$, so the amount of gelatin needed is $4\pi(0.7)^2 = 6.16$ cm^2.

25. a. The volume of a cylinder with radius r and height h is $\pi r^2 h$. So the amount of grain that can be stored is $\pi(6)^2 23.2 \approx 2623.86$ m^3.

  b. The surface area of a cylinder is $2\pi(rh + r^2)$, so the amount of paint needed is $2\pi\left([15.3]4.5 + [15.3]^2\right) \approx 1903.43$ square inches.

27. a. The surface area of a box with height h, length l, and width w is $S = 2hl + 2hw + 2wl$. Since $l = 20$ and $w = 16$, we have the formula $S = 2h(20) + 2h(16) + 2(16)(20)$ which simplifies to $S = 72h + 640$.

  b. Solve for h in the equation:

$1216 = 72h + 640$ Subtract 640 from both sides.

 $576 = 72h$ Divide both sides by 72.

  $8 = h$

The height can be no greater than 8 inches

Exercise A.4

1. a. The second coordinate value of the highest point on the graph is the highest temperature and the second coordinate value of the lowest point on the graph is the lowest temperature. These values are approximately 7°F for the high temperature and –19°F for the low temperature.

b. From the graph we notice that the temperature at noon and at 3 P.M. is about 5°F. Between these times the temperature is greater than 5°F. At 9 A.M. the temperature is –5°F and from midnight to 9 A.M. the temperature is below –5°F. The temperature is also –5°F at approximately 7 P.M. and from 7 P.M to midnight the temperature is below –5°F. Therefore, the temperature is below –5°F from 7 P.M. to 9 A.M.

c. At 7 A.M. the temperature was approximately –10°F. At 1 P.M. the temperature was approximately 6°F. The temperature was approximately 0°F at 10 A.M. and 5 P.M. The temperature was approximately –12°F at about 6 A.M. and 10 P.M.

d. From 3 A.M. to 6 A.M. the temperature increased from approximately –19°F to –12°F, a change of 7°F. Between 9 A.M. and noon the temperature increased from approximately –5°F to 5°F, a change of 10°F. From 6 P.M. to 9 P.M. the temperature decreased from approximately –1°F to –10°F, a change of 9°F.

e. The greatest increase was from 9 A.M. to noon where the temperature increased about 10°F in 3 hours. The greatest decrease was from 6 P.M. to 9 P.M. when the temperature dropped about 9°F in 3 hours.

3. a. From the graph, the gas mileage at 43 mph is approximately 27 miles per gallon.

b. A gas mileage of 34 miles per gallon is achieved around 51 mph.

c. The best gas mileage is achieved at around 70 mph. The gas mileage would most likely drop at some point since driving at high speeds requires a high engine temperature, and at some point really hot engines run inefficiently.

d. Other factors to consider are driving conditions (e.g., rain, severe heat or cold) and the maintenance level of the engine (e.g., is it in tune, does it have the proper amount of oil, etc.).

5. a. The only time the car's speed was 0 during the trip was after approximately 12 minutes. This has to be when the car stopped at a traffic signal.

b. During the first 38 minutes, the car's speed fluctuated between 0 and 25 mph, indicative of city traffic.

c. The car traveled on the freeway from approximately 40 minutes into the journey until 50 minutes, since this was the only period where the car traveled at speeds in the 50 to 55 mph range.

7. a. From the graph: $(-3, -2)$, $(1, 6)$, $(-2, 0)$, and $(0, 4)$.

b. Algebraic verification:
$2(-3) + 4 = -6 + 4 = -2$
$2(1) + 4 = 2 + 4 = 6$
$2(-2) + 4 = -4 + 4 = 0$
$2(0) + 4 = 0 + 4 = 4$

9. a. From the graph: $(-2, 6)$, $(2, 6)$, $(\pm 1, 3)$, and $(0, 2)$.

b. Algebraic verification:
$(-2)^2 + 2 = 4 + 2 = 6$
$(2)^2 + 2 = 4 + 2 = 6$
$(-1)^2 + 2 = 1 + 2 = 3$
$(1)^2 + 2 = 1 + 2 = 3$
$(0)^2 + 2 = 0 + 2 = 2$

11. a. From the graph: $\left(-1, -\dfrac{1}{2}\right)$, $\left(\dfrac{1}{2}, -2\right)$, $\left(4, \dfrac{1}{3}\right)$, and $(0, -1)$.

b. Algebraic verification:
$$\frac{1}{-1-1} = -\frac{1}{2}$$
$$\frac{1}{\frac{1}{2}-1} = \frac{1}{-\frac{1}{2}} = -2$$
$$\frac{1}{4-1} = \frac{1}{3}$$
$$\frac{1}{0-1} = \frac{1}{-1} = -1$$

13. a. From the graph: $(-2, -8)$, $\left(\dfrac{1}{2}, \dfrac{1}{8}\right)$, $(0, 0)$, and $(-1, -1)$.

b. Algebraic verification:
$(-2)^3 = 8$
$\left(\dfrac{1}{2}\right)^3 = \dfrac{1}{8}$
$(0)^3 = 0$
$(-1)^3 = -1$

341

Exercise A.5

1. a. $b^4 \cdot b^5 = b^{4+5} = b^9$ by the first law of exponents.

   b. $b^2 \cdot b^8 = b^{2+8} = b^{10}$ by the first law of exponents.

   c. $(q^3)(q)(q^5) = q^{3+1+5} = q^9$ by the first law of exponents.

   d. $(p^2)(p^4)(p^4) = p^{2+4+4} = p^{10}$ by the first law of exponents.

3. a. $2^7 \cdot 2^2 = 2^{7+2} = 2^9 = 512$ by the first law of exponents.

   b. $6^5 \cdot 6^3 = 6^{5+3} = 6^8 = 1,679,616$ by the first law of exponents.

   c. $\dfrac{2^9}{2^4} = 2^{9-4} = 2^5 = 32$ by the second law of exponents.

   d. $\dfrac{8^6}{8^2} = 8^{6-2} = 8^4 = 4096$ by the second law of exponents.

5. a. $(6x)^3 = 6^3 x^3 = 216x^3$ by the fourth law of exponents.

   b. $(3y)^4 = 3^4 y^4 = 81y^4$ by the fourth law of exponents.

   c. $(2t^3)^5 = (2^5)(t^3)^5$ Apply fourth law.
   $= 2^5 t^{3 \cdot 5}$ Apply third law.
   $= 32t^{15}$

   d. $(6s^2)^2 = (6^2)(s^2)^2$ Apply fourth law.
   $= 6^2 s^{2 \cdot 2}$ Apply third law.
   $= 36s^4$

7. a. $\left(\dfrac{h^2}{m^3}\right)^4 = \dfrac{(h^2)^4}{(m^3)^4}$ Apply fifth law.

$\qquad\qquad = \dfrac{h^{2\cdot4}}{m^{3\cdot4}}$ Apply third law.

$\qquad\qquad = \dfrac{h^8}{m^{12}}$

b. $\left(\dfrac{n^3}{k^4}\right)^8 = \dfrac{(n^3)^8}{(k^4)^8}$ Apply fifth law.

$\qquad\qquad = \dfrac{n^{3\cdot8}}{k^{4\cdot8}}$ Apply third law.

$\qquad\qquad = \dfrac{n^{24}}{k^{32}}$

c. $(-4a^2b^4)^4 = (-4)^4(a^2)^4(b^4)^4$ Apply fourth law.

$\qquad\qquad = 4^4 a^{2\cdot4} b^{4\cdot4}$ Apply third law.

$\qquad\qquad = 256 a^8 b^{16}$

d. $(-5ab^8)^3 = (-5)^3 a^3 (b^8)^3$ Apply fourth law.

$\qquad\qquad = (-5)^3 a^3 b^{8\cdot3}$ Apply third law.

$\qquad\qquad = -125 a^3 b^{24}$

9. a. $w + w = 2w$ Combine like terms.

b. $w(w) = w^{1+1} = w^2$ by the first law of exponents.

11. a. $4z^2 - 6z^2 = -2z^2$ Combine like terms.

b. $4z^2(-6z^2) = -24z^{2+2} = -24z^4$ by the first law of exponents.

13. a. $4p^2 + 3p^3$ Cannot be simplified (unlike terms).

b. $4p^2(3p^3) = 12p^2 p^3 = 12p^{2+3} = 12p^5$ by the first law of exponents.

15. a. $3^9 \cdot 3^8 = 3^{9+8} = 3^{17} = 129,140,163$ by the first law of exponents.

b. $3^9 + 3^8 = 19,683 + 6561 = 26,244$

17. a. $(4y)(-6y) = (-24)(y \cdot y) = -24y^{1+1} = -24y^2$

b. $(-4z)(-8z) = (32)(z \cdot z) = 32z^{1+1} = 32z^2$

19. a. $\quad -4x(3xy)(xy^3) = -12(x \cdot x \cdot x)(y \cdot y^3) = -12x^{1+1+1}y^{1+3} = -12x^3y^4$

b. $\quad -5x^2(2xy)(5x^2) = (-5 \cdot 2 \cdot 5)(x^2 \cdot x \cdot x^2)(y) = -50x^{2+1+2}y = -50x^5y$

21. a. Simplify numerical coefficients and apply second law:

$$\frac{2a^3b}{8a^4b^5} = \frac{a^3b}{4a^4b^5} = \frac{1}{4a^{4-3}b^{5-1}} = \frac{1}{4ab^4}$$

b. Simplify numerical coefficients and apply second law:

$$\frac{8a^2b}{12a^5b^3} = \frac{2a^2b}{3a^5b^3} = \frac{2}{3a^{5-2}b^{3-1}} = \frac{2}{3a^3b^2}$$

23. a. Simplify numerical coefficients, apply first law in numerator, and apply second law:

$$\frac{-15bc(b^2c)}{-3b^3c^4} = \frac{5bc(b^2c)}{b^3c^4} = \frac{5(b \cdot b^2)(c \cdot c)}{b^3c^4} = \frac{5b^{1+2}c^{1+1}}{b^3c^4} = \frac{5b^3c^2}{b^3c^4} = \frac{5}{c^{4-2}} = \frac{5}{c^2}$$

b. Simplify numerical coefficients, apply first law in numerator, and apply second law:

$$\frac{-25c(c^2d^2)}{-5c^8d^2} = \frac{5c(c^2d^2)}{c^8d^2} = \frac{5(c \cdot c^2)(d^2)}{c^8d^2} = \frac{5c^{1+2}d^2}{c^8d^2} = \frac{5c^3d^2}{c^8d^2} = \frac{5}{c^{8-3}} = \frac{5}{c^5}$$

25. a. $b^3(b^2)^5 = b^3(b^{2\cdot5})\quad$ Apply third law.

$\qquad\qquad = b^3(b^{10})$

$\qquad\qquad = b^{10+3} \qquad$ Apply first law.

$\qquad\qquad = b^{13}$

b. $b(b^4)^6 = b(b^{4\cdot6})\quad$ Apply third law.

$\qquad\qquad = b(b^{24})$

$\qquad\qquad = b^{1+24} \qquad$ Apply first law.

$\qquad\qquad = b^{25}$

27. a. $(2x^3y)^2(xy^3)^4 = \left(2^2(x^3)^2y^2\right)\left(x^4(y^3)^4\right)$ Apply fourth law.

$$= \left(2^2(x^{3\cdot2})y^2\right)\left(x^4(y^{3\cdot4})\right)$$ Apply third law.

$$= (2^2x^6y^2)(x^4y^{12})$$

$$= 2^2(x^6\cdot x^4)(y^2\cdot y^{12})$$ Rearrange factors.

$$= 2^2x^{6+4}y^{2+12}$$ Apply first law.

$$= 4x^{10}y^{14}$$

b. $(3xy^2)^3(2x^2y^2)^2 = \left(3^3x^3(y^2)^3\right)\left(2^2(x^2)^2(y^2)^2\right)$ Apply fourth law.

$$= (27x^3y^{2\cdot3})(4x^{2\cdot2}y^{2\cdot2})$$ Apply third law.

$$= (27x^3y^6)(4x^4y^4)$$

$$= (27\cdot4)(x^3\cdot x^4)(y^6\cdot y^4)$$ Rearrange factors.

$$= 108x^{3+4}y^{6+4}$$ Apply first law.

$$= 108x^7y^{10}$$

29. a. $\left(\dfrac{-2x}{3y^2}\right)^3 = \left(\dfrac{(-2)^3x^3}{3^3(y^2)^3}\right)$ Apply fifth law.

$$= \frac{-8x^3}{27y^{2\cdot3}}$$ Apply third law.

$$= \frac{-8x^3}{27y^6}$$

b. $\left(\dfrac{-x^2}{2y}\right)^4 = \left(\dfrac{(-1)^4(x^2)^4}{2^4y^4}\right)$ Apply fifth law.

$$= \frac{x^{2\cdot4}}{16y^4}$$ Apply third law.

$$= \frac{x^8}{16y^4}$$

Exercise A.5

31. a. $\dfrac{(xy)^2(-x^2y)^3}{(x^2y^2)^2} = \dfrac{(x^2y^2)\left((-1)^3(x^2)^3y^3\right)}{(x^2)^2(y^2)^2}$ Apply fourth law.

$$= \dfrac{(x^2y^2)\left(-(x^{2\cdot3})y^3\right)}{x^{2\cdot2}y^{2\cdot2}}$$ Apply third law.

$$= \dfrac{(x^2y^2)(-x^6y^3)}{x^4y^4}$$

$$= \dfrac{-(x^2\cdot x^6)(y^2\cdot y^3)}{x^4y^4}$$ Rearrange factors.

$$= \dfrac{-(x^{2+6})(y^{2+3})}{x^4y^4}$$ Apply first law.

$$= -\dfrac{x^8y^5}{x^4y^4}$$

$$= -x^{8-4}y^{5-4}$$ Apply second law.

$$= -x^4y$$

b. $\dfrac{(-x)^2(-x^2)^4}{(x^2)^3} = \dfrac{(-1)^2x^2(-1)^4(x^2)^4}{(x^2)^3}$ Apply fourth law.

$$= \dfrac{x^2x^{2\cdot4}}{x^{2\cdot3}}$$ Apply third law.

$$= \dfrac{x^2x^8}{x^6}$$

$$= \dfrac{x^{2+8}}{x^6}$$ Apply first law.

$$= \dfrac{x^{10}}{x^6}$$

$$= x^{10-6}$$ Apply second law.

$$= x^4$$

Exercise A.6

1. binomial; degree 3

3. monomial; degree 4

5. trinomial; degree 2

7. trinomial; degree 3

9. **a.** polynomial

    b. This is not a polynomial since there is a variable in a denominator.

    c. This is not a polynomial since there is a variable under a radical.

    d. polynomial

11. **a.** This is not a polynomial since there are variables in the denominator.

    b. This is not a polynomial since there is a variable in an exponent.

    c. This is not a polynomial since there is a variable under a radical.

    d. polynomial

13. $P = x^3 - 3x^2 + x + 1$

    a. $P = (2)^3 - 3(2)^2 + (2) + 1 = 8 - 3 \cdot 4 + 2 + 1 = 8 - 12 + 2 + 1 = -1$

    b. $P = (-2)^3 - 3(-2)^2 + (-2) + 1 = -8 - 3 \cdot 4 - 2 + 1 = -8 - 12 - 2 + 1 = -21$

    c. $P = (2b)^3 - 3(2b)^2 + (2b) + 1 = 2^3 b^3 - 3 \cdot 2^2 b^2 + 2b + 1 = 8b^3 - 12b^2 + 2b + 1$

15. $Q = t^2 + 3t + 1$

    a. $Q = \left(\frac{1}{2}\right)^2 + 3\left(\frac{1}{2}\right) + 1 = \frac{1}{4} + \frac{3}{2} + 1 = \frac{1}{4} + \frac{6}{4} + \frac{4}{4} = \frac{11}{4}$

    b. $Q = \left(-\frac{1}{3}\right)^2 + 3\left(-\frac{1}{3}\right) + 1 = \frac{1}{9} - 1 + 1 = \frac{1}{9}$

    c. $Q = (-w)^2 + 3(-w) + 1 = (-1)^2 w^2 - 3w + 1 = w^2 - 3w + 1$

17. $R = 3z^4 - 2z^2 + 3$

    a. $R = 3(1.8)^4 - 2(1.8)^2 + 3 = 3(10.4976) - 2(3.24) + 3$
         $= 31.4928 - 6.48 + 3 = 28.0128$

    b. $R = 3(-2.6)^4 - 2(-2.6)^2 + 3 = 3(45.6976) - 2(6.76) + 3$
         $= 137.0928 - 13.52 + 3 = 126.5728$

    c. $R = 3(k - 1)^4 - 2(k - 1)^2 + 3$

19. $N = a^6 - a^5$

    a. $N = (-1)^6 - (-1)^5 = 1 - (-1) = 1 + 1 = 2$

    b. $N = (-2)^6 - (-2)^5 = 64 - (-32) = 64 + 32 = 96$

    c. $N = \left(\dfrac{m}{3}\right)^6 - \left(\dfrac{m}{3}\right)^5 = \dfrac{m^6}{3^6} - \dfrac{m^5}{3^5} = \dfrac{m^6}{3^6} - \dfrac{3m^5}{3 \cdot 3^5} = \dfrac{m^6}{3^6} - \dfrac{3m^5}{3^6} = \dfrac{m^6 - 3m^5}{729}$

21. $4y(x - 2y) = 4y(x) + 4y(-2y) = 4xy - 8y^2$

23. $-6x(2x^2 - x + 1) = -6x(2x^2) - 6x(-x) - 6x(1) = -12x^3 + 6x^2 - 6x$

25. $a^2b(3a^2 - 2ab - b) = a^2b(3a^2) + a^2b(-2ab) + a^2b(-b) = 3a^4b - 2a^3b^2 - a^2b^2$

27. $2x^2y^3(4xy^4 - 2xy - 3x^3y^2) = 2x^2y^3(4xy^4) + 2x^2y^3(-2xy) + 2x^2y^3(-3x^3y^2)$
$$= 8x^3y^7 - 4x^3y^4 - 6x^5y^5$$

29. $(n + 2)(n + 8) = n^2 + 8n + 2n + 16 = n^2 + 10n + 16$

31. $(r + 5)(r - 2) = r^2 - 2r + 5r - 10 = r^2 + 3r - 10$

33. $(2z + 1)(z - 3) = 2z^2 - 6z + z - 3 = 2z^2 - 5z - 3$

35. $(4r + 3s)(2r - s) = 8r^2 - 4rs + 6rs - 3s^2 = 8r^2 + 2rs - 3s^2$

37. $(2x - 3y)(3x - 2y) = 6x^2 - 4xy - 9xy + 6y^2 = 6x^2 - 13xy + 6y^2$

39. $(3t - 4s)(3t + 4s) = 9t^2 + 12st - 12st - 16s^2 = 9t^2 - 16s^2$

41. $(2a^2 + b^2)(a^2 - 3b^2) = 2a^4 - 6a^2b^2 + a^2b^2 - 3b^4 = 2a^4 - 5a^2b^2 - 3b^4$

43. $4x^2z + 8xz = (4xz)(x) + (4xz)(2) = 4xz(x + 2)$

45. $3n^4 - 6n^3 + 12n^2 = (3n^2)(n^2) + (3n^2)(-2n) + 3n^2(4) = 3n^2(n^2 - 2n + 4)$

47. $15r^2s + 18rs^2 - 3r = (3r)(5rs) + (3r)(6s^2) + (3r)(-1) = 3r(5rs + 6s^2 - 1)$

49. $3m^2n^4 - 6m^3n^3 + 14m^3n^2 = (m^2n^2)(3n^2) + (m^2n^2)(-6mn) + (m^2n^2)(14m)$
$$= m^2n^2(3n^2 - 6mn + 14m)$$

51. $15a^4b^3c^4 - 12a^2b^2c^5 + 6a^2b^3c^4$
$$= (3a^2b^2c^4)(5a^2b) + (3a^2b^2c^4)(-4c) + (3a^2b^2c^4)(2b)$$
$$= 3a^2b^2c^4(5a^2b - 4c + 2b)$$

53. $a(a+3)+b(a+3)=(a+3)(a+b)$

55. $y(y-2)-3x(y-2)=(y-2)(y-3x)$

57. $4(x-2)^2-8x(x-2)^3=4(x-2)^2[1-2x(x-2)]$
$$=4(x-2)^2(1-2x^2+4x)$$
$$=4(x-2)^2(-2x^2+4x+1)$$
$$=-4(x-2)^2(2x^2-4x-1)$$

59. $x(x-5)^2-x^2(x-5)^3=x(x-5)^2[1-x(x-5)]$
$$=x(x-5)^2(1-x^2+5x)$$
$$=x(x-5)^2(-x^2+5x+1)$$
$$=-x(x-5)^2(x^2-5x-1)$$

61. $3m-2n=-(-3m+2n)=-(2n-3m)$

63. $-2x+2=-2(x-1)$

65. $-ab-ac=-a(b+c)$

67. $2x-y+3z=-(-2x+y-3z)$

69. Since $2+3=5$ and $2\cdot3=6$, $x^2+5x+6=(x+2)(x+3)$.

71. Since $(-4)+(-3)=-7$ and $(-4)(-3)=12$, $y^2-7y+12=(y-3)(y-4)$.

73. $x^2-6-x=x^2-x-6$
Since $2\cdot3=6$ and $2+(-3)=-1$, $x^2-x-6=(x-3)(x+2)$

75. $2x^2+3x-2$;
2 factors as $1\cdot2$ and -2 factors as $-1\cdot2$ or $1\cdot(-2)$. Since the constant term is negative, the middle signs of the two factors will be opposites. So the possibilities are:
$(2x-1)(x+2)$,
$(2x+1)(x-2)$,
$(x-1)(2x+2)$, and
$(x+1)(2x-2)$.
From checking the middle terms of the product, $2x^2+3x-2=(2x-1)(x+2)$.

77. $7x + 4x^2 - 2 = 4x^2 + 7x - 2$;

·4 factors as $1 \cdot 4$ or $2 \cdot 2$; –2 factors as $-1 \cdot 2$ or $1 \cdot (-2)$. Since the constant term is negative, the middle signs of the two factors will be opposites. So the possibilities are:

$(x - 1)(4x + 2)$,
$(x + 1)(4x - 2)$,
$(4x - 1)(x + 2)$,
$(4x + 1)(x - 2)$,
$(2x - 1)(2x + 2)$, and
$(2x + 1)(2x - 2)$.

From checking the middle terms of the product, $4x^2 + 7x - 2 = (4x - 1)(x + 2)$.

79. $9y^2 - 21y - 8$; Since the constant term is negative, the middle signs of the two factors will be opposites. By trial and error with the factors of 9 and 8, we have
$9y^2 - 21y - 8 = (3y + 1)(3y - 8)$.

81. $10u^2 - 3 - u = 10u^2 - u - 3$; Since the constant term is negative, the middle signs of the two factors will be opposites. By trial and error with the factors of 10 and 3, we have $10u^2 - u - 3 = (2u + 1)(5u - 3)$.

83. $21x^2 - 43x - 14$; Since the constant term is negative, the middle signs of the two factors will be opposites. By trial and error with the factors of 21 and 14, we have $21x^2 - 43x - 14 = (7x + 2)(3x - 7)$.

85. $5a + 72a^2 - 12 = 72a^2 + 5a - 12$; Since the constant term is negative, the middle signs of the two factors will be opposites. By trial and error with the factors of 72 and 12, we have $72a^2 + 5a - 12 = (9a + 4)(8a - 3)$.

87. $12 - 53x + 30x^2 = 30x^2 - 53x + 12$; Since the coefficient of x is negative and the constant term is positive, the middle signs of both factors will be negative. By trial and error with the factors of 30 and 12, we have $30x^2 - 53x + 12 = (2x - 3)(15x - 4)$.

89. First, write in descending order and factor out the greatest common factor:
$-30t - 44 + 54t^2 = 54t^2 - 30t - 44 = 2(27t^2 - 15t - 22)$. Since the constant term, (-22) is negative, the middle signs of the two factors will be opposites. By trial and error with the factors of 27 and 22, we have $2(27t^2 - 15t - 22) = 2(3t + 2)(9t - 11)$.

91. $3x^2 - 7ax + 2a^2$; Since the last term is positive and the middle term is negative, the middle signs of both factors will be negative. By trial and error on the factors of 3 and 2, we have $3x^2 - 7ax + 2a^2 = (x - 2a)(3x - a)$.

93. $15x^2 - 4xy - 4y^2$; Since the last term is negative, the middle signs of the two factors will be opposites. By trial and error on the factors of 15 and 4, we have
$15x^2 - 4xy - 4y^2 = (3x - 2y)(5x + 2y)$.

95. Write in descending powers of u: $18u^2 + 20v^2 - 39uv = 18u^2 - 39uv + 20v^2$. Since the last term is positive and the middle term is negative, the middle signs of both factors will be negative. By trial and error on the factors of 18 and 20, we have
$18u^2 - 39uv + 20v^2 = (3u - 4v)(6u - 5v)$.

97. Write in descending powers of a: $12a^2 - 14b^2 - 13ab = 12a^2 - 13ab - 14b^2$; Since the last term is negative, the middle signs of the two factors will be opposites. By trial and error on the factors of 12 and 14, we have
$12a^2 - 13ab - 14b^2 = (3a + 2b)(4a - 7b)$.

99. $10a^2b^2 - 19ab + 6$; Since the last term is positive and the middle sign is negative, the middle signs of the two factors will both be negative. By trial and error on the factors of 10 and 6, we have $10a^2b^2 - 19ab + 6 = (2ab - 3)(5ab - 2)$.

101. Begin by factoring out the greatest common factor of 2:
$56x^2y^2 - 2xy - 4 = 2(28x^2y^2 - xy - 2)$. Since the last term is negative, the middle signs of the two factors will be opposites. By trial and error on the factors of 28 and 2, we have $2(28x^2y^2 - xy - 2) = 2(7xy - 2)(4xy + 1)$.

103. Write in descending order: $22a^2z^2 - 21 - 19az = 22a^2z^2 - 19az - 21$. Since the last term is negative, the middle signs of the two factors will be opposites. By trial and error on the factors of 22 and 21, we have $22a^2z^2 - 19az - 21 = (11az + 7)(2az - 3)$.

105. $(x + 3)^2 = (x + 3)(x + 3) = x^2 + 3x + 3x + 9 = x^2 + 6x + 9$

107. $(2y - 5)^2 = (2y - 5)(2y - 5) = 4y^2 - 10y - 10y + 25 = 4y^2 - 20y + 25$

109. $(x + 3)(x - 3) = x^2 - 3x + 3x - 9 = x^2 - 9$

111. $(3t - 4s)(3t + 4s) = 9t^2 + 12st - 12st - 16s^2 = 9t^2 - 16s^2$

113. $(5a - 2b)(5a - 2b) = 25a^2 - 10ab - 10ab + 4b^2 = 25a^2 - 20ab + 4b^2$

115. $(8xz + 3)(8xz + 3) = 64x^2z^2 + 24xz + 24xz + 9 = 64x^2z^2 + 48xz + 9$

117. $x^2 - 25 = x^2 - 5^2 = (x + 5)(x - 5)$

119. $x^2 - 24x + 144 = x^2 - 2(1)(12)x + 12^2 = (x - 12)(x - 12) = (x - 12)^2$

Exercise A.6

121. $x^2 - 4y^2 = x^2 - (2y)^2 = (x + 2y)(x - 2y)$

123. $4x^2 + 12x + 9 = (2x)^2 + 2(2)(3)x + 3^2 = (2x + 3)(2x + 3) = (2x + 3)^2$

125. $9u^2 - 30uv + 25v^2 = (3u)^2 - 2(3u)(5v) + (5v)^2 = (3u - 5v)(3u - 5v) = (3u - 5v)^2$

127. $4a^2 - 25b^2 = (2a)^2 - (5b)^2 = (2a + 5b)(2a - 5b)$

129. $x^2 y^2 - 81 = (xy)^2 - 9^2 = (xy + 9)(xy - 9)$

131. $9x^2 y^2 + 6xy + 1 = (3xy)^2 + 2(3xy)(1) + (1)^2 = (3xy + 1)(3xy + 1) = (3xy + 1)^2$

133. $16x^2 y^2 - 1 = (4xy)^2 - 1^2 = (4xy + 1)(4xy - 1)$

135. $(x + 2)^2 - y^2 = [(x + 2) + y][(x + 2) - y] = (x + 2 + y)(x + 2 - y)$

Exercise A.7

1. There are nine numbers. Write them in increasing order:
   $1, 2, 3, 4, 4, 4, 5, 6, 7$

   The mean is equal to
   $$\frac{1+2+3+4+4+4+5+6+7}{9} = 4.$$

   The median is 4, since it is the middle data value when they are listed in increasing order.

   The mode is 4, since it is the score with the highest frequency.

3. There are seven numbers. Write them in increasing order.
   $27, 44, 66, 77, 87, 93, 110$

   The mean is equal to
   $$\frac{27+44+66+77+87+93+110}{7}$$
   $$= \frac{504}{7} = 72$$

   The median is 77, since it is the middle data value when they are listed in increasing order.

   Because no data value occurs more often than the rest, there is no mode.

5. **a.** There are 7 players. The sum of the seven salaries (in thousands) is 1351. Dividing by 7 gives $\frac{1351}{7} = 193$. Therefore, the mean of the salaries is $193,000. The middle value is 10, so the median of the salaries is $10,000. The data value with highest frequency is 7, so the mode of the salaries is $7000.

   b. Let x be Player A's new salary, in thousands. Then the mean is
   $$\frac{x+100+20+10+7+7+7}{7}$$
   $$= \frac{x+151}{7}$$
   For the mean salary to be $100,000, let the above expression equal 100 and solve for x:
   $$\frac{x+151}{7} = 100$$
   $$x+151 = 700$$
   $$x = 549$$
   Player A's new salary should be $549,000 (as compared to the old salary of $1,200,000).

7. If the median of the 7 numbers is 10, then three numbers are at or below 10 and three are at or above 10. Two of the numbers are 5 and the mode is 20, so all three numbers above 10 must be 20. Let x be the missing number. Note that this number must be below 10. The mean is 0, so solve:

$$\frac{x+5+5+10+20+20+20}{7} = 0$$

$$\frac{x+80}{7} = 0$$

$$x+80 = 0$$

$$x = -80$$

Thus, the seven numbers are: $-80, 5, 5, 10, 20, 20, 20$.

9. There are many possible answers. Say there are 5 numbers. Let the median be 10. This is less than the mean, and the mean is less than the mode. Choose a number greater than 10 to be the mode, say 15. So the numbers are ?, ?, 10, 15, 15. Since the mean is between 10 and 15, choose two different numbers who are less than 10 by no more than 5 units. For example, choose 8 and 9. Now we have 8, 9, 10, 15, 15. The mean of this data set is

$$\frac{8+9+10+15+15}{5} = 11.4 \text{, so the}$$

median (10) is less than the mean (11.4) which is in turn less than the mode (15).

11. 1, 2, 3, 4, 4, 4, 5, 6, 7
The lowest and highest scores are 1 and 7. The median is 4, since this is the middle of the data values. The lower quartile is 2.5, since this is the midpoint between the two middle values (2 and 3) of the four numbers below the median. The upper quartile is 5.5, since is this the midpoint between the two middle values (5 and 6) of the four numbers above the median. The five number summary is 1, 2.5, 4, 5.5, 7.

13. 27, 44, 66, 77, 87, 93, 110
The lowest and highest scores are 27 and 110. The median is 77, since this is the middle of the data values. The lower quartile is 44, since this is the middle of the three numbers below the median. The upper quartile is 93, since is this the middle of the three numbers above the median. The five number summary is 27, 44, 77, 93, 110.

15. More students received a score of 7 than any other score, so 7 is the mode. Because of the large number of scores in this problem, it is helpful to use our graphing calculator to finish the problem . In your STAT menu, enter the scores in L_1 and the frequencies in L_2 as follows:

L1	⬛	L3	2
0	1	------	
2	1		
4	3		
5	3		
6	8		
7	14		
8	12		

$L2 = \langle 1,1,3,3,8,1\ldots$

L1	L2	L3	1
5	3		
6	8		
7	14		
8	12		
9	6		
10	2		

$L1(10) =$

Select "1-Var Stats" from the CALC menu within the STAT menu, and then L_1, L_2 (which can be found using *2nd* 1, *comma*, and *2nd* 2.)

```
1-Var Stats
 x̄=6.9
 Σx=345
 Σx²=2555
 Sx=1.887120688
 σx=1.868154169
↓n=50
```

```
1-Var Stats
↑n=50
 minX=0
 Q₁=6
 Med=7
 Q₃=8
 maxX=10
■
```

Thus, the mean is 6.9, the mode is 7, and the five number summary is 0, 6, 7, 8, 10.

17. a. range: $98 - 78 = 70$

   b. The upper quartile is 81.

   c. 73 is the median, so 50% of the class scored at or below 73.

   d. 81 is the upper quartile, so 25% of the class scored at or above 81.

19. a. The difference between extreme values in the class in Exercise 17 ("your section") is $98 - 28 = 70$, while the difference between extremes in the class for Exercise 19 (the "other section") is $94 - 3 = 91$, so the other section has a wider range of scores.

   b. In your section, 75% of the scores were at or below 81, so at least 75% of the scores were at or below 82. In the other section, 82 is the median, so 50% of the scores were at or below 82. Hence your section had the larger fraction of scores at or below 82.

   c. In your section, we know that 25% of the scores were at or above 81 and in the other section, we know that 25% of the scores were at or above 91, so 25% of the scores were above 90. So the other section had the larger fraction of scores above 90.

   d. The other section did better overall than your section, since the other section had a higher percentage scoring above 74, above 82, and above 91.

21. When the mode exists, it must be one of the values in the data set. There are many possible data sets where the mean and median are not in the set. For example, 0, 0, 0, 1, 5, 6 has a mean of 2 and a median of 0.5.

Appendix B: The Number System

Exercise B.1

1. $-\dfrac{5}{8}$ is a rational number since it is a quotient of integers. It is also a real number.

3. $\sqrt{8} = 2.828427\ldots$ is an irrational number since the decimal is non-repeating. It is also a real number.

5. -36 is an integer, a rational number, and a real number.

7. 0 is a whole number, an integer, a rational number, and a real number.

9. $13.\overline{289}$ is a rational number since it is a repeating decimal, so can be written as a quotient of integers. It is also a real number.

11. $2\pi = 6.2831853\ldots$ is an irrational number since the decimal is non-repeating. It is also a real number.

13. $\dfrac{3}{8} = 0.375$; terminating decimal

15. $\dfrac{2}{7} = 0.\overline{285714}$; repeats the pattern 285714

17. $\dfrac{7}{16} = 0.4375$; terminating decimal

19. $\dfrac{11}{13} = 0.\overline{846153}$; repeats the pattern 846153

Exercise B.2

1. $(11-4i)-(-2-8i) = 11-4i+2+8i = 13+4i$

3. $(2.1+5.6i)+(-1.8i-2.9) = 2.1+5.6i-1.8i-2.9 = -0.8+3.8i$

5. $5i(2-4i) = 10i-20i^2 = 10i-20(-1) = 20+10i$

7. $(4-i)(-6+7i) = -24+28i+6i-7i^2 = -24+28i+6i-7(-1)$
    $= -24+28i+6i+7 = -17+34i$

9. $\left(7+i\sqrt{3}\right)^2 = \left(7+i\sqrt{3}\right)\left(7+i\sqrt{3}\right) = 49+7i\sqrt{3}+7i\sqrt{3}+3i^2 = 49+14i\sqrt{3}+3(-1)$
    $= 49+14i\sqrt{3}-3 = 46+14i\sqrt{3}$

11. $\left(7+i\sqrt{3}\right)\left(7-i\sqrt{3}\right) = 49-7i\sqrt{3}+7i\sqrt{3}-3i^2 = 49-3i^2 = 49-3(-1) = 49+3 = 52$

13. z^2+9

    a. $(3i)^2+9 = 9i^2+9 = 9(-1)+9 = -9+9 = 0$

    b. $(-3i)^2+9 = 9i^2+9 = 9(-1)+9 = -9+9 = 0$

15. x^2-2x+2

    a. $(1-i)^2-2(1-i)+2 = (1-i)(1-i)-2(1-i)+2 = 1-i-i+i^2-2+2i+2$
        $= 1+i^2 = 1+(-1) = 0$

    b. $(1+i)^2-2(1+i)+2 = (1+i)(1+i)-2(1+i)+2 = 1+i+i+i^2-2-2i+2$
        $= 1+i^2 = 1+(-1) = 0$

17. $2y^2-y+2$

    a. $2(2-i)^2-(2-i)+2 = 2(2-i)(2-i)-(2-i)+2$
        $= 2(4-4i+i^2)-2+i+2 = 8-8i+2i^2-2+i+2 = 8-7i+2(-1) = 6-7i$

    b. $2(-2-i)^2-(-2-i)+2 = 2(-2-i)(-2-i)-(-2-i)+2$
        $= 2(4+4i+i^2)+2+i+2 = 8+8i+2i^2+2+i+2 = 12+9i+2(-1) = 10+9i$

19. To rationalize the denominator, multiply the numerator and denominator by i.
    $$\frac{12+3i}{-3i} = \frac{12+3i}{-3i}\cdot\frac{i}{i} = \frac{12i+3i^2}{-3i^2} = \frac{12i+3(-1)}{-3(-1)} = \frac{-3+12i}{3} = -1+4i$$

21. To rationalize the denominator, multiply the numerator and denominator by the conjugate of the denominator, $2 - i$.

$$\frac{10 + 15i}{2 + i} = \frac{(10 + 15i)(2 - i)}{(2 + i)(2 - i)} = \frac{20 - 10i + 30i - 15i^2}{4 - i^2} = \frac{20 + 20i - 15(-1)}{4 - (-1)}$$

$$= \frac{20 + 20i + 15}{4 + 1} = \frac{35 + 20i}{5} = 7 + 4i$$

23. The conjugate of the denominator is $2 + 5i$, so

$$\frac{5i}{2 - 5i} = \frac{5i(2 + 5i)}{(2 - 5i)(2 + 5i)} = \frac{10i + 25i^2}{4 - 25i^2} = \frac{10i + 25(-1)}{4 - 25(-1)} = \frac{-25 + 10i}{29} = -\frac{25}{29} + \frac{10}{29}i$$

25. The conjugate of the denominator is $\sqrt{3} - i$, so

$$\frac{\sqrt{3}}{\sqrt{3} + i} = \frac{\sqrt{3}(\sqrt{3} - i)}{(\sqrt{3} + i)(\sqrt{3} - i)} = \frac{\sqrt{9} - i\sqrt{3}}{\sqrt{9} - i^2} = \frac{3 - i\sqrt{3}}{3 - (-1)} = \frac{3 - i\sqrt{3}}{4} = \frac{3}{4} - \frac{\sqrt{3}}{4}i$$

27. The conjugate of the denominator is $1 + i\sqrt{5}$, so

$$\frac{1 + i\sqrt{5}}{1 - i\sqrt{5}} = \frac{(1 + i\sqrt{5})(1 + i\sqrt{5})}{(1 - i\sqrt{5})(1 + i\sqrt{5})} = \frac{1 + i\sqrt{5} + i\sqrt{5} + i^2\sqrt{25}}{1 - i^2\sqrt{25}} = \frac{1 + 2i\sqrt{5} + (-1)(5)}{1 - (-1)(5)}$$

$$= \frac{-4 + 2i\sqrt{5}}{6} = -\frac{2}{3} + \frac{\sqrt{5}}{3}i$$

29. The conjugate of the denominator is $2 + 3i$, so

$$\frac{3 + 2i}{2 - 3i} = \frac{(3 + 2i)(2 + 3i)}{(2 - 3i)(2 + 3i)} = \frac{6 + 9i + 4i + 6i^2}{4 - 9i^2} = \frac{6 + 13i + 6(-1)}{4 - 9(-1)} = \frac{13i}{13} = i$$

31. $(2z + 7i)(2z - 7i) = 4z^2 - 14zi + 14zi - 49i^2 = 4z^2 - 49i^2 = 4z^2 - 49(-1) = 4z^2 + 49$

33. $[x + (3 + i)][x + (3 - i)] = x^2 + (3 - i)x + (3 + i)x + (3 + i)(3 - i)$
$= x^2 + 3x - ix + 3x + ix + 9 - i^2 = x^2 + 6x + 9 - (-1) = x^2 + 6x + 10$

35. $[v - (4 + i)][v - (4 - i)] = v^2 - v(4 - i) - v(4 + i) + (4 + i)(4 - i)$
$= v^2 - 4v + iv - 4v - iv + 16 - i^2 = v^2 - 8v + 16 - (-1) = v^2 - 8v + 17$

37. $\sqrt{x - 5}$ is real when $x - 5 \geq 0$ and imaginary when $x - 5 < 0$. Hence, $\sqrt{x - 5}$ is real for $x \geq 5$ and imaginary for $x < 5$.

39. a. $i^6 = (i^2)^3 = (-1)^3 = -1$; Alternate method: $i^6 = i^2 \cdot i^4 = (-1)(1) = -1$

b. $i^{12} = (i^2)^6 = (-1)^6 = 1$; Alternate method: $i^{12} = (i^4)^3 = (1)^3 = 1$

c. $i^{15} = (i^2)^7 \cdot i = (-1)^7 i = -i$; Alternate method: $i^{15} = (i^4)^3 \cdot i^2 \cdot i = (1)^3(-1) \cdot i = -i$

d. $i^{102} = (i^2)^{51} = (-1)^{51} = -1$; Alternate method: $i^{102} = (i^4)^{25} \cdot i^2 = (1)^{25}(-1) = -1$